WATER WAVE MECHANICS FOR ENGINEERS AND SCIENTISTS

ADVANCED SERIES ON OCEAN ENGINEERING

Series Editor-in-Chief
Philip L- F Liu
Cornell University, USA

Advanced Series on Ocean Engineering — Volume 2

WATER WAVE MECHANICS FOR ENGINEERS AND SCIENTISTS

Robert G. Dean
University of Florida

Robert A. Dalrymple
University of Delaware

World Scientific
Singapore • New Jersey • London • Hong Kong

Published by

World Scientific Publishing Co. Pte. Ltd.

P O Box 128, Farrer Road, Singapore 912805

USA office: Suite 1B, 1060 Main Street, River Edge, NJ 07661

UK office: 57 Shelton Street, Covent Garden, London WC2H 9HE

Library of Congress Cataloging-in-Publication Data

Dean, Robert G. (Robert George), 1930–
   Water wave mechanics for engineers and scientists / Robert G. Dean
and Robert A. Dalrymple.
       p.   cm.
   Includes bibliographical references and index.
   ISBN 9810204205. -- ISBN 9810204213 (pbk.)
   1. Water waves.   2. Fluid mechanics.   I. Dalrymple, Robert A.,
1945–   .   II. Title
TC172.D4   1991
627'.042--dc20   90-27331
                                    CIP

First published in 1984 by Prentice Hall, Inc.

Copyright © 1991 by World Scientific Publishing Co. Pte. Ltd.

Reprinted in 1992, 1993, 1994, 1995, 1998, 2000.

Printed in Singapore.

To Phyllis and Candice

Contents

4

ENGINEERING WAVE PROPERTIES   **78**

5

LONG WAVES   **131**

6

7

8

9

Preface

The initial substantive interest in and contributions to water wave mechanics date from more than a century ago, beginning with the analysis of linear wave theory by Airy in 1845 and continuing with higher order theories by Stokes in 1847, long wave theories by Boussinesq in 1872, and limiting wave heights by Michell in 1893 and McCowan in 1894.

Following that half-century of pioneering developments, research continued at a relatively slow pace until the amphibious landings in the Second World War emphasized the need for a much better understanding of wave initiation and growth due to winds, the conservative and dissipative transformation mechanisms occurring from the source area to the shoaling, and the breaking processes at the shore. The largely unsuccessful attempt to utilize portable and floating breakwaters in the surprise amphibious landing at Normandy, France, stimulated interest in wave interaction with fixed and floating objects.

After the Second World War, the activity in water wave research probably would have subsided without the rather explosive growth in ocean-related engineering in scientific, industrial, and military activities. From the 1950s to the 1980s, offshore drilling and production of petroleum resources progressed from water depths of approximately 10 meters to over 300 meters, platforms for the latter being designed for wave heights on the order of 25 meters and costing in excess of $700,000,000 (U.S.). The financial incentives of well-planned and comprehensive studies of water wave phenomena became much greater. Laboratory studies as well as much more expensive field programs were required to validate design methodology and to provide a better basis for describing the complex and nonlinear directional seas. A second and substantial impetus to nearshore research on water waves has been the interest in coastal erosion, an area still only poorly understood. For example, although the momentum flux concepts were systematized by Longuet-Higgins and Stewart and applied to a number of relevant problems in the 1960s, the usual (spilling wave) assumption of the wave height inside the surf zone being proportional to the water depth avoids the important matter of the distribution of the applied longshore stress across the surf zone. This can only be reconciled through careful laboratory and field measurements of wave breaking. Wave energy provides another example. In the last two decades remote sensing has indicated the potential of defining synoptic measures of wave intensity over very wide areas, with the associated benefits to shipping efficiency. Simple calculations of the magnitudes of the "standing crop" of wave energy have stimulated many scientists and engineers to devise ingenious mechanisms to harvest this energy. Still, these mechanisms must operate in a harsh environment known for its long-term corrosive and fouling effects and the high-intensity forces during severe storms.

The problem of quantifying the wave climate, understanding the interaction of waves with structures and/or sediment, and predicting the associated responses of interest underlies almost every problem in coastal and ocean engineering. It is toward this goal that this book is directed. Although the book is intended for use primarily as a text at the advanced undergraduate or first-year graduate level, it is hoped that it will serve also as a reference and will assist one to learn the field through self-study. Toward these objectives, each chapter concludes with a number of problems developed to illustrate by application the material presented. The references included should aid the student and the practicing engineer to extend their knowledge further.

The book is comprised of twelve chapters. Chapter 1 presents a number of common examples illustrating the wide range of water wave phenomena, many of which can be commonly observed. Chapter 2 offers a review of potential flow hydrodynamics and vector analysis. This material is presented for the sake of completeness, even though it will be familiar to many readers. Chapter 3 formulates the linear water wave theory and develops the simplest two-dimensional solution for standing and progressive waves. Chapter 4 extends the solutions developed in Chapter 3 to many features of engineering relevance, including kinematics, pressure fields, energy, shoaling, refraction, and diffraction. Chapter 5 investigates long wave phenomena, such as kinematics, seiching, standing and progressive waves with friction, and long waves including geostrophic forces and storm surges. Chapter 6 explores various wavemaker problems, which are relevant to problems of wave tank and wave basin design and to problems of damping of floating bodies. The utility of spectral analysis to combine many elemental solutions is explored in Chapter 7. In this manner a complex sea comprising a spectrum of frequencies and, at each frequency, a continuum of directions can be represented. Chapter 8 examines the problem of wave forces on structures. A slight modification of the problem of two-dimensional idealized flow about a cylinder yields the well-known Morison equation. Both drag- and inertia-dominant systems are discussed, including methods for data analysis, and some field data are presented. This chapter concludes with a brief description of the Green's function representation for calculating the forces on large bodies. Chapter 9 considers the effects of waves propagating over seabeds which may be porous, viscous, and/or compressible and at which frictional effects may occur in the bottom boundary layer. Chapter 10 develops a number of nonlinear (to second order in wave height) results that, somewhat surprisingly, may be obtained from linear wave theory. These results, many of which are of engineering concern, include mass transport, momentum flux, set-down and set-up of the mean water level, mean pressure under a progressive wave, and the "microseisms," in-phase pressure fluctuations that occur under two-dimensional standing waves. Chaper 11 introduces the perturbation method to develop and solve various nonlinear wave theories, including the Stokes second order theory, and the solitary and cnoidal wave theories. The procedure for developing numerical wave theories to high order is described, as are the analytical and physical validities of theories. Finally, Chapter 12 presents a number of water wave experiments (requiring only simple instrumentation) that the authors have found useful for demonstrating the theory and introducing the student to wave experimentation, specifically methodology, instrumentation, and frustrations.

Each chapter is dedicated to a scientist who contributed importantly to this field. Brief biographies were gleaned from such sources as *The Dictionary of National Biography* (United Kindom scientists; Cambridge University Press), *Dictionary of Scientific Biography* (Charles Scribner's Sons, New York), *Neue Deutsche Biographie* (Helmholtz; Duncker and Humblot, Berlin) and *The London Times* (Havelock). These productive and influential individuals are but a few of those who have laid the foundations of our present-day knowledge; however, the biographies illustrate the level of effort and intensity of those people and their eras, through which great scientific strides were made.

The authors wish to acknowledge the stimulating discussions and inspiration provided by many of their colleagues and former professors. In particular, Professors R. O. Reid, B. W. Wilson, A. T. Ippen, and C. L. Bretschneider were central in introducing the authors to the field. Numerous focused discussions with M. P. O'Brien have crystallized understanding of water wave phenomena and their effects on sediment transport. Drs. Todd L. Walton and Ib A. Svendsen provided valuable reviews of the manuscript, as have a number of students who have taken the Water Wave Mechanics course at the University of Delaware. Mrs. Sue Thompson deserves great praise for her cheerful disposition and faultless typing of numerous drafts of the manuscript, as does Mrs. Connie Weber, who managed final revision.

Finally the general support and encouragement provided by the University of Delaware is appreciated.

Robert G. Dean
Robert A. Dalrymple

1

Introduction to Wave Mechanics

Dedication
SIR HORACE LAMB

Sir Horace Lamb (1849–1934) is best known for his extremely thorough and well-written book, *Hydrodynamics*, which first appeared in 1879 and has been reprinted numerous times. It still serves as a compendium of useful information as well as the source for a great number of papers and books. If this present book has but a small fraction of the appeal of *Hydrodynamics*, the authors would be well satisfied.

Sir Horace Lamb was born in Stockport, England in 1849, educated at Owens College, Manchester, and then Trinity College, Cambridge University, where he studied with professors such as J. Clerk Maxwell and G. G. Stokes. After his graduation, he lectured at Trinity (1822–1825) and then moved to Adelaide, Australia, to become Professor of Mathematics.

After ten years, he returned to Owens College (part of Victoria University of Manchester) as Professor of Pure Mathematics; he remained until 1920.

Professor Lamb was noted for his excellent teaching and writing abilities. In response to a student tribute on the occasion of his eightieth birthday, he replied: "I did try to make things clear, first to myself...and then to my students, and somehow make these dry bones live."

His research areas encompassed tides, waves, and earthquake properties as well as mathematics.

1.1 INTRODUCTION

Rarely can one find a body of water open to the atmosphere that does not have waves on its surface. These waves are a manifestation of forces acting on the fluid tending to deform it against the action of gravity and surface tension, which together act to maintain a level fluid surface. Thus it requires a force of some kind, such as would be caused by a gust of wind or a falling stone impacting on the water, to create waves. Once these are created, gravitational and surface tension forces are activated that allow the waves to propagate, in the same manner as tension on a string causes the string to vibrate, much to our listening enjoyment.

Waves occur in all sizes and forms, depending on the magnitude of the forces acting on the water. A simple illustration is that a small stone and a large rock create different-size waves after impacting on water. Further, different speeds of impact create different-size waves, which indicates that the pressure forces acting on the fluid surface are important, as well as the magnitude of the displaced fluid. The gravitational attraction of the moon, sun, and other astronomical bodies creates the longest known water waves, the tides. These waves circle halfway around the earth from end to end and travel with tremendous speeds. The shortest waves can be less than a centimeter in length. The length of the wave gives one an idea of the magnitude of the forces acting on the waves. For example, the longer the wave, the more important gravity (comprised of the contributions from the earth, the moon, and the sun) is in relation to surface tension.

The importance of waves cannot be overestimated. Anything that is near or in a body of water is subject to wave action. At the coast, this can result in the movement of sand along the shore, causing erosion or damage to structures during storms. In the water, offshore oil platforms must be able to withstand severe storms without destruction. At present drilling depths exceeding 300 m, this requires enormous and expensive structures. On the water, all ships are subjected to wave attack, and countless ships have foundered due to waves which have been observed to be as large as 34 m in height. Further, any ship moving through water creates a pressure field and, hence, waves. These waves create a significant portion of the resistance to motion enountered by the ships.

1.2 CHARACTERISTICS OF WAVES

The important parameters to describe waves are their length and height, and the water depth over which they are propagating. All other parameters, such as wave-induced water velocities and accelerations, can be determined theoretically from these quantities. In Figure 1.1, a two-dimensional schematic of a wave propagating in the x direction is shown. The length of the wave,

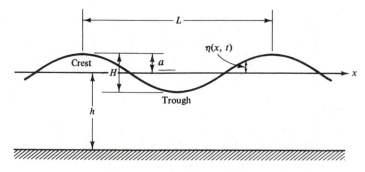

Figure 1.1   Wave characteristics.

L, is the horizontal distance between two successive wave crests, or the high points on a wave, or alternatively the distance between two wave troughs. The wave length will be shown later to be related to the water depth h and wave period T, which is the time required for two successive crests or troughs to pass a particular point. As the wave, then, must move a distance L in time T, the speed of the wave, called the celerity, C, is defined as $C = L/T$. While the wave form travels with celerity C, the water that comprises the wave does not translate in the direction of the wave.

The coordinate axis that will be used to describe wave motion will be located at the still water line, $z = 0$. The bottom of the water body will be at $z = -h$.

Waves in nature rarely appear to look exactly the same from wave to wave, nor do they always propagate in the same direction. If a device to measure the water surface elevation, η, as a function of time was placed on a platform in the middle of the ocean, it might obtain a record such as that shown in Figure 1:2. This sea can be seen to be a superposition of a large number of sinusoids going in different directions. For example, consider the two sine waves shown in Figure 1.3 and their sum. It is this superposition of sinusoids that permits the use of Fourier analysis and spectral techniques to be used in describing the sea. Unfortunately, there is a great amount of randomness in the sea, and statistical techniques need to be brought to bear. Fortunately, very large waves or, alternatively, waves in shallow water appear

Figure 1.2   Example of a possible recorded wave form.

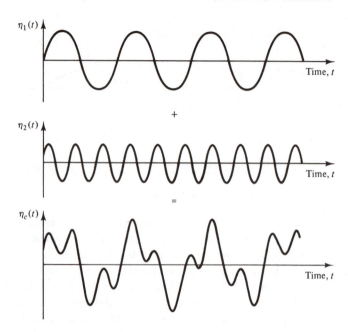

Figure 1.3 Complex wave form resulting as the sum of two sinusoids.

to be more regular than smaller waves or those in deeper water, and not so random. Therefore, in these cases, each wave is more readily described by one sinusoid, which repeats itself periodically. Realistically, due to shallow water nonlinearities, more than one sinusoid, all of the same phase, are necessary; however, using one sinusoid has been shown to be reasonably accurate for some purposes. It is this surprising accuracy and ease of application that have maintained the popularity and the widespread usage of so-called linear, or small-amplitude, wave theory. The advantages are that it is easy to use, as opposed to more complicated nonlinear theories, and lends itself to superposition and other complicated manipulations. Moreover, linear wave theory is an effective stepping-stone to some nonlinear theories. For this reason, this book is directed primarily to linear theory.

1.3 HISTORICAL AND PRESENT LITERATURE

The field of water wave theory is over 150 years old and, of course, during this period of time numerous books and articles have been written about the subject. Perhaps the most outstanding is the seminal work of Sir Horace Lamb. His *Hydrodynamics* has served as a source book since its original publication in 1879.

Other notable books with which the reader should become acquainted are R. L. Wiegel's *Oceanographical Engineering* and A. T. Ippen's *Estuary and Coastline Hydrodynamics.* These two books, appearing in the 1960s, provided the education of many of the practicing coastal and ocean engineers of today.

The authors also recommend for further studies on waves the book by G. B. Witham entitled *Linear and Nonlinear Waves*, from which a portion of Chapter 11 is derived, and the article "Surface Waves," by J. V. Wehausen and E. V. Laitone, in the *Handbuch der Physik.*

In terms of articles, there are a number of journals and proceedings that will provide the reader with more up-to-date material on waves and wave theory and its applications. These include the American Society of Civil Engineers' *Journal of Waterway, Port, Coastal and Ocean Division,* the *Journal of Fluid Mechanics,* the *Proceedings of the International Coastal Engineering Conferences,* the *Journal of Geophysical Research, Coastal Engineering, Applied Ocean Research,* and the *Proceedings of the Offshore Technology Conference.*

2

A Review of Hydrodynamics and Vector Analysis

Dedication
LEONHARD EULER

Leonhard Euler (1707–1783), born in Basel, Switzerland, was one of the earliest practitioners of applied mathematics, developing with others the theory of ordinary and partial differential equations and applying them to the physical world. The most frequent use of his work here is the use of the Euler equations of motion, which describe the flow of an inviscid fluid.

In 1722 he graduated from the University of Basel with a degree in Arts. During this time, however, he attended the lectures of Johan I. Bernoulli (Daniel Bernoulli's father), and turned to the study of mathematics. In 1723 he received a master's level degree in philosophy and began to teach in the philosophy department. In 1727 he moved to St. Petersburg, Russia, and to the St. Petersburg Academy of Science, where he worked in physiology and mathematics and succeeded Daniel Bernoulli as Professor of Physics in 1731.

In 1741 he was invited to work in the Berlin Society of Sciences (founded by Leibniz). Some of his work there was applied as opposed to theoretical. He worked on the hydraulic works of Frederick the Great's summer residence as well as in ballistics, which was of national interest. In Berlin he published 380 works related to mathematical physics in such areas as geometry, optics, electricity, and magnetism. In 1761 he published his monograph, "Principia motus fluidorum," which put forth the now-familiar Euler and continuity equations.

He returned to St. Petersburg in 1766 after a falling-out with

Frederick the Great and began to depend on coauthors for a number of his works, as he was going blind. He died there in 1783.

In mathematics, Euler was responsible for introducing numerous notations: for example, $i \equiv \sqrt{-1}$, e for base of the natural log, and the finite difference Δx.

2.1 INTRODUCTION

In order to investigate water waves most effectively, a reasonably good background in fluid dynamics and mathematics is helpful. Although it is anticipated that the reader has this background, a review of the essential derivations and equations is offered here as a refresher and to acquaint the reader with the notation to be used throughout the book.

A mathematical tool that will be used often is the Taylor series. Mathematically, it can be shown that if a continuous function $f(x, y)$ of two independent variables x and y is known at, say, x equal to x_0, then it can be approximated at another location on the x axis, $x_0 + \Delta x$, by the Taylor series.

$$f(x_0 + \Delta x, y) = f(x_0, y) + \frac{\partial f(x_0, y)}{\partial x}\Delta x + \frac{\partial^2 f(x_0, y)}{\partial x^2}\frac{(\Delta x)^2}{2!} \qquad (2.1)$$

$$+ \cdots + \frac{\partial^n f(x_0, y)}{\partial x^n}\frac{(\Delta x)^n}{n!} + \cdots$$

where the derivatives of $f(x, y)$ are all taken at $x = x_0$, the location for which the function is known. For very small values of Δx, the terms involving $(\Delta x)^n$, where $n > 1$, are very much smaller than the first two terms on the right-hand side of the equation and often in practice can be neglected. If $f(x, y)$ varies linearly with x, for example, $f(x, y) = y^2 + mx + b$, truncating the Taylor series to two terms involves no error, for all values of Δx.[1] Through the use of the Taylor series, it is possible to develop relationships between fluid properties at two closely spaced locations.

2.2 REVIEW OF HYDRODYNAMICS

2.2.1 Conservation of Mass

In a real fluid, mass must be conserved; it cannot be created or destroyed. To develop a mathematical equation to express this concept, consider a very small cube located with its center at x, y, z in a Cartesian coordinate system as shown in Figure 2.1. For the cube with sides Δx, Δy, and

---

[1]In fact, for any nth-order function, the expression (2.1) is exact as long as $(n + 1)$ terms in the series are obtained.

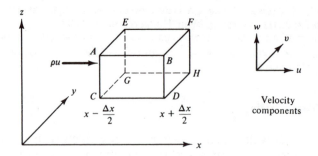

Figure 2.1   Reference cube in a fluid.

Δz, the rate at which fluid mass flows into the cube across the various faces must equal the sum of the rate of mass accumulation in the cube and the mass fluxes out of the faces.

Taking first the x face at $x - \Delta x/2$, the rate at which the fluid mass flows in is equal to the velocity component in the x direction times the area through which it is crossing, all multiplied times the density of the fluid, ρ. Therefore, the mass inflow rate at $x - \Delta x/2$, or side $ACEG$, is

$$\rho(x - \frac{\Delta x}{2}, y, z)u(x - \frac{\Delta x}{2}, y, z)\, \Delta y\, \Delta z \tag{2.2}$$

where the terms in parentheses denote the coordinate location.

This mass flow rate can be related to that at the center of the cube by the truncated Taylor series, keeping in mind the smallness of the cube,

$$\rho(x - \frac{\Delta x}{2}, y, z)u(x - \frac{\Delta x}{2}, y, z)\, \Delta y\, \Delta z \tag{2.3}$$

$$= \left[\rho(x, y, z)u(x, y, z) - \frac{\partial(\rho u)}{\partial x}\frac{\Delta x}{2} + \cdots \right]\Delta y\, \Delta z$$

For convenience, the coordinates of ρ and u at the center of the cube will not be shown hereafter. The mass flow rate out of the other x face, at $x + \Delta x/2$, face $BDFH$, can also be represented by the Taylor series,

$$\left[\rho u + \frac{\partial(\rho u)}{\partial x}\frac{\Delta x}{2} + \cdots \right]\Delta y\, \Delta z \tag{2.4}$$

By subtracting the mass flow rate out from the mass flow rate in, the net flux of mass into the cube in the x direction is obtained, that is, the rate of mass accumulation in the x direction:

$$-\frac{\partial(\rho u)}{\partial x}\Delta x\, \Delta y\, \Delta z + O(\Delta x)^4 \tag{2.5}$$

where the term $O(\Delta x)^4$ denotes terms of higher order, or power, than $(\Delta x)^3$

and is stated as "order of $(\Delta x)^4$." This term is a result of neglected higher-order terms in the Taylor series and implicitly assumes that Δx, Δy, and Δz are the same order of magnitude. If the procedure is followed for the y and z directions, their contributions will also be obtained. The net rate of mass accumulation inside the control volume due to flux across all six faces is

$$-\left[\frac{\partial(\rho u)}{\partial x} + \frac{\partial(\rho v)}{\partial y} + \frac{\partial(\rho w)}{\partial z}\right]\Delta x\,\Delta y\,\Delta z + O(\Delta x)^4 \tag{2.6}$$

Let us now consider this accumulation of mass to occur for a time increment Δt and evaluate the increase in mass within the volume. The mass of the volume at time t is $\rho(t)\,\Delta x\,\Delta y\,\Delta z$ and at time $(t + \Delta t)$ is $\rho(t + \Delta t)\,\Delta x\,\Delta y\,\Delta z$. The increase in mass is therefore

$$[\rho(t + \Delta t) - \rho(t)]\,\Delta x\,\Delta y\,\Delta z = \left[\frac{\partial\rho}{\partial t}\,\Delta t + O(\Delta t)^2\right]\Delta x\,\Delta y\,\Delta z \tag{2.7}$$

where $O(\Delta t)^2$ represents the higher-order terms in the Taylor series. Since mass must be conserved, this increase in mass must be due to the net inflow rate [Eq. (2.6)] occurring over a time increment Δt, that is,

$$\left[\frac{\partial\rho}{\partial t}\,\Delta t + O(\Delta t)^2\right]\Delta x\,\Delta y\,\Delta z$$

$$= -\left[\frac{\partial(\rho u)}{\partial x} + \frac{\partial(\rho v)}{\partial y} + \frac{\partial(\rho w)}{\partial z}\right]\Delta x\,\Delta y\,\Delta z\,\Delta t + O(\Delta x)^4\,\Delta t \tag{2.8}$$

Dividing both sides by $\Delta x\,\Delta y\,\Delta z\,\Delta t$ and allowing the time increment and size of the volume to approach zero, the following *exact* equation results:

$$\frac{\partial\rho}{\partial t} + \frac{\partial\rho u}{\partial x} + \frac{\partial\rho v}{\partial y} + \frac{\partial\rho w}{\partial z} = 0 \tag{2.9}$$

By expanding the product terms, a different form of the continuity equation can be derived.

$$\frac{1}{\rho}\left(\frac{\partial\rho}{\partial t} + u\frac{\partial\rho}{\partial x} + v\frac{\partial\rho}{\partial y} + w\frac{\partial\rho}{\partial z}\right) + \frac{\partial u}{\partial x} + \frac{\partial v}{\partial y} + \frac{\partial w}{\partial z} = 0 \tag{2.10}$$

Recalling the definition for the total derivative from the calculus, the term within brackets can be seen to be the total derivative[2] of $\rho(x, y, z, t)$ with respect to time, $D\rho/Dt$ or $d\rho/dt$, given $u = dx/dt$, $v = dy/dt$, and $w = dz/dt$. The first term is then $(1/\rho)(d\rho/dt)$ and is related to the change in pressure through the bulk modulus E of the fluid, where

$$E \equiv \rho\frac{dp}{d\rho} \tag{2.11}$$

[2]This is discussed later in the chapter.

where dp is the incremental change in pressure, causing the compression of the fluid. Thus

$$\frac{1}{\rho}\frac{d\rho}{dt} = \frac{1}{E}\frac{dp}{dt} \qquad (2.12)$$

For water, $E = 2.07 \times 10^9\ \text{Nm}^{-2}$, a very large number. For example, a $1 \times 10^6\ \text{Nm}^{-2}$ increase in pressure results in a 0.05% change in density of water. Therefore, it will be assumed henceforth that water is incompressible.

From Eq. (2.10), the conservation of mass equation for an *incompressible* fluid can be stated simply as

$$\boxed{\frac{\partial u}{\partial x} + \frac{\partial v}{\partial y} + \frac{\partial w}{\partial z} = 0} \qquad (2.13)$$

which must be true at every location in the fluid. This equation is also referred to as the *continuity* equation, and the flow field satisfying Eq. (2.13) is termed a "nondivergent flow." Referring back to the cube in Figure 2.1, this equation requires that if there is a change in the flow in a particular direction across the cube, there must be a corresponding flow change in another direction, to ensure no fluid accumulation in the cube.

Example 2.1

An example of an incompressible flow is accelerating flow into a corner in two dimensions, as shown in Figure 2.2 The velocity components are $u = -Axt$ and $w = Azt$. To determine if it is an incompressible flow, substitute the velocity components into the continuity equation, $-At + At = 0$. Therefore, it is incompressible.

2.2.2 Surface Stresses on a Particle

The motion of a fluid particle is induced by the forces that act on the particle. These forces are of two types, as can be seen if we again refer to the fluid cube that was utilized in the preceding section. Surface forces include pressure and shear stresses which act on the surface of the volume. Body forces, on the other hand, act throughout the volume of the cube. These forces

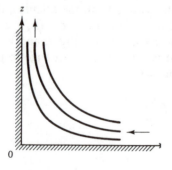

Figure 2.2   Fluid flow in a corner. Flow is tangent to solid lines.

include gravity, magnetic, and other forces that act directly on each individual particle in the volume under consideration.

All of these forces which act on the cube of fluid will cause it to move as predicted by Newton's second law, $\mathbf{F} = m\mathbf{a}$, for a volume of constant mass m. This law, which relates the resultant forces on a body to its resultant acceleration $\mathbf{a}$, is a vector equation, being made up of forces and accelerations in the x, y, and z coordinate directions, and therefore all forces for convenience must be resolved into their components.

Hydrostatic pressure.   By definition, a fluid is a substance distinguished from solids by the fact that it deforms continuously under the action of *shear* stresses. This deformation occurs by the fluid's flowing. Therefore, for a still fluid, there are no shear stresses and the normal stresses or forces must balance each other, $\mathbf{F} = 0$. Normal (perpendicular) stresses must be present because we know that a fluid column has a weight and this weight must be supported by a pressure times the area of the column. Using this static force balance, we will show first that the pressure is the same in all directions (i.e., a scalar) and then derive the hydrostatic pressure relationship.

For a container of fluid, as illustrated in Figure 2.3a, the only forces that act are gravity and hydrostatic pressure. If we first isolate a stationary prism of fluid with dimensions Δx, Δz, Δl [$= \sqrt{(\Delta x)^2 + (\Delta z)^2}$], we can examine the force balance on it. We will only consider the x and z directions for now; the forces in the y direction do not contribute to the x direction.

On the left side of the prism, there is a pressure force acting in the positive x direction, $p_x \, \Delta z \, \Delta y$. On the diagonal face, there must be a balanc-

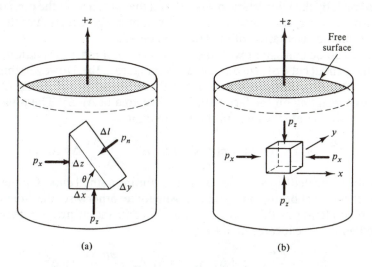

(a)   (b)

Figure 2.3   Hydrostatic pressures on (a) a prism and (b) a cube.

ing component of p_n, which yields the following form of Newton's second law:

$$p_x \, \Delta z \, \Delta y = p_n \sin \theta \, \Delta l \, \Delta y \qquad (2.14)$$

In the vertical direction, the force balance yields

$$p_z \, \Delta x \, \Delta y = p_n \cos \theta \, \Delta l \, \Delta y + \tfrac{1}{2} \rho g \, \Delta z \, \Delta x \, \Delta y \qquad (2.15)$$

where the second term on the right-hand side corresponds to the weight of the prism, which also must be supported by the vertical pressure force. From the geometry of the prism, $\sin \theta = \Delta z/\Delta l$ and $\cos \theta = \Delta x/\Delta l$, and after substitution we have

$$p_x = p_n$$

$$p_z = p_n + \tfrac{1}{2} \rho g \, \Delta z$$

If we let the prism shrink to zero, then

$$p_x = p_z = p_n$$

which indicates that the pressures in the x-z plane are the same at a point irrespective of the orientation of the prism's diagonal face, since the final equations do not involve the angle θ. This result would still be valid, of course, if the prism were oriented along the y axis, and thus we conclude at a point,

$$p_x = p_y = p_z \qquad (2.16)$$

or, the pressure at a point is independent of direction. An important point to notice is that the pressure is not a vector; it is a scalar and thus has no direction associated with it. Any surface immersed in a fluid will have a force exerted on it by the hydrostatic pressure, and the force acts in the direction of the normal, or perpendicular to the surface; that is, the direction of the force depends on the orientation of the face considered.

Now, to be consistent with the conservation of mass derivation, let us examine a small cube of size Δx, Δy, Δz (see Figure 2.3b). However, this time we will not shrink the cube to a point. On the left-hand face at $x - \Delta x/2$ there is a pressure acting on the face with a surface area of $\Delta y \, \Delta z$. The total force tending to accelerate the cube in the $+x$ direction is

$$p\left(x - \frac{\Delta x}{2}, y, z \right) \Delta y \, \Delta z = p(x, y, z) \, \Delta y \, \Delta z - \frac{\partial p}{\partial x} \frac{\Delta x}{2} \Delta y \, \Delta z + \cdots \quad (2.17)$$

where the truncated Taylor series is used, assuming a small cube. On the other x face, there must be an equal and opposite force; otherwise, the cube would have to accelerate in this direction. The force in the minus x direction is exerted on the face located at $x + \Delta x/2$.

$$p\left(x + \frac{\Delta x}{2}, y, z \right) \Delta y \, \Delta z = p \, \Delta y \, \Delta z + \frac{\partial p}{\partial x} \frac{\Delta x}{2} \Delta y \, \Delta z \qquad (2.18)$$

Equating the two forces yields

$$\frac{\partial p}{\partial x} = 0 \tag{2.19}$$

For the y direction, a similar result is obtained,

$$\frac{\partial p}{\partial y} = 0$$

In the vertical, z, direction the force acting upward is

$$p\left(x, y, z - \frac{\Delta z}{2}\right) \Delta x\, \Delta y = \left(p - \frac{\partial p}{\partial z}\frac{\Delta z}{2}\right) \Delta x\, \Delta y \tag{2.20}$$

which must be equal to the pressure force acting downward, and the weight of the cube, $\rho g\, \Delta x\, \Delta y\, \Delta z$, where g is the acceleration of gravity.

Summing these forces yields

$$+\frac{\partial p}{\partial z}\frac{\Delta z}{2} \Delta x\, \Delta y = -\frac{\partial p}{\partial z}\frac{\Delta z}{2} \Delta x\, \Delta y - \rho g\, \Delta x\, \Delta y\, \Delta z \tag{2.21}$$

or dividing by the volume of the small cube, we have

$$\frac{\partial p}{\partial z} = -\rho g \tag{2.22}$$

Integrating the three partial differential equations for the pressure results in the hydrostatic pressure equation

$$p = -\rho g z + C \tag{2.23}$$

Evaluating the constant C at the free surface, $z = 0$, where $p = 0$ (gage pressure),

$$p = -\rho g z \tag{2.24}$$

The pressure increases linearly with increasing depth into the fluid.[3]

The *buoyancy* force is just a result of the hydrostatic pressure acting over the surface of a body. In a container of fluid, imagine a small sphere of fluid that could be denoted by some means such as dye. The spherical boundaries of this fluid would be acted upon by the hydrostatic pressure, which would be greater at the bottom of the sphere, as it is deeper there, than at the top of the sphere. The sphere does not move because the pressure difference supports the weight of the sphere. Now, if we could remove the fluid sphere and replace it with a sphere of lesser density, the same pressure forces would exist at its surface, yet the weight would be less and therefore the hydrostatic force would push the object upward. Intuitively, we would say

---

[3]Note that z is negative into the fluid and therefore Eq. (2.24) does yield positive pressure underwater.

that the buoyancy force due to the fluid pressure is equal to the weight of the fluid displaced by the object. To examine this, let us look again at the force balance in the z direction, Eq. (2.21):

$$-\frac{\partial p}{\partial z}\,\Delta z\,\Delta x\,\Delta y = \rho g\,\Delta x\,\Delta y\,\Delta z = \rho g\,\Delta V = dF_z \qquad (2.25)$$

which states that the net force in the z direction for the incremental area $\Delta x\,\Delta y$ equals the weight of the incremental volume of fluid delimited by that area. There is no restriction on the size of the cube due to the linear variation of hydrostatic pressure.

If we now integrate the pressure force over the surface of the object, we obtain

$$F_{\text{buoyancy}} = \rho g V \qquad (2.26)$$

The buoyancy force is equal to the weight of the fluid displaced by the object, as discovered by Archimedes in about 250 B.C., and is in the positive z (vertical) direction (and it acts through the center of gravity of the displaced fluid).

Shear stresses. Shear stresses also act on the surface; however, they differ from the pressure in that they are not isotropic. Shear stresses are caused by forces acting tangentially to a surface; they are always present in a real flowing fluid and, as pressures, have the units of force per unit area.

If we again examine our small volume (see Figure 2.4), we can see that there are three possible stresses for each of the six faces of the cube; two shear stresses and a normal stress, perpendicular to the face. Any other arbitrarily oriented stress can always be expressed in terms of these three. On the x face at $x + \Delta x/2$ which will be designated the positive x face, the stresses are σ_{xx}, τ_{xy}, and τ_{xz}. The notation convention for stresses is that the first subscript

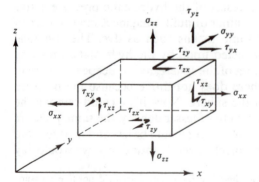

Figure 2.4   Shear and normal stresses on a fluid cube.

refers to the axis to which the face is perpendicular and the second to the direction of the stress. For a positive face, the stresses point in the positive axes directions. For the negative x face at $x - \Delta x/2$, the stresses are again σ_{xx}, τ_{xy}, and τ_{xz}, but they point in the direction of *negative* x, y, and z, respectively.[4] Although these stresses have the same designation as those in the positive x face, in general they will differ in magnitude. In fact, it is the difference in magnitude that leads to a net force on the cube and a corresponding acceleration.

There are nine stresses that are exerted on the cube faces. Three of these stresses include the pressure, as the normal stresses are written as

$$\sigma_{xx} = -p + \tau_{xx}$$

$$\sigma_{yy} = -p + \tau_{yy} \tag{2.27}$$

$$\sigma_{zz} = -p + \tau_{zz}$$

where
$$p \equiv -\left(\frac{\sigma_{xx} + \sigma_{yy} + \sigma_{zz}}{3} \right)$$

for both still and flowing fluids. It is possible, however, to show that some of the shear stresses are identical. To do this we use Newton's second law as adapted to moments and angular momentum. If we examine the moments about the z axis, we have

$$M_z = I_z \dot{\omega}_z \tag{2.28}$$

where M_z is the sum of the moments about the z axis, I_z is the moment of inertia, and $\dot{\omega}_z$ is the z component of the angular acceleration of the body. The moments about an axis through the center of the cube, parallel to the z axis, can be readily identified if a slice is taken through the fluid cube perpendicularly to the z axis. This is shown in Figure 2.5. Considering moments about the center of the element and positive in the clockwise direction, Eq. (2.28) is written, in terms of the stresses existing at the center of

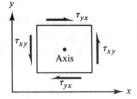

Figure 2.5 Shear stresses contributing to moments about the z-axis. Note that τ_{xy}, τ_{yx} are functions of x and y.

[4]Can you identify the missing stresses on the $(y - \Delta y/2)$ face and orient them correctly?

the cube,

$$\left(\tau_{yx} - \frac{\partial \tau_{yx}}{\partial y}\frac{\Delta y}{2}\right)\Delta x\,\Delta z\,\frac{\Delta y}{2} + \left(\tau_{yx} + \frac{\partial \tau_{yx}}{\partial y}\frac{\Delta y}{2}\right)\Delta x\,\Delta z\,\frac{\Delta y}{2} \qquad (2.29)$$

$$- \left(\tau_{xy} + \frac{\partial \tau_{xy}}{\partial x}\frac{\Delta x}{2}\right)\Delta y\,\Delta z\,\frac{\Delta x}{2} - \left(\tau_{xy} - \frac{\partial \tau_{xy}}{\partial x}\frac{\Delta x}{2}\right)\Delta y\,\Delta z\,\frac{\Delta x}{2} = I_z\dot{\omega}_z$$

Reducing the equation leaves

$$\tau_{yx}\,\Delta x\,\Delta y\,\Delta z - \tau_{xy}\,\Delta x\,\Delta y\,\Delta z = \tfrac{1}{12}\rho[\Delta x\,\Delta y\,\Delta z\,(\Delta x^2 + \Delta y^2)]\dot{\omega}_z \qquad (2.30)$$

For a nonzero difference, on the left-hand side, as the cube is taken to be smaller and smaller, the acceleration $\dot{\omega}_z$ must become greater, as the moment of inertia involves terms of length to the fifth power, whereas the stresses involve only the length to the third power. Therefore, in order that the angular acceleration of the fluid particle not unrealistically be infinite as the cube reduces in size, we conclude that $\tau_{yx} = \tau_{xy}$ (i.e., the two shear stresses must be equal). Further, similar logic will show that $\tau_{xz} = \tau_{zx}$, $\tau_{yz} = \tau_{zy}$. Therefore, there are only six unknown stresses (σ_{xx}, τ_{xy}, τ_{xz}, τ_{yz}, σ_{yy}, and σ_{zz}) on the element. These stresses depend on parameters such as fluid viscosity and fluid turbulence and will be discussed later.

2.2.3 The Translational Equations of Motion

For the x direction, Newton's second law is, again, $\Sigma F_x = ma_x$, where a_x is the particle acceleration in the x direction. By definition $a_x = du/dt$, where u is the velocity in the x direction. This velocity, however, is a function of space and time, $u = u(x, y, z, t)$; therefore, its total derivative is

$$\frac{du}{dt} = \frac{\partial u}{\partial t} + \frac{\partial u}{\partial x}\frac{dx}{dt} + \frac{\partial u}{\partial y}\frac{dy}{dt} + \frac{\partial u}{\partial z}\frac{dz}{dt} \qquad (2.31)$$

or, since dx/dt is u, and so forth,

$$\frac{du}{dt} = \frac{\partial u}{\partial t} + u\frac{\partial u}{\partial x} + v\frac{\partial u}{\partial y} + w\frac{\partial u}{\partial z} \qquad (2.32)$$

This is the total acceleration and will be denoted as Du/Dt. The derivative is composed of two types of terms, the local acceleration, $\partial u/\partial t$, which is the change of u observed at a point with time, and the convective acceleration terms

$$u\frac{\partial u}{\partial x} + v\frac{\partial u}{\partial y} + w\frac{\partial u}{\partial z}$$

which are the changes of u that result due to the motion of the particle. For

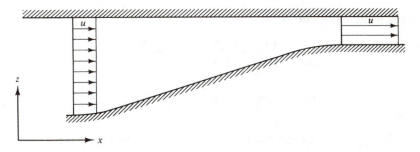

Figure 2.6   Acceleration of flow through a convergent section.

example, if we follow a water particle in a steady flow (i.e., a flow which is independent of time so that $\partial u/\partial t = 0$) into a transition section as shown in Figure 2.6, it is clear that the fluid accelerates. The important terms applicable to the figure are the $u\,\dfrac{\partial u}{\partial x}$ and the $w\,\dfrac{\partial u}{\partial z}$ terms.

The equation of motion in the x direction can now be formulated:

$$\Sigma\,F_x = m\frac{Du}{Dt}$$

From Figure 2.4, the surface forces can be obtained on the six faces via the truncated Taylor series

$$\left(\sigma_{xx} + \frac{\partial \sigma_{xx}}{\partial x}\frac{\Delta x}{2}\right)\Delta y\,\Delta z - \left(\sigma_{xx} - \frac{\partial \sigma_{xx}}{\partial x}\frac{\Delta x}{2}\right)\Delta y\,\Delta z + \left(\tau_{yx} + \frac{\partial \tau_{yx}}{\partial y}\frac{\Delta y}{2}\right)\Delta x\,\Delta z$$

$$-\left(\tau_{yx} - \frac{\partial \tau_{yx}}{\partial y}\frac{\Delta y}{2}\right)\Delta x\,\Delta z + \left(\tau_{zx} + \frac{\partial \tau_{zx}}{\partial z}\frac{\Delta z}{2}\right)\Delta x\,\Delta y \qquad (2.33)$$

$$-\left(\tau_{zx} - \frac{\partial \tau_{zx}}{\partial z}\frac{\Delta z}{2}\right)\Delta x\,\Delta y + \rho\,\Delta x\,\Delta y\,\Delta z X = \rho\,\Delta x\,\Delta y\,\Delta z\,\frac{Du}{Dt}$$

The capital X denotes any body force per unit mass acting in the x direction. Combining terms and dividing by the volume of the cube yields

$$\rho\frac{Du}{Dt} = \frac{\partial \sigma_{xx}}{\partial x} + \frac{\partial \tau_{yx}}{\partial y} + \frac{\partial \tau_{zx}}{\partial z} + \rho X \qquad (2.34)$$

or

$$\frac{Du}{Dt} = -\frac{1}{\rho}\frac{\partial p}{\partial x} + \frac{1}{\rho}\left(\frac{\partial \tau_{xx}}{\partial x} + \frac{\partial \tau_{yx}}{\partial y} + \frac{\partial \tau_{zx}}{\partial z}\right) + X \qquad (2.35)$$

and, by exactly similar developments, the equations of motion are obtained for the y and z directions:

$$\frac{Dv}{Dt} = -\frac{1}{\rho}\frac{\partial p}{\partial y} + \frac{1}{\rho}\left(\frac{\partial \tau_{xy}}{\partial x} + \frac{\partial \tau_{yy}}{\partial y} + \frac{\partial \tau_{zy}}{\partial z}\right) + Y \qquad (2.36)$$

$$\frac{Dw}{Dt} = -\frac{1}{\rho}\frac{\partial p}{\partial z} + \frac{1}{\rho}\left(\frac{\partial \tau_{xz}}{\partial x} + \frac{\partial \tau_{yz}}{\partial y} + \frac{\partial \tau_{zz}}{\partial z}\right) + Z \qquad (2.37)$$

To apply the equations of motion for a fluid particle, it is necessary to know something about stresses in a fluid. The most convenient assumption, one that is reasonably valid for most problems in water wave mechanics, is that the shear stresses are zero, which results in the Euler equations. Expressing the body force per unit mass as $-g$ in the z direction and zero in the x and y directions, we have

$$\frac{Du}{Dt} = -\frac{1}{\rho}\frac{\partial p}{\partial x} \qquad (2.38a)$$

$$\frac{Dv}{Dt} = -\frac{1}{\rho}\frac{\partial p}{\partial y} \qquad \text{the Euler equations} \qquad (2.38b)$$

$$\frac{Dw}{Dt} = -\frac{1}{\rho}\frac{\partial p}{\partial z} - g \qquad (2.38c)$$

In many real flow cases, the flow is turbulent and shear stresses are influenced by the turbulence and thus the previous stress terms must be retained. If the flow is laminar, that is there is no turbulence in the fluid, the stresses are governed by the Newtonian shear stress relationship and the accelerations are governed by

$$\frac{Du}{Dt} = -\frac{1}{\rho}\frac{\partial p}{\partial x} + \frac{\mu}{\rho}\left(\frac{\partial^2 u}{\partial x^2} + \frac{\partial^2 u}{\partial y^2} + \frac{\partial^2 u}{\partial z^2}\right) + X \qquad (2.39a)$$

$$\frac{Dv}{Dt} = -\frac{1}{\rho}\frac{\partial p}{\partial y} + \frac{\mu}{\rho}\left(\frac{\partial^2 v}{\partial x^2} + \frac{\partial^2 v}{\partial y^2} + \frac{\partial^2 v}{\partial z^2}\right) + Y \qquad (2.39b)$$

$$\frac{Dw}{Dt} = -\frac{1}{\rho}\frac{\partial p}{\partial z} + \frac{\mu}{\rho}\left(\frac{\partial^2 w}{\partial x^2} + \frac{\partial^2 w}{\partial y^2} + \frac{\partial^2 w}{\partial z^2}\right) + Z \qquad (2.39c)$$

and μ is the dynamic (molecular) viscosity of the fluid. Often μ/ρ is replaced by ν, defined as the kinematic viscosity.

For turbulent flows, where the velocities and pressure fluctuate about mean values due to the presence of eddies, these equations are modified to describe the mean and the fluctuating quantities separately, in order to facilitate their use. We will not, however, be using these turbulent forms of the equations directly.

2.3 REVIEW OF VECTOR ANALYSIS

Throughout the book, vector algebra will be used to facilitate proofs and minimize required algebra; therefore, the use of vectors and vector analysis is reviewed briefly below.

In a three-dimensional Cartesian coordinate system, a reference system (x, y, z) as has been used before can be drawn (see Figure 2.7). For each coordinate direction, there is a unit vector, that is, a line segment of unit length oriented such that it is directed in the corresponding coordinate direction. These unit vectors are defined as $(\mathbf{i}, \mathbf{j}, \mathbf{k})$ in the (x, y, z) directions. The boldface type denotes vector quantities. Any vector with orientation and a length can be expressed in terms of unit vectors. For example, the vector $\mathbf{a}$ can be represented as

$$\mathbf{a} = a_x\mathbf{i} + a_y\mathbf{j} + a_z\mathbf{k} \tag{2.40}$$

where a_x, a_y, and a_z are the projections of $\mathbf{a}$ on the x, y, and z axes.

2.3.1 The Dot Product

The dot (or inner or scalar) product is defined as

$$\mathbf{a} \cdot \mathbf{b} = |a|\,|b|\,\cos\theta \tag{2.41}$$

where the absolute value sign refers to the magnitude or length of the vectors and θ refers to the angle between them. For the unit vectors, the following identities readily follow:

$$\mathbf{i} \cdot \mathbf{i} = 1$$
$$\mathbf{i} \cdot \mathbf{j} = 0$$
$$\mathbf{i} \cdot \mathbf{k} = 0 \tag{2.42}$$
$$\mathbf{j} \cdot \mathbf{j} = 1$$
$$\mathbf{j} \cdot \mathbf{k} = 0$$
$$\mathbf{k} \cdot \mathbf{k} = 1$$

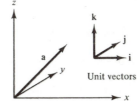

Figure 2.7   Unit vectors in a Cartesian coordinate system.

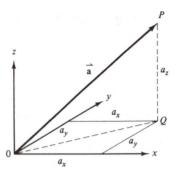

Figure 2.8 Projections of vector **a**.

These rules are commutative, also, so that reversing the order of the operation does not alter the results. For instance,

$$\mathbf{i} \cdot \mathbf{j} = \mathbf{j} \cdot \mathbf{i} \tag{2.43}$$

or $\mathbf{a} \cdot \mathbf{b} = \mathbf{b} \cdot \mathbf{a}$. Consider taking a dot product of the vector with itself.

$$\mathbf{a} \cdot \mathbf{a} = (a_x\mathbf{i} + a_y\mathbf{j} + a_z\mathbf{k}) \cdot (a_x\mathbf{i} + a_y\mathbf{j} + a_z\mathbf{k}) \tag{2.44}$$

$$= a_x^2 + a_y^2 + a_z^2$$

A graphical interpretation of $\mathbf{a} \cdot \mathbf{a}$ can be obtained from Figure 2.8, where the magnitude of vector **a** is the length $\overline{OP}$. From the Pythagorean theorem, $\overline{OP}^2 = \overline{OQ}^2 + \overline{PQ}^2$. But $\overline{PQ}$ is just a_z and $\overline{OQ}^2 = a_x^2 + a_y^2$. Therefore, $\overline{OP}^2 = a_x^2 + a_y^2 + a_z^2$. Therefore, the magnitude of vector **a** can be written as

$$|\mathbf{a}| = \overline{OP} = \sqrt{\mathbf{a} \cdot \mathbf{a}} \tag{2.45}$$

The quantity $\mathbf{a} \cdot \mathbf{b}$ as shown before is a scalar quantity; that is, it has a magnitude, but no direction (therefore, it is not a vector). Another way to express $\mathbf{a} \cdot \mathbf{b}$ is

$$\mathbf{a} \cdot \mathbf{b} = |a| \, |b| \cos \theta = a_x b_x + a_y b_y + a_z b_z \tag{2.46}$$

Note that if $\mathbf{a} \cdot \mathbf{b}$ is zero, but neither a or b is the zero vector, defined as $(0\mathbf{i} + 0\mathbf{j} + 0\mathbf{k})$, then $\cos \theta$ must be zero; the vectors are perpendicular to one another.

An important use of the dot product is in determining the projection of a vector onto another vector. For example, the projection of vector **a** onto the x axis is $\mathbf{a} \cdot \mathbf{i}$. In general, the projection of **a** onto the **b** vector direction would be $\mathbf{a} \cdot \mathbf{b}/|\mathbf{b}|$.

2.3.2 The Cross Product

The cross product (or outer, or vector product) is a *vector quantity* which is defined as $\mathbf{a} \times \mathbf{b} = |a| \, |b| \sin \theta$, but with a direction perpendicular to the plane of **a** and **b** according to the right-hand rule. For the unit vectors,

$$\mathbf{i} \times \mathbf{i} = \mathbf{j} \times \mathbf{j} = \mathbf{k} \times \mathbf{k} = 0; \qquad \mathbf{i} \times \mathbf{j} = \mathbf{k}, \quad \mathbf{j} \times \mathbf{k} = \mathbf{i}, \quad \mathbf{k} \times \mathbf{i} = \mathbf{j} \tag{2.47}$$

but this rule is not commutative. So, for example, $\mathbf{j} \times \mathbf{i} = -\mathbf{k}$. A convenient method for evaluating the cross product of two vectors is to use a determinant form:

$$\mathbf{a} \times \mathbf{b} = \begin{vmatrix} \mathbf{i} & \mathbf{j} & \mathbf{k} \\ a_x & a_y & a_z \\ b_x & b_y & b_z \end{vmatrix} = (a_y b_z - a_z b_y)\mathbf{i} + (a_z b_x - a_x b_z)\mathbf{j} + (a_x b_y - a_y b_x)\mathbf{k} \tag{2.48}$$

If neither $\mathbf{a}$ nor $\mathbf{b}$ is the zero vector, yet their cross product is zero, then from the definition of the cross product, the two vectors must be parallel, or in other words, they are collinear vectors.

2.3.3 The Vector Differential Operator and the Gradient

Consider a scalar field in space; for example, this might be the temperature $T(x, y, z)$ in a room. Because of uneven heating, it is logical to expect that the temperature will vary both with height and horizontal distance into the room. If the temperature and spatial gradients at one point are known, the truncated three-dimensional Taylor series can be used to estimate the temperature at a small distance $d\mathbf{r}$ ($= dx\mathbf{i} + dy\mathbf{j} + dz\mathbf{k}$) away.

$$T(x + \Delta x, y + \Delta y, z + \Delta z) \tag{2.49}$$

$$= T(x, y, z) + \frac{\partial T(x, y, z)}{\partial x} \Delta x + \frac{\partial T(x, y, z)}{\partial y} \Delta y + \frac{\partial T(x, y, z)}{\partial z} \Delta z$$

The last three terms in this expression may be written as the dot product of two vectors:

$$\left(\frac{\partial T}{\partial x}\mathbf{i} + \frac{\partial T}{\partial y}\mathbf{j} + \frac{\partial T}{\partial z}\mathbf{k} \right) \cdot (\Delta x\mathbf{i} + \Delta y\mathbf{j} + \Delta z\mathbf{k}) \tag{2.50}$$

The first term is defined as the gradient of the temperature and the second is the differential vector $\Delta \mathbf{r}$.

The gradient or gradient vector is often written as grad T or ∇T, and can be further broken down to

$$\nabla T = \left(\mathbf{i}\frac{\partial}{\partial x} + \mathbf{j}\frac{\partial}{\partial y} + \mathbf{k}\frac{\partial}{\partial z} \right) T(x, y, z) \tag{2.51}$$

where the first term on the right-hand side is defined as the vector differential operator ∇, and the second, of course, is just the scalar temperature.

The gradient always indicates the direction of maximum change of a scalar field[5] and can be used to indicate perpendicular, or *normal*, vectors to

---

[5] The total differential $dT = \nabla T \cdot d\mathbf{r} = |\nabla T| \, |d\mathbf{r}| \cos \theta$. The maximum value occurs when $d\mathbf{r}$ is in the direction of $|\nabla \mathbf{T}|$.

a surface. For example, if the temperature in a room was stably stratified, the temperature would be solely a function of elevation in the room, or $T(x, y, z) = T(z)$. If we move horizontally across the room to a new point, the change in temperature would be zero, as we have moved along a surface of constant temperature. Therefore,

$$dT = \frac{\partial T}{\partial x}\,dx + \frac{\partial T}{\partial y}\,dy + \frac{\partial T}{\partial z}\,dz = 0 \tag{2.52}$$

where

$$\frac{\partial T}{\partial x} = \frac{\partial T}{\partial y} = 0, \qquad \Delta \mathbf{r} = dx\mathbf{i} + dy\mathbf{j} + 0\mathbf{k} \tag{2.53}$$

or

$$\nabla T \cdot \Delta \mathbf{r} = 0 \tag{2.54}$$

which means, using the definition of the dot product, that ∇T is perpendicular to the surface of constant temperature. The unit normal vector will be defined here as the vector $\mathbf{n}$, having a magnitude of 1 and directed perpendicular to the surface. For this example,

$$\mathbf{n} = \frac{\nabla T}{|\nabla T|} \tag{2.55}$$

or

$$\mathbf{n} = 0\mathbf{i} + 0\mathbf{j} + 1\mathbf{k} = \mathbf{k}$$

2.3.4 The Divergence

If the vector differential operator is applied to a vector using a dot product rather than to a scalar, as in the gradient, we have the divergence

$$\nabla \cdot \mathbf{a} = \left(\mathbf{i}\frac{\partial}{\partial x} + \mathbf{j}\frac{\partial}{\partial y} + \mathbf{k}\frac{\partial}{\partial z} \right) \cdot (a_x\mathbf{i} + a_y\mathbf{j} + a_z\mathbf{k}) \tag{2.56}$$

$$= \frac{\partial a_x}{\partial x} + \frac{\partial a_y}{\partial y} + \frac{\partial a_z}{\partial z}$$

We have already seen this operator in the continuity equation, Eq. (2.10), which can be rewritten as

$$\frac{1}{\rho}\frac{D\rho}{Dt} + \nabla \cdot \mathbf{u} = 0 \tag{2.57}$$

where $\mathbf{u}$ is the velocity vector, $\mathbf{u} = \mathbf{i}u + \mathbf{j}v + \mathbf{k}w$,

$$\nabla \cdot \mathbf{u} = \frac{\partial u}{\partial x} + \frac{\partial v}{\partial y} + \frac{\partial w}{\partial z} \tag{2.58}$$

For an incompressible fluid, for which $(1/\rho) (D\rho/Dt)$ is equal to zero, the divergence of the velocity is also zero, and therefore the fluid is divergence-less. Another useful result may be obtained by taking the divergence of a gradient,

$$\nabla \cdot \nabla T = \left(\mathbf{i} \frac{\partial}{\partial x} + \mathbf{j} \frac{\partial}{\partial y} + \mathbf{k} \frac{\partial}{\partial z} \right) \cdot \left(\mathbf{i} \frac{\partial T}{\partial x} + \mathbf{j} \frac{\partial T}{\partial y} + \mathbf{k} \frac{\partial T}{\partial z} \right)$$

$$= \frac{\partial^2 T}{\partial x^2} + \frac{\partial^2 T}{\partial y^2} + \frac{\partial^2 T}{\partial z^2} \tag{2.59}$$

$$= \nabla^2 T$$

Del squared (∇^2) is known as the Laplacian operator, named after the famous French mathematician Laplace (1749–1827).[6]

2.3.5 The Curl

If the vector differential operator is applied to a vector using the cross product, then the *curl* of the vector results.

$$\nabla \times \mathbf{a} = \left(\mathbf{i} \frac{\partial}{\partial x} + \mathbf{j} \frac{\partial}{\partial y} + \mathbf{k} \frac{\partial}{\partial z} \right) \times (a_x \mathbf{i} + a_y \mathbf{j} + a_z \mathbf{k}) \tag{2.60}$$

Carrying out the cross product, which can be done by evaluating the following determinant, yields

$$\nabla \times \mathbf{a} = \begin{vmatrix} \mathbf{i} & \mathbf{j} & \mathbf{k} \\ \dfrac{\partial}{\partial x} & \dfrac{\partial}{\partial y} & \dfrac{\partial}{\partial z} \\ a_x & a_y & a_z \end{vmatrix} = \left(\frac{\partial a_z}{\partial y} - \frac{\partial a_y}{\partial z} \right) \mathbf{i} + \left(\frac{\partial a_x}{\partial z} - \frac{\partial a_z}{\partial x} \right) \mathbf{j} \tag{2.61}$$

$$+ \left(\frac{\partial a_y}{\partial x} - \frac{\partial a_x}{\partial y} \right) \mathbf{k}$$

As we will see later, the curl of a velocity vector is a measure of the rotation in the velocity field.

As an example of the curl operator, let us determine the divergence of the curl of **a**.

$$\nabla \cdot (\nabla \times \mathbf{a}) = \left(\mathbf{i} \frac{\partial}{\partial x} + \mathbf{j} \frac{\partial}{\partial y} + \mathbf{k} \frac{\partial}{\partial z} \right) \cdot \left[\left(\frac{\partial a_z}{\partial y} - \frac{\partial a_y}{\partial z} \right) \mathbf{i} \right.$$

$$\left. + \left(\frac{\partial a_x}{\partial z} - \frac{\partial a_z}{\partial x} \right) \mathbf{j} + \left(\frac{\partial a_y}{\partial x} - \frac{\partial a_x}{\partial y} \right) \mathbf{k} \right] = 0$$

[6]Chapter 3 is dedicated to Laplace.

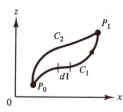

Figure 2.9 Integration paths between two points.

This is an *identity* for any vector that has continuous first and second derivatives.

2.3.6 Line Integrals

In Figure 2.9, two points are shown in the $(x\text{-}y)$ plane, P_0 and P_1. Over this plane the vector $\mathbf{a}(x, y)$ exists. Consider the integral from P_0 to P_1 of the projection of the vector $\mathbf{a}$ on the contour line C_1. We will denote this integral as F:

$$F \equiv \oint_{1 \, P_0}^{P_1} \mathbf{a} \cdot d\mathbf{l} \tag{2.62}$$

It is anticipated that should we have chosen contour C_2, a different value of the integral would have resulted. The question is whether constraints can be prescribed on the nature of $\mathbf{a}$ such that it makes no difference whether we go from P_0 to P_1 on contour C_1 or C_2.

If Eq. (2.62) were rewritten as

$$F = \oint_{1 \, P_0}^{P_1} dF$$

where dF is the exact differential of F, then F would be equal to $F(P_1) - F(P_0)$; that is, it is only a function of the end points of the integration. Therefore, if we can require that $\mathbf{a} \cdot d\mathbf{l}$ be of the form dF, independence of path should ensue. Now,

$$\mathbf{a} \cdot d\mathbf{l} = a_x \, dx + a_z \, dz \quad \text{for two dimensions, as} \quad d\mathbf{l} = dx\mathbf{i} + dz\mathbf{k}$$

and the total differential of F is

$$dF = \frac{\partial F}{\partial x} dx + \frac{\partial F}{\partial z} dz = \nabla F \cdot d\mathbf{l} \tag{2.63}$$

By equating $\mathbf{a} \cdot d\mathbf{l}$ with dF, we see that independence of path requires, in two dimensions,

$$a_x = \frac{\partial F}{\partial x} \quad \text{and} \quad a_z = \frac{\partial F}{\partial z} \quad \text{or} \quad \mathbf{a} = \nabla F \tag{2.64}$$

If this is true for a_x and a_z, it follows that

$$\frac{\partial a_x}{\partial z} - \frac{\partial a_z}{\partial x} = 0 \tag{2.65}$$

as

$$\frac{\partial^2 F}{\partial z \, \partial x} - \frac{\partial^2 F}{\partial x \, \partial z} = 0$$

Therefore, in summary, independence of path of the line integral requires that Eq. (2.65) be satisfied. For three dimensions it can be shown that this condition requires that the curl of **a** must be zero.

Example 2.2

What is the value of

$$F = \oint \mathbf{a} \cdot d\mathbf{l}$$

if $\nabla \times \mathbf{a} = 0$ and where the $\oint$ indicates a complete circuit around the closed contour composed of C_1 and C_2? Do this by parts.

Solution.

$$F = \oint_{P_0}^{P_1} \mathbf{a} \cdot d\mathbf{l} + \oint_{P_1}^{P_0} \mathbf{a} \cdot d\mathbf{l} = F(P_1) - F(P_0) + F(P_0) - F(P_1) = 0$$

Alternatively, note that by Stokes's theorem, the integral can be cast into another form:

$$F = \oint \mathbf{a} \cdot d\mathbf{l} = \int \int (\nabla \times \mathbf{a}) \cdot \mathbf{n} \, ds$$

where ds is a surface element contained within the perimeter of $C_1 + C_2$, and **n** is an outward unit normal to ds. Therefore, if $\nabla \times \mathbf{a}$ is zero, $F = 0$.

2.3.7 Velocity Potential

Instead of discussing the vector **a**, let us consider **u**, the vector velocity, given by

$$\mathbf{u}(x, y, z, t) = u\mathbf{i} + v\mathbf{j} + w\mathbf{k} \tag{2.66}$$

Now, let us define the value of the line integral of **u** as $-\phi$:

$$-\phi = \oint_1 \mathbf{u} \cdot d\mathbf{l} = \oint_1 (u \, dx + v \, dy + w \, dz) \tag{2.67}$$

The quantity $\mathbf{u} \cdot d\mathbf{l}$ is a measure of the fluid velocity in the direction of the contour at each point. Therefore, $-\phi$ is related to the product of the velocity and length along the path between the two points P_0 and P_1. The minus sign is a matter of definitional convenience; quite often in the literature it is not present.

For the value of ϕ to be independent of path, that is, for the flow rate between P_0 and P_1 to be the same no matter how the integration is carried out, the terms in the integral must be an exact differential $d\phi$, and therefore

$$u = -\frac{\partial \phi}{\partial x} \tag{2.68a}$$

$$v = -\frac{\partial \phi}{\partial y} \tag{2.68b}$$

$$w = -\frac{\partial \phi}{\partial z} \tag{2.68c}$$

To ensure that this scalar function ϕ exists, the curl of the velocity vector must be zero:

$$\nabla \times \mathbf{u} = 0 = \left(\frac{\partial w}{\partial y} - \frac{\partial v}{\partial z}\right)\mathbf{i} + \left(\frac{\partial u}{\partial z} - \frac{\partial w}{\partial x}\right)\mathbf{j} + \left(\frac{\partial v}{\partial x} - \frac{\partial u}{\partial y}\right)\mathbf{k} \tag{2.69}$$

The curl of the velocity vector is referred to as the vorticity ω.

The velocity vector $\mathbf{u}$ can therefore be conveniently represented as

$$\mathbf{u} = -\nabla \phi \tag{2.70}$$

That is, we can express the vector quantity by the gradient of a scalar function ϕ for a flow with no vorticity. Further $\mathbf{u}$ flows "downhill," that is, in the direction of decreasing ϕ.[7] If $\phi (x, y, z, t)$ is known over all space, then u, v, and w can be determined. Note that ϕ has the units of length squared divided by time.

Let us examine more closely the line integral of the velocity component along the contour. If we consider the closed path from P_0 to P_1 and then back again, we know, from before, that the integral is zero.

$$\oint \mathbf{u} \cdot d\mathbf{l} = 0 \tag{2.71}$$

which means that if, for example, the path taken from P_0 to P_1 and back again were circular, no fluid would travel this circular path. Therefore, we expect no rotation of the fluid in circles if the curl of the velocity vector is zero.

To examine this irrotationality concept more fully, consider the average rate of rotation of a pair of orthogonal axes drawn on the small water mass

[7]This is the reason for the minus sign in the defintion of ϕ.

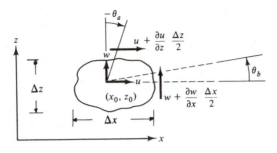

Figure 2.10

shown in Figure 2.10. Denoting the positive rotation in the counterclockwise direction, the average rate of rotation of the axes will be given by Eq. (2.72).

$$\theta = \frac{\theta_b + \theta_a}{2} \tag{2.72}$$

Now if u and w are known at (x_0, z_0), the coordinates of the center of the fluid mass, then at the edges of the mass the velocities are approximated as

$$u\left(x_0, z_0 + \frac{\Delta z}{2}\right) = u(x_0, z_0) + \frac{\partial u(x_0, z_0)}{\partial z} \frac{\Delta z}{2}$$

and

$$w\left(x_0 + \frac{\Delta x}{2}, z_0\right) = w(x_0, z_0) + \frac{\partial w(x_0, z_0)}{\partial x} \frac{\Delta x}{2}$$

Now the angular velocity of the z axis can be expressed as

$$\dot{\theta}_a = -\frac{u(x_0, z_0 + \Delta z/2) - u(x_0, z_0)}{\Delta z/2} = -\frac{\partial u}{\partial z}$$

and similarly for $\dot{\theta}_b$:

$$\dot{\theta}_b = \frac{\partial w}{\partial x}$$

The average rate of rotation is therefore

$$\dot{\theta} = \frac{\dot{\theta}_b + \dot{\theta}_a}{2} = \frac{1}{2}\left(\frac{\partial w}{\partial x} - \frac{\partial u}{\partial z}\right) \tag{2.73}$$

Therefore, the **j** component of the curl of the velocity vector is equal to twice the rate of rotation of the fluid particles, or $\nabla \times \mathbf{u} = 2\dot{\theta} \equiv \boldsymbol{\omega}$, where $\boldsymbol{\omega}$ is the fluid vorticity.

A mechanical analog to irrotational and rotational flows can be depicted by considering a carnival Ferris wheel. Under normal operating

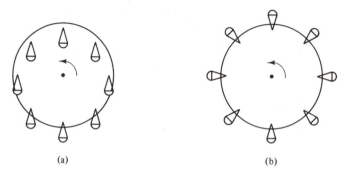

Figure 2.11 (a) Irrotational motion of chairs on a Ferris wheel; (b) rotational motion of the chairs.

conditions the chairs do not rotate; they always have the same orientation with respect to the earth (see Figure 2.11a). As far as the occupants are concerned, this is irrotational motion. If, on the other hand, the cars were fixed rigidly to the Ferris wheel, we would have, first, rotational motion (Figure 2.11b) and then perhaps a castastrophe.

For an inviscid and incompressible fluid, where the Euler equations are valid, there are only normal stresses (pressures) acting on the surface of a fluid particle; since the shear stresses are zero, there are no stresses to impart a rotation on a fluid particle. Therefore, in an inviscid fluid, a nonrotating particle remains nonrotating. However, if an initial vorticity exists in the fluid, the vorticity remains constant. To see this, we write the Euler equations in vector form:

$$\frac{D\mathbf{u}}{Dt} = -\frac{1}{\rho}\nabla p - g\mathbf{k} \qquad (2.74)$$

Taking the curl of this equation and substituting $\nabla \times \mathbf{u} = \boldsymbol{\omega}$ and $\nabla \times \nabla p = 0$ (identically), we have

$$\frac{D\boldsymbol{\omega}}{Dt} = 0 \qquad (2.75)$$

Therefore, there can be no change in the vorticity or the rotation of the fluid with time. This theory is due to Lord Kelvin (1869).[8]

2.3.8 Stream Function

For the velocity potential, we defined ϕ as (minus) the line integral of the velocity vector projected onto the line element; let us now define the line integral composed of the velocity component *perpendicular* to the line

[8]Chapter 5 is dedicated to Lord Kelvin.

element in two dimensions.

$$\psi = \oint_{C_1}^{P_1}{}_{P_0} \mathbf{u} \cdot \mathbf{n} \, dl \tag{2.76}$$

where $dl = |d\mathbf{l}|$. Consideration of the integrand above will demonstrate that ψ represents the amount of fluid crossing the line C_1 between points P_0 and P_1. The unit vector $\mathbf{n}$ is perpendicular to the path of integration C_1.

To determine the unit normal vector $\mathbf{n}$, it is necessary to find a normal vector $\mathbf{N}$ such that

$$\mathbf{N} \cdot d\mathbf{l} = 0$$

or $$N_x \, dx + N_z \, dz = 0$$

This is always true if

$$N_x = -dz \quad \text{and} \quad N_z = dx$$

It would have been equally valid to take $N_x = dz$ and $N_z = -dx$; however, this would have resulted in $\mathbf{N}$ directed to the right along the path of integration instead of the left.

To find the unit normal $\mathbf{n}$, it remains only to normalize $\mathbf{N}$.

$$\mathbf{n} = \frac{\mathbf{N}}{|\mathbf{N}|} = \frac{-dz\mathbf{i} + dx\mathbf{k}}{\sqrt{dx^2 + dz^2}} = \frac{-dz\mathbf{i} + dx\mathbf{k}}{dl}$$

The integral can thus be written as

$$\psi = \oint_{C_1}^{P_1}{}_{P_0} (-u \, dz + w \, dx) \tag{2.77}$$

For independence of path, so that the flow between P_0 and P_1 will be measured the same way no matter which way we connect the points, the integrand must be an exact differential, $d\psi$. This requires that

$$w = \frac{\partial\psi}{\partial x}; \quad u = -\frac{\partial\psi}{\partial z} \tag{2.78}$$

and thus the condition for independence of path [Eq. (2.65)] is

$$\frac{\partial w}{\partial z} + \frac{\partial u}{\partial x} = 0 \tag{2.79}$$

which is the two-dimensional form of the continuity equation. Therefore, for two-dimensional incompressible flow, a stream function exists and if we know its functional form, we know the velocity vector.

In general, there can be no stream function for *three-dimensional* flows, with the exception of axisymmetric flows. However, the velocity potential exists in any three-dimensional flow that is irrotational.

Note that the flow rate (per unit width) between points P_0 and P_1 is measured by the difference between $\psi(P_1)$ and $\psi(P_0)$. If an arbitrary constant is added to both values of the stream function, the flow rate is not affected.

2.3.9 Streamline

A streamline is defined as a line that is everywhere tangent to the velocity vector, or, on a streamline, $\mathbf{u} \cdot \mathbf{n} = 0$, where $\mathbf{n}$ is the normal to the streamline. From the earlier section,

$$\mathbf{u} \cdot \mathbf{n} = -u\,dz + w\,dx = 0 \qquad \text{or} \qquad \frac{dx}{u} = \frac{dz}{w} \quad \text{or} \quad \frac{dz}{dx} = \frac{w}{u} \qquad (2.80)$$

along a streamline. These are the equations for a streamline in two dimensions. Streamlines are a physical concept and therefore must also exist in all three-dimensional flows and all compressible flows.

From the definition of the stream function in two-dimensional flows, $\partial\psi/\partial l = 0$ on a streamline, and therefore the stream function, when it exists, is a constant along a streamline. This leads to the result $\nabla\psi \cdot dl = 0$ along a streamline, and therefore the gradient of ψ is perpendicular to the streamlines and in the direction normal to the velocity vector.

2.3.10 Relationship between Velocity Potential and Stream Function

For a three-dimensional flow, the velocity field may be determined from a velocity potential ϕ if the fluid is *irrotational*. For some three-dimensional flows and all two-dimensional flows for which the fluid is *incompressible*, a stream function ψ exists. Each is a measure of the flow rate between two points: in either the normal or transverse direction. For two-dimensional incompressible fluid flow, which is irrotational, both the stream function and the velocity potential exist and must be related through the velocity components.

The streamline, or line of constant stream function, and the lines of constant velocity potential are perpendicular, as can be seen from the fact that their gradients are perpendicular:

$$\nabla\phi \cdot \nabla\psi = 0$$

as

$$\left(\frac{\partial\phi}{\partial x}\mathbf{i} + \frac{\partial\phi}{\partial z}\mathbf{k} \right) \cdot \left(\frac{\partial\psi}{\partial x}\mathbf{i} + \frac{\partial\psi}{\partial z}\mathbf{k} \right) =$$

$$(-u\mathbf{i} - w\mathbf{k}) \cdot (+w\mathbf{i} - u\mathbf{k}) = \qquad (2.81)$$

$$-uw + uw = 0$$

The primary advantage of either the stream function or the velocity potential is that they are scalar quantities from which the velocity vector field can be obtained. As one can easily imagine, it is far easier to work with scalar rather than with vector functions.

Often, the stream function or the velocity potential is known and the other is desired. To obtain one from the other, it is necessary to relate the two. Recalling the definition of the velocity components

$$u = -\frac{\partial \phi}{\partial x} = -\frac{\partial \psi}{\partial z}$$

$$w = -\frac{\partial \phi}{\partial z} = \frac{\partial \psi}{\partial x}$$

we have

$$\frac{\partial \phi}{\partial x} = \frac{\partial \psi}{\partial z} \tag{2.82a}$$

$$\frac{\partial \phi}{\partial z} = -\frac{\partial \psi}{\partial x} \tag{2.82b}$$

These relationships are called the Cauchy–Riemann conditions and enable the hydrodynamicist to utilize the powerful techniques of complex variable analysis. See for example, Milne-Thomson (1949).

Example 2.3

For the following velocity potential, determine the corresponding stream function.

$$\phi(x, z, t) = (-3x + 5z) \cos \frac{2\pi t}{T}$$

This velocity potential represents a to-and-fro motion of the fluid with the streamlines slanted with respect to the origin as shown in Figure 2.12. The velocity components are

$$u = -\frac{\partial \phi}{\partial x} = 3 \cos \frac{2\pi t}{T}$$

$$w = -\frac{\partial \phi}{\partial z} = -5 \cos \frac{2\pi t}{T}$$

Solution. From the Cauchy–Riemann conditions

$$-\frac{\partial \psi}{\partial z} = 3 \cos \frac{2\pi t}{T}$$

or, integrating,

$$\psi(x, z, t) = -3z \cos \frac{2\pi t}{T} + G_1(x, t)$$

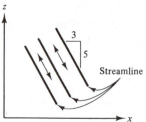

Figure 2.12

Note that because we integrated a partial differential, the unknown quantity that results is a function of both x and t. For the vertical velocity,

$$\frac{\partial \psi}{\partial x} = -5 \cos \frac{2\pi t}{T}$$

or

$$\psi(x, z, t) = -5x \cos \frac{2\pi t}{T} + G_2(z, t)$$

Comparing these two equations, which must be the same stream function, it is apparent that

$$\psi(x, z, t) = -(5x + 3z) \cos \frac{2\pi t}{T} + G(t)$$

The quantity $G(t)$ is a constant with regard to the space variables x and z and can, in fact, vary with time. This time dependency, due to $G(t)$, has no bearing whatsoever on the flow field; hence $G(t)$ can be set equal to zero without affecting the flow field.

2.4 CYLINDRICAL COORDINATES

The most appropriate coordinate system to describe a particular problem usually is that for which constant values of a coordinate most nearly conform to the boundaries or response variables in the problem. Therefore, for the case of circular waves, which might be generated when a stone is dropped into a pond, it is not convenient to use Cartesian coordinates to describe the problem, but cylindrical coordinates. These coordinates are (r, θ, z), which are shown in Figure 2.13. The transformation between coordinates depends on these equations, $x = r \cos \theta$, $y = r \sin \theta$, and $z = z$. For a velocity potential defined in terms of (r, θ, z), the velocity components are

$$v_r = -\frac{\partial \phi}{\partial r} \tag{2.83a}$$

$$v_\theta = -\frac{1}{r} \frac{\partial \phi}{\partial \theta} \tag{2.83b}$$

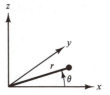

Figure 2.13 Relationship between
Cartesian and cylindrical coordinate
systems r and θ lie in the x-y plane.

$$v_z = -\frac{\partial \phi}{\partial z} \tag{2.83c}$$

As noted previously, the stream function exists only for those three-dimensional flows which are axisymmetric. The stream function for an axisymmetric flow in cylindrical coordinates is called the "Stokes" stream function. The derivation of this stream function is presented in numerous references, however this form is not used extensively in wave mechanics and therefore will not be discussed further here.

2.5 THE BERNOULLI EQUATION

The Bernoulli equation is simply an integrated form of Euler equations of motion and provides a relationship between the pressure field and kinematics, and will be useful later. Retaining our assumptions of irrotational motion and an incompressible fluid, the governing equations of motion in the fluid for the x-z plane are the Euler equations, Eqs. (2.38).

$$\frac{\partial u}{\partial t} + u\frac{\partial u}{\partial x} + w\frac{\partial u}{\partial z} = -\frac{1}{\rho}\frac{\partial p}{\partial x} \tag{2.84a}$$

$$\frac{\partial w}{\partial t} + u\frac{\partial w}{\partial x} + w\frac{\partial w}{\partial z} = -\frac{1}{\rho}\frac{\partial p}{\partial z} - g \tag{2.84b}$$

Substituting in the two-dimensional irrotationality condition [Eq. (2.69)],

$$\frac{\partial u}{\partial z} = \frac{\partial w}{\partial x} \tag{2.85}$$

the equations can be rewritten as

$$\frac{\partial u}{\partial t} + \frac{\partial(u^2/2)}{\partial x} + \frac{\partial(w^2/2)}{\partial x} = -\frac{1}{\rho}\frac{\partial p}{\partial x} \tag{2.86}$$

$$\frac{\partial w}{\partial t} + \frac{\partial(u^2/2)}{\partial z} + \frac{\partial(w^2/2)}{\partial z} = -\frac{1}{\rho}\frac{\partial p}{\partial z} - g \tag{2.87}$$

Now, since a velocity potential exists for the fluid, we have

$$u = -\frac{\partial \phi}{\partial x}; \qquad w = -\frac{\partial \phi}{\partial z} \tag{2.88}$$

Therefore, if we substitute these definitions into Eqs. (2.86) and (2.87), we get

$$\frac{\partial}{\partial x}\left[-\frac{\partial \phi}{\partial t} + \frac{1}{2}(u^2 + w^2) + \frac{p}{\rho} \right] = 0 \tag{2.89a}$$

$$\frac{\partial}{\partial z}\left[-\frac{\partial \phi}{\partial t} + \frac{1}{2}(u^2 + w^2) + \frac{p}{\rho} \right] = -g \tag{2.89b}$$

where it has been assumed that the density is uniform throughout the fluid. Integrating the x equation yields

$$-\frac{\partial \phi}{\partial t} + \frac{1}{2}(u^2 + w^2) + \frac{p}{\rho} = C'(z, t) \tag{2.90}$$

where, as indicated, the constant of integration $C'(z, t)$ varies only with z and t. Integrating the z equation yields

$$-\frac{\partial \phi}{\partial t} + \frac{1}{2}(u^2 + w^2) + \frac{p}{\rho} = -gz + C(x, t) \tag{2.91}$$

Examining these two equations, which have the same quantity on the left-hand sides, shows clearly that

$$C'(z, t) = -gz + C(x, t)$$

Thus C cannot be a function of x, as neither C' nor (gz) depend on x. Therefore, $C'(z, t) = -gz + C(t)$. The resulting equation is

$$\boxed{-\frac{\partial \phi}{\partial t} + \frac{1}{2}(u^2 + w^2) + \frac{p}{\rho} + gz = C(t)} \tag{2.92}$$

The steady-state form of this equation, the integrated form of the equations of motion, is called the Bernoulli equation, which is valid throughout the fluid. In this book we will refer to Eq. (2.92) as the unsteady form of the Bernoulli equation or, for brevity, as simply the Bernoulli equation. The function $C(t)$ is referred to as the Bernoulli term and is a constant for steady flows.

The Bernoulli equation can also be written as

$$-\frac{\partial \phi}{\partial t} + \frac{p}{\rho} + \frac{1}{2}\left[\left(\frac{\partial \phi}{\partial x}\right)^2 + \left(\frac{\partial \phi}{\partial z}\right)^2 \right] + gz = C(t) \tag{2.93}$$

which interrelates the fluid pressure, particle elevation, and velocity potential. Between any two points in the fluid of known elevation and velocity potential, pressure differences can be obtained by this equation; for example, for points A and B at elevations z_A and z_B, the pressure at A is

$$p_A = p_B - \frac{\rho}{2}\left[\left(\frac{\partial\phi}{\partial x}\right)^2 + \left(\frac{\partial\phi}{\partial z}\right)^2\right]_A + \frac{\rho}{2}\left[\left(\frac{\partial\phi}{\partial x}\right)^2 + \left(\frac{\partial\phi}{\partial z}\right)^2\right]_B \tag{2.94}$$

$$+ \rho\left[\left(\frac{\partial\phi}{\partial t}\right)_A - \left(\frac{\partial\phi}{\partial t}\right)_B\right] + \rho g(z_B - z_A)$$

Notice that the Bernoulli constant is the same at both locations and thus dropped out of the last equation. [Another method to eliminate the constant is to absorb it into the velocity potential. Starting with Eq. (2.93) for the Bernoulli equation, we can define a function $f(t)$ such that

$$\frac{\partial f(t)}{\partial t} = C(t)$$

Therefore, the Bernoulli equation can be written as

$$-\frac{\partial(\phi + f(t))}{\partial t} + \frac{1}{2}\left[\left(\frac{\partial\phi}{\partial x}\right)^2 + \left(\frac{\partial\phi}{\partial z}\right)^2\right] + \frac{p}{\rho} + gz = 0 \tag{2.95}$$

Now, if we define $\phi'(x, z, t) = \phi(x, z, t) + f(t)$,[9]

$$-\frac{\partial\phi'}{\partial t} + \frac{1}{2}\left[\left(\frac{\partial\phi'}{\partial x}\right)^2 + \left(\frac{\partial\phi'}{\partial z}\right)^2\right] + gz + \frac{p}{\rho} = 0 \tag{2.96}$$

Often we will use the ϕ' form of the velocity potential, or, equivalently, we will take the Bernoulli constant as zero.] For three-dimensional flows, Eq. (2.96) would be modified only by the addition of $(1/2)(\partial\phi/\partial y)^2$ on the left-hand side.

In the following paragraphs a form of the Bernoulli equation will be derived for two-dimensional steady flow in which the density is uniform and the shear stresses are zero; however, in contrast to the previous case, the results apply to *rotational* flow fields (i.e., *the velocity potential does not exist*). In Figure 2.14 the velocity vector at a point on a streamline is shown, as is a coordinate system, **s** and **n**, in the streamline tangential and normal directions.

By definition of a streamline, at A a tangential velocity exists, u_s, but there is no normal velocity to the streamline u_n. Referring to Eq. (2.84), the steady-state form of the equation of motion for a particle at A would be

---

[9] The kinematics associated with $\phi'(x, z, t)$ are exactly the same as $\phi(x, z, t)$, as can be shown easily by the reader.

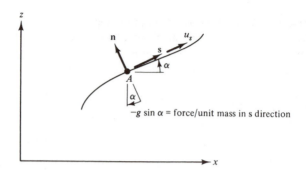

Figure 2.14 Definition sketch for derivation of steady-state two-dimensional Bernoulli equation for rotational flows.

written as

$$u_s \frac{\partial u_s}{\partial s} = -\frac{1}{\rho} \frac{\partial p}{\partial s} - g \sin \alpha \qquad (2.97)$$

where $\sin \alpha$ accounts for the fact that the streamline coordinate system is inclined with respect to the horizontal plane. From the figure, $\sin \alpha = dz/ds$, and therefore the equation of motion is

$$\frac{\partial}{\partial s} \left(\frac{u_s^2}{2} + \frac{p}{\rho} + gz \right) = 0$$

where again we have assumed the density ρ to be a constant along the streamline. Integrating along the streamline, we have

$$\frac{u_s^2}{2} + \frac{p}{\rho} + gz = C(\psi) \qquad (2.98)$$

This is nearly the familiar form of the Bernoulli equation, except that the time-dependent term resulting from the local acceleration is not present due to the assumption of steady flow and also, the Bernoulli constant is a function of the streamline on which we integrated the equation. In contrast to the Bernoulli equation for an ideal flow, in this case we cannot apply the Bernoulli equation everywhere, only at points along the same streamline.

REFERENCES

MILNE-THOMSON, L. M., *Theoretical Hydrodynamics*, 4th ed., The Macmillan Co., N.Y., 1960.

PROBLEMS

2.1 Consider the following transition section:

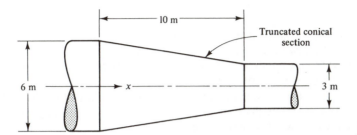

(a) The flow from left to right is constant at $Q = 12\pi$ m³/s. What is the total acceleration of a water particle in the x direction at $x = 5$ m? Assume that the water is incompressible and that the x component of velocity is uniform across each cross section.

(b) The flow of water from *right* to *left* is given by

$$Q(t) = \pi t^2$$

Calculate the total acceleration at $x = 5$ m for $t = 2.0$ s. Make the same assumptions as in part (a).

2.2 Consider the following transition section:

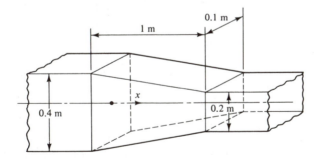

(a) If the flow of water from left to right is constant at $Q = .1$ m³/s, what is the total acceleration of a water particle at $x = 0.5$ m? Assume that the water is incompressible and that the x component of velocity is uniform across each cross section.

(b) The flow of water from *right* to *left* is expressed by

$$Q(t) = t^2/100$$

Calculate the total acceleration at $x = 0.5$ m for $t = 4.48$ s. Make the same assumptions as in part (a).

2.3 The velocity potential for a particular two-dimensional flow field in which the density is uniform is

$$\phi = (-3x + 5z) \cos \frac{2\pi}{T} t$$

where the z axis is oriented vertically upward.
(a) Is the flow irrotational?
(b) Is the flow nondivergent? If so, derive the stream function and sketch any two streamlines for $t = T/8$.

2.4 If the water (assumed inviscid) in the U-tube is displaced from its equilibrium position, it will oscillate about this position with its natural period. Assume that the displacement of the surface is

$$\eta(t) = A \cos \frac{2\pi}{T} t$$

where the amplitude A is 10 cm and the natural period T is 8 s. What will be the pressure at a distance 20 cm below the instantaneous water surface for $\eta = +10$, 0, and -10 cm? Assume that $g = 980$ cm/s^2 and $\rho = 1$ g/cm^3.

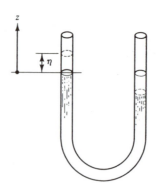

2.5 Suppose that we measure the mass density ρ at a fixed point (x, y, z) as a function of time and observe the following:

From this information alone, is it possible to determine whether the flow is nondivergent?

2.6 Derive the following equation for an inviscid fluid and a nondivergent steady flow:

$$-g - \frac{1}{\rho}\frac{\partial p}{\partial z} = \frac{\partial(uw)}{\partial x} + \frac{\partial(vw)}{\partial y} + \frac{\partial(w^2)}{\partial z}$$

2.7 Expand the following expression so that gradients of *products* of scalar functions do not appear in the result:

$$\nabla(\phi\psi f)$$

where ϕ, ψ, and f are scalar functions.

2.8 The velocity components in a two-dimensional flow of an inviscid fluid are

$$u = \frac{Kx}{x^2 + z^2}$$

$$w = \frac{Kz}{x^2 + z^2}$$

(a) Is the flow nondivergent?
(b) Is the flow irrotational?
(c) Sketch the two streamlines passing through points A and B, where the coordinates of these points are:

$$\text{Point } A: \quad x = 1, z = 1$$
$$\text{Point } B: \quad x = 1, z = 2$$

2.9 For a particular fluid flow, the velocity components u, v, and w in the x, y, and z directions, respectively, are

$$u = x + 8y + 6tz + t^4$$

$$v = 8x - 7y + 6z$$

$$w = 12x + 6y + 12z \cos\frac{2\pi t}{T} + t^2$$

(a) Are there any times for which the flow is nondivergent? If so, when?
(b) Are there any times for which the flow is irrotational? If so, when?
(c) Develop the expression for the pressure gradient in the vertical (z) direction as a function of space and time.

2.10 The stream function for an inviscid fluid flow is

$$\psi = Ax^2zt$$

where x, z, $t \geq 0$.
(a) Sketch the streamlines $\psi = 0$ and $\psi = 6A$ for $t = 3$ s.
(b) For $t = 5$ s, what are the coordinates of the point where the streamline slope dz/dx is -5 for the particular streamline $\psi = 100A$?
(c) What is the pressure gradient at $x = 2$, $z = 5$ and at time $t = 3$ s? $A = 1.0$, $\rho = 1.0$.

2.11 Develop expressions for $\sinh x$ and $\cosh x$ for small values of x, using the Taylor series expansion.

2.12 The pressures $p_A(t)$ and $p_B(t)$ act on the massless pistons containing the inviscid, incompressible fluid in the horizontal tube shown below. Develop an expression for the velocity of the fluid as a function of time $\rho = 1$ gm/cm^3.

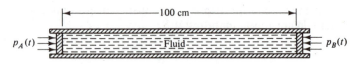

Note:

$$p_A(t) = C_A \sin \sigma t$$

$$p_B(t) = C_B \sin (\sigma t + \alpha)$$

where

$$\sigma = 0.5 \text{ rad/s}$$

$$\alpha = \frac{\pi}{2}$$

$$C_A = C_B = 10 \text{dyn/cm}^3$$

2.13 An early experimenter of waves and other two-dimensional fluid motions closely approximating irrotational flows noted that at an impermeable horizontal boundary, the gradient of horizontal velocity in the vertical direction is always zero. Is this finding in accordance with hydrodynamic fundamentals? If so, prove your answer.

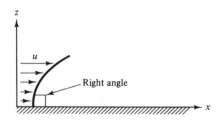

3

Small-Amplitude Water Wave Theory Formulation and Solution

Dedication

PIERRE SIMON LAPLACE

Pierre Simon Laplace (1749–1827) is well known for the equation that bears his name. The Laplace equation is one of the most ubiquitous equations of mathematical physics (the Helmholtz, the diffusion, and the wave equation being others); it appears in electrostatics, hydrodynamics, groundwater flow, thermostatics, and other fields.

As had Euler, Laplace worked in a great variety of areas, applying his knowledge of mathematics to physical problems. He has been called the Newton of France.

He was born in Beaumont-en-Auge, Normandy, France, and educated at Caen (1765–1767). In 1768 he became Professor of Mathematics at the École Militaire in Paris. Later he moved to the École Normale, also in Paris.

Napoleon appointed him Minister of the Interior in 1799, and he became a Count in 1806 and a Marquis in 1807, the same year that he assumed the presidency of the French Academy of Sciences.

A large portion of Laplace's research was devoted to astronomy. He wrote on the orbital motion of the planets and celestial mechanics and on the stability of the solar system. He also developed the hypothesis that the solar system coalesced out of a gaseous nebula.

In other areas of physics, he developed the theory of tides which bears his name, worked with Lavoisier on specific heat of solids, studied capillary action, surface tension, and electric theory, and with Legendre, introduced partial differential equations into the study of probability. He also developed and applied numerous solutions (potential functions) of the Laplace equation.

3.1 INTRODUCTION

Real water waves propagate in a viscous fluid over an irregular bottom of varying permeability. A remarkable fact, however, is that in most cases the main body of the fluid motion is nearly irrotational. This is because the viscous effects are usually concentrated in thin "boundary" layers near the surface and the bottom. Since water can also be considered reasonably incompressible, a velocity potential and a stream function should exist for waves. To simplify the mathematical analysis, numerous other assumptions must and will be made as the development of the theory proceeds.

3.2 BOUNDARY VALUE PROBLEMS

In formulating the small-amplitude water wave problem, it is useful to review, in very general terms, the structure of boundary value problems, of which the present problem of interest is an example. Numerous classical

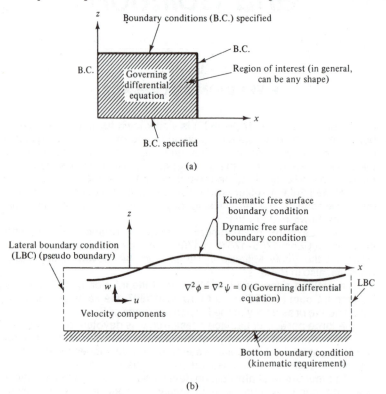

(a)

(b)

Figure 3.1 (a) General structure of two-dimensional boundary value problems. (*Note*: The number of boundary conditions required depends on the order of the differential equation.) (b) Two-dimensional water waves specified as a boundary value problem.

problems of physics and most analytical problems in engineering may be posed as boundary value problems; however, in some developments, this may not be apparent.

The formulation of a boundary value problem is simply the expression in mathematical terms of the physical situation such that a unique solution exists. This generally consists of first establishing a region of interest and specifying a differential equation that must be satisfied within the region (see Figure 3.1a). Often, there are an infinite number of solutions to the differential equation and the remaining task is selecting the one or more solutions that are relevant to the physical problem under investigation. This selection is effected through the boundary conditions, that is, rejecting those solutions that are not compatible with these conditions.

In addition to the *spatial* (or geometric) boundary conditions, there are temporal boundary conditions which specify the state of the variable of interest at some point in time. This temporal condition is termed an "initial condition." If we are interested in water waves, which are periodic in space, then we might specify, for example, that the waves are propagating in the positive x direction and that at $t = 0$, the wave crest is located at $x = 0$.

In the following development of linear water wave theory, it will be helpful to relate each major step to the general structure of boundary value problems discussed previously. Figure 3.1b presents the region of interest, the governing differential equations, and indicates in a general manner the important boundary conditions.

3.2.1 The Governing Differential Equation

With the assumption of irrotational motion and an incompressible fluid, a velocity potential exists which should satisfy the continuity equation

$$\nabla \cdot \mathbf{u} = 0 \tag{3.1a}$$

or

$$\nabla \cdot \nabla \phi = 0 \tag{3.1b}$$

As was shown in Chapter 2, the divergence of a gradient leads to the Laplace equation, which must hold throughout the fluid.

$$\nabla^2 \phi = \frac{\partial^2 \phi}{\partial x^2} + \frac{\partial^2 \phi}{\partial y^2} + \frac{\partial^2 \phi}{\partial z^2} = 0 \tag{3.2}$$

The Laplace equation occurs frequently in many fields of physics and engineering and numerous solutions to this equation exist (see, e.g., the book by Bland, 1961), and therefore it is necessary to select only those which are applicable to the particular water wave motion of interest.

In addition, for flows that are nondivergent and irrotational, the Laplace equation also applies to the stream function. The incompressibility or, equivalently, the nondivergent condition for two dimensions guarantees the existence of a stream function, from which the velocities under the wave can be determined. Substituting these velocities into the irrotationality condition again yields the Laplace equation, except for the stream function this time,

$$\frac{\partial w}{\partial x} - \frac{\partial u}{\partial z} = 0 \tag{3.3a}$$

or

$$\nabla^2 \psi = \frac{\partial^2 \psi}{\partial x^2} + \frac{\partial^2 \psi}{\partial z^2} = 0 \tag{3.3b}$$

This equation must hold throughout the fluid. If the motion had been rotational, yet frictionless, the governing equation would be

$$\nabla^2 \psi = \omega \tag{3.4}$$

where ω is the vorticity.

A few comments on the velocity potential and the stream function may help in obtaining a better understanding for later applications. First, as mentioned earlier, the velocity potential can be defined for both two and three dimensions, whereas the definition of the stream function is such that it can only be defined for three dimensions if the flow is symmetric about an axis (in this case although the flow occurs in three dimensions, it is mathematically two-dimensional). It therefore follows that the stream function is of greatest use in cases where the wave motion occurs in one plane. Second, the Laplace equation is linear; that is, it involves no products and thus has the interesting and valuable property of superposition; that is, if ϕ_1 and ϕ_2 each satisfy the Laplace equation, then $\phi_3 = A\phi_1 + B\phi_2$ also will solve the equation, where A and B are arbitrary constants. Therefore, we can add and subtract solutions to build up solutions applicable for different problems of interest.

3.2.2 Boundary Conditions

Kinematic boundary conditions.   At any boundary, whether it is fixed, such as the bottom, or free, such as the water surface, which is free to deform under the influence of forces, certain physical conditions must be satisfied by the fluid velocities. These conditions on the water particle kinematics are called *kinematic boundary conditions*. At any surface or fluid interface, it is clear that there must be no flow across the interface; otherwise, there would be no interface. This is most obvious in the case of an impermeable fixed surface such as a sheet pile seawall.

The mathematical expression for the kinematic boundary condition may be derived from the equation which describes the surface that constitutes the boundary. Any fixed or moving surface can be expressed in terms of a mathematical expression of the form $F(x, y, z, t) = 0$. For example, for a stationary sphere of fixed radius a, $F(x, y, z, t) = x^2 + y^2 + z^2 - a^2 = 0$. If the surface varies with time, as would the water surface, then the total derivative of the surface with respect to time would be zero on the surface. In other words, if we move with the surface, it does not change.

$$\frac{DF(x, y, z, t)}{Dt} = 0 = \frac{\partial F}{\partial t} + u \frac{\partial F}{\partial x} + v \frac{\partial F}{\partial y} + w \frac{\partial F}{\partial z} \bigg|_{\text{on } F(x,y,z,t) = 0} \tag{3.5a}$$

or

$$-\frac{\partial F}{\partial t} = \mathbf{u} \cdot \nabla F = \mathbf{u} \cdot \mathbf{n} |\nabla F| \tag{3.5b}$$

where the unit vector normal to the surface has been introduced as $\mathbf{n} = \nabla F/|\nabla F|$.

Rearranging the kinematic boundary condition results:

$$\boxed{\mathbf{u} \cdot \mathbf{n} = \frac{-\partial F/\partial t}{|\nabla F|} \qquad \text{on } F(x, y, z, t) = 0} \tag{3.6}$$

where

$$|\nabla F| = \sqrt{\left(\frac{\partial F}{\partial x}\right)^2 + \left(\frac{\partial F}{\partial y}\right)^2 + \left(\frac{\partial F}{\partial z}\right)^2}$$

This condition requires that the component of the fluid velocity normal to the surface be related to the local velocity of the surface. If the surface does not change with time, then $\mathbf{u} \cdot \mathbf{n} = 0$; that is, the velocity component normal to the surface is zero.

Example 3.1

Fluid in a U-tube has been forced to oscillate sinusoidally due to an oscillating pressure on one leg of the tube (see Figure 3.2). Develop the kinematic boundary condition for the free surface in leg A.

Solution.   The still water level in the U-tube is located at $z = 0$. The motion of the free surface can be described by $z = \eta(t) = a \cos t$, where a is the amplitude of the variation of η.

If we examine closely the motion of a fluid particle at the surface (Figure 3.2b), as the surface drops, with velocity w, it follows that the particle has to move with the speed of the surface or else the particle leaves the surface. The same is true for a rising surface. Therefore, we would postulate on physical grounds that

$$w = \frac{d\eta}{dt}\bigg|_{z=\eta(t)}$$

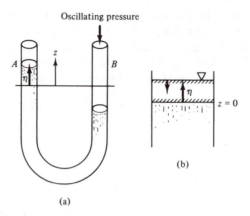

Figure 3.2   (a) Oscillating flow in a U-tube; (b) details of free surface.

where $d\eta/dt$ = the rate of rise or fall of the surface. To ensure that this is formally correct, we follow the equation for the kinematic boundary condition, Eq. (3.6), where $F(z, t) = z - \eta(t) = 0$. Therefore,

$$\mathbf{u} \cdot \mathbf{n} = \frac{\partial \eta}{\partial t}$$

where $\mathbf{n} = 0\mathbf{i} + 0\mathbf{j} + 1\mathbf{k}$, directed vertically upward and $\mathbf{u} = u\mathbf{i} + v\mathbf{j} + w\mathbf{k}$, and carrying out the scalar product, we find that

$$w = \frac{\partial \eta}{\partial t}$$

which is the same as obtained previously, when we realize that $d\eta/dt = \partial\eta/\partial t$, as η is only a function of time.

The Bottom Boundary Condition (BBC).   In general, the lower boundary of our region of interest is described as $z = -h(x)$ for a two-dimensional case where the origin is located at the still water level and h represents the depth. If the bottom is impermeable, we expect that $\mathbf{u} \cdot \mathbf{n} = 0$, as the bottom does not move with time. (For some cases, such as earthquake motions, obviously the time dependency of the bottom must be included.)

The surface equation for the bottom is $F(x, z) = z + h(x) = 0$. Therefore,

$$\mathbf{u} \cdot \mathbf{n} = 0 \tag{3.7}$$

where

$$\mathbf{n} = \frac{\nabla F}{|\nabla F|} = \frac{\dfrac{dh}{dx}\mathbf{i} + 1\mathbf{k}}{\sqrt{(dh/dx)^2 + 1}} \tag{3.8}$$

Carrying out the dot product and multiplying through by the square root, we have

$$u \frac{dh}{dx} + w = 0 \qquad \text{on } z = -h(x) \tag{3.9a}$$

or

$$w = -u \frac{dh}{dx} \qquad \text{on } z = -h(x) \tag{3.9b}$$

For a horizontal bottom, then, $w = 0$ on $z = -h$. For a sloping bottom, we have

$$\frac{w}{u} = -\frac{dh}{dx} \tag{3.10}$$

Referring to Figure 3.3, it is clear that the kinematic condition states that the flow at the bottom is tangent to the bottom. In fact, we could treat the bottom as a streamline, as the flow is everywhere tangential to it. The bottom boundary condition, Eq. (3.7), also applies directly to flows in three dimensions in which h is $h(x, y)$.

Kinematic Free Surface Boundary Condition (KFSBC).   The free surface of a wave can be described as $F(x, y, z, t) = z - \eta(x, y, t) = 0$, where $\eta(x, y, t)$ is the displacement of the free surface about the horizontal plane, $z = 0$. The kinematic boundary condition at the free surface is

$$\mathbf{u} \cdot \mathbf{n} = \frac{\partial \eta / \partial t}{\sqrt{(\partial \eta / \partial x)^2 + (\partial \eta / \partial y)^2 + 1}} \qquad \text{on } z = \eta(x, y, t) \tag{3.11a}$$

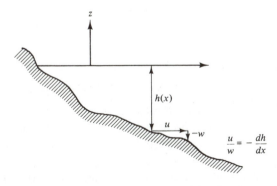

Figure 3.3   Illustration of bottom boundary condition for the two-dimensional case.

where

$$\mathbf{n} = \frac{-\dfrac{\partial \eta}{\partial x}\,\mathbf{i} - \dfrac{\partial \eta}{\partial y}\,\mathbf{j} + 1\mathbf{k}}{\sqrt{(\partial \eta/\partial x)^2 + (\partial \eta/\partial y)^2 + 1}} \qquad (3.11b)$$

Carrying out the dot product yields

$$w = \frac{\partial \eta}{\partial t} + u\,\frac{\partial \eta}{\partial x} + v\,\frac{\partial \eta}{\partial y}\Bigg|_{\text{on } z=\eta(x,y,t)} \qquad (3.11c)$$

This condition, the KFSBC, is a more complicated expression than that obtained for (1), the U-tube, where the flow was normal to the surface and (2) the bottom, where the flow was tangential. In fact, inspection of Eq. (3.11c) will verify that the KFSBC is a combination of the other two conditions, which are just special cases of this more general type of condition.[1]

Dynamic Free Surface Boundary Condition. The boundary conditions for fixed surfaces are relatively easy to prescribe, as shown in the preceding section, and they apply on the known surface. A distinguishing feature of fixed (in space) surfaces is that they can support pressure variations. However, surfaces that are "free," such as the air–water interface, cannot support variations in pressure[2] across the interface and hence must respond in order to maintain the pressure as uniform. A second boundary condition, termed a dynamic boundary condition, is thus required on any *free* surface or interface, to prescribe the pressure distribution pressures on this boundary. An interesting effect of the displacement of the free surface is that the position of the upper boundary is not known a priori in the water wave problem. This aspect causes considerable difficulty in the attempt to obtain accurate solutions that apply for large wave heights (Chapter 11).

As the dynamic free surface boundary condition is a requirement that the pressure on the free surface be uniform along the wave form, the Bernoulli equation [Eq. (2.92)] with p_η = constant is applied on the free surface, $z = \eta(x, t)$,

$$-\frac{\partial \phi}{\partial t} + \frac{1}{2}\,(u^2 + w^2) + \frac{p_\eta}{\rho} + gz = C(t) \qquad (3.12)$$

where p_η is a constant and usually taken as gage pressure, $p_\eta = 0$.

Conditions at "Responsive" Boundaries. As noted previously, an additional condition must be imposed on those boundaries that can respond to spatial or temporal variations in pressure. In the case of wind blowing across

---

[1]The reader is urged to develop the general kinematic free surface boundary condition for a wave propagating in the x direction alone.

[2]Neglecting surface tension.

a water surface and generating waves, if the pressure relationship were known, the Bernoulli equation would serve to couple that wind field with the kinematics of the wave. The wave and wind field would be interdependent and the wave motion would be termed "coupled." If the wave were driven by, but did not affect the applied surface pressure distribution, this would be a case of "forced" wave motion and again the Bernoulli equation would serve to express the boundary condition. For the simpler case that is explored in some detail in this chapter, the pressure will be considered to be uniform and hence a case of "free" wave motion exists. Figure 3.4 depicts various degrees of coupling between the wind and wave fields.

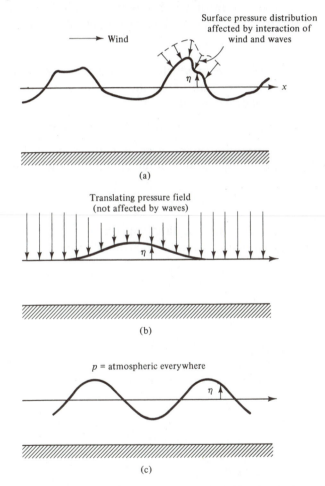

Figure 3.4   Various degrees of air-water boundary interaction and coupling to atmospheric pressure field: (a) coupled wind and waves; (b) forced waves due to moving pressure field; (c) free waves—not affected by pressure variations at air-water interface.

The boundary condition for free waves is termed the "dynamic free surface boundary condition" (DFSBC), which the Bernoulli equation expresses as Eq. (3.13) with a uniform surface pressure p_η:

$$-\frac{\partial\phi}{\partial t} + \frac{p_\eta}{\rho} + \frac{1}{2}\left[\left(\frac{\partial\phi}{\partial x}\right)^2 + \left(\frac{\partial\phi}{\partial z}\right)^2\right] + gz = C(t), \qquad z = \eta(x, t) \qquad (3.13)$$

where p_η is a constant and usually taken as gage pressure, $p_\eta = 0$.

If the wave lengths are very short (on the order of several centimeters), the surface is no longer "free." Although the pressure is uniform above the water surface, as a result of the surface curvature, a nonuniform pressure will occur within the water immediately below the surface film. Denoting the coefficient of surface tension as σ', the tension per unit length T is simply

$$T = \sigma' \qquad (3.14)$$

Consider now a surface for which a curvature exists as shown in Figure 3.5. Denoting p as the pressure under the free surface, a free-body force analysis in the vertical direction yields

$$T\left[-\sin\alpha|_x + \sin\alpha|_{x+\Delta x}\right] + (p - p_\eta)\,\Delta x + \text{terms of order } \Delta x^2 = 0$$

in which the approximation $\partial\eta/\partial x \approx \sin\alpha$ will be made. Expanding by Taylor's series and allowing the size of the element to shrink to zero yields

$$p = p_\eta - \sigma'\frac{\partial^2\eta}{\partial x^2} \qquad (3.15)$$

Thus for cases in which surface tension forces are important, the dynamic free surface boundary condition is modified to

$$-\frac{\partial\phi}{\partial t} + \frac{p_\eta}{\rho} - \frac{\sigma'}{\rho}\frac{\partial^2\eta}{\partial x^2} + \frac{1}{2}\left[\left(\frac{\partial\phi}{\partial x}\right)^2 + \left(\frac{\partial\phi}{\partial z}\right)^2\right] + gz = C(t), \qquad z = \eta(x, t) \quad (3.16)$$

which will be of use in our later examination of capillary water waves.

Lateral Boundary Conditions.   At this stage boundary conditions have been discussed for the bottom and upper surfaces. In order to complete specification of the boundary value problem, conditions must also be speci-

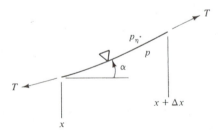

Figure 3.5   Definition sketch for surface element.

fied on the remaining lateral boundaries. There are several situations that must be considered.

If the waves are propagating in one direction (say the x direction), conditions are two-dimensional and then "no-flow" conditions are appropriate for the velocities in the y direction. The boundary conditions to be applied in the x direction depend on the problem under consideration. If the wave motion results from a prescribed disturbance of, say, an object at $x = 0$, which is the classical wavemaker problem, then at the object, the usual kinematic boundary condition is expressed by Figure 3.6a.

Consider a vertical paddle acting as a wavemaker in a wave tank. If the displacement of the paddle may be described as $x = S(z, t)$, the kinematic boundary condition is

$$\mathbf{u} \cdot \mathbf{n} = \frac{\partial S(z, t)}{\partial t} \Big/ \sqrt{1 + \left(\frac{\partial S}{\partial z}\right)^2}$$

where

$$\mathbf{n} = \frac{1\mathbf{i} - \dfrac{\partial S}{\partial z}\mathbf{k}}{\sqrt{1 + (\partial S/\partial z)^2}}$$

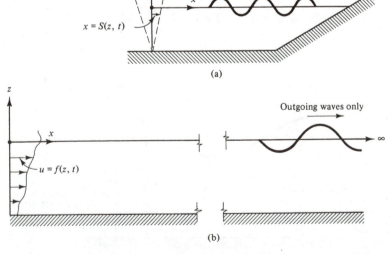

(a)

(b)

Figure 3.6 (a) Schematic of wavemaker in a wave tank; (b) radiation condition for wavemaker problem for region unbounded in x direction.

or, carrying out the dot product,

$$u - w \frac{\partial S}{\partial z} = \frac{\partial S}{\partial t} \bigg|_{\text{on } x=S(z,t)} \qquad (3.17)$$

which, of course, requires that the fluid particles at the moving wall follow the wall.

Two different conditions occur at the other possible lateral boundaries: at a fixed beach as shown at the right side of Figure 3.6a, where a kinematic condition would be applied, or as in Figure 3.6b, where a "radiation" boundary condition is applied which requires that only outgoing waves occur at infinity. This precludes incoming waves which would not be physically meaningful in a wavemaker problem.

For waves that are periodic in space and time, the boundary condition is expressed as a periodicity condition,

$$\phi(x, t) = \phi(x + L, t) \qquad (3.18a)$$

$$\phi(x, t) = \phi(x, t + T) \qquad (3.18b)$$

where L is the wave length and T is the wave period.

3.3 SUMMARY OF THE TWO-DIMENSIONAL PERIODIC WATER WAVE BOUNDARY VALUE PROBLEM

The governing second-order differential equation for the fluid motion under a periodic two-dimensional water wave is the Laplace equation, which holds throughout the fluid domain consisting of one wave, shown in Figure 3.7.

$$\nabla^2 \phi = 0, \qquad 0 < x < L, \qquad -h < z < \eta \qquad (3.19)$$

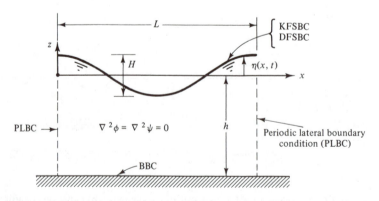

Figure 3.7 Boundary value problem specification for periodic water waves.

At the bottom, which is assumed to be horizontal, a no-flow condition applies (BBC):

$$w = 0 \qquad \text{on } z = -h \qquad (3.20a)$$

or

$$-\frac{\partial \phi}{\partial z} = 0 \qquad \text{on } z = -h \qquad (3.20b)$$

At the free surface, two conditions must be satisfied. The KFSBC, Eq. (3.11c),

$$-\frac{\partial \phi}{\partial z} = \frac{\partial \eta}{\partial t} - \frac{\partial \phi}{\partial x}\frac{\partial \eta}{\partial x} \qquad \text{on } z = \eta(x, t) \qquad (3.11c)$$

The DFSBC, Eq. (3.13), with $p_\eta = 0$,

$$-\frac{\partial \phi}{\partial t} + \frac{1}{2}\left[\left(\frac{\partial \phi}{\partial x}\right)^2 + \left(\frac{\partial \phi}{\partial z}\right)^2\right] + g\eta = C(t) \qquad \text{on } z = \eta(x, t) \qquad (3.13)$$

Finally, the periodic lateral boundary conditions apply in both time and space, Eqs. (3.18).

$$\phi(x, t) = \phi(x + L, t) \qquad (3.18a)$$

$$\phi(x, t) = \phi(x, t + T) \qquad (3.18b)$$

3.4 SOLUTION TO LINEARIZED WATER WAVE BOUNDARY VALUE PROBLEM FOR A HORIZONTAL BOTTOM

In this section a solution is developed for the boundary value problem representing waves that are periodic in space and time propagating over a horizontal bottom. This requires solution of the Laplace equation with the boundary conditions as expressed by Eqs. (3.19), (3.20b), (3.11c), (3.13), and (3.18).

3.4.1 Separation of Variables

A convenient method for solving some linear partial differential equations is called separation of variables. The assumption behind its use is that the solution can be expressed as a product of terms, each of which is a function of only one of the independent variables. For our case,

$$\phi(x, z, t) = X(x) \cdot Z(z) \cdot T(t) \qquad (3.21)$$

where $X(x)$ is some function that depends only on x, the horizontal coordinate, $Z(z)$ depends only on z, and $T(t)$ varies only with time. Since we know that ϕ must be periodic in time by the lateral boundary conditions, we can specify $T(t) = \sin \sigma t$. To find σ, the angular frequency of the wave, we utilize the periodic boundary condition, Eq. (3.18b).

$$\sin \sigma t = \sin \sigma(t + T)$$

or

$$\sin \sigma t = \sin \sigma t \cos \sigma T + \cos \sigma t \sin \sigma T$$

which is true for $\sigma T = 2\pi$ or $\sigma = 2\pi/T$. Equally as likely, we could have chosen $\cos \sigma t$ or some combination of the two: $A \cos \sigma t + B \sin \sigma t$. Since the equations to be solved will be linear and superposition is valid, we can defer generalizing the solution in time until after the solution components have been obtained and discussed. The velocity potential now takes the form

$$\phi(x, z, t) = X(x) \cdot Z(z) \cdot \sin \sigma t \qquad (3.22)$$

Substituting into the Laplace equation, we have

$$\frac{d^2 X(x)}{dx^2} \cdot Z(z) \cdot \sin \sigma t + X(x) \cdot \frac{d^2 Z(z)}{dz^2} \cdot \sin \sigma t = 0$$

Dividing through by ϕ gives us

$$\frac{1}{X} \frac{\partial^2 X}{\partial x^2} + \frac{1}{Z} \frac{\partial^2 Z}{\partial z^2} = 0 \qquad (3.23)$$

Clearly, the first term of this equation depends on x alone, while the second term depends only on z. If we consider a variation in z in Eq. (3.23) holding x constant, the second term could conceivably vary, whereas the first term could not. This would give a nonzero sum in Eq. (3.23) and thus the equation would not be satisfied. The only way that the equation would hold is if each term is equal to the same constant except for a sign difference, that is,

$$\frac{d^2 X(x)/dx^2}{X(x)} = -k^2 \qquad (3.24a)$$

$$\frac{d^2 Z(z)/dz^2}{Z(z)} = +k^2 \qquad (3.24b)$$

The fact that we have assigned a minus constant to the x term is not of importance, as we will permit the separation constant k to have an imaginary value in this problem and in general the separation constant can be complex.

Equations (3.24) are now ordinary differential equations and may be solved separately. Three possible cases may now be examined depending on the nature of k; these are for k real, $k = 0$, and k a pure imaginary number. Table 3.1 lists the separate cases. (Note that if k consisted of both a real and an imaginary part, this could imply a change of wave height with distance, which may be valid for cases of waves propagating with damping or wave growth by wind.)

TABLE 3.1   Possible Solutions to the Laplace Equation, Based on Separation of Variables

Character of k, the Separation Constant	Ordinary Differential Equations	Solutions
Real $k^2 > 0$	$\dfrac{d^2X}{dx^2} + k^2X = 0$ $\dfrac{d^2Z}{dz^2} - k^2Z = 0$	$X(x) = A\cos kx + B\sin kx$ $Z(z) = Ce^{kz} + De^{-kz}$
$k = 0$	$\dfrac{d^2X}{dx^2} = 0$ $\dfrac{d^2Z}{dz^2} = 0$	$X(x) = Ax + B$ $Z(z) = Cz + D$
Imaginary $k^2 < 0, k = i\,|k|$ $|k|$ = magnitude of k	$\dfrac{d^2X}{dx^2} - |k|^2X = 0$ $\dfrac{d^2Z}{dz^2} + |k|^2Z = 0$	$X(x) = Ae^{|k|x} + Be^{-|k|x}$ $Z(z) = C\cos|k|z + D\sin|k|z$

3.4.2 Application of Boundary Conditions

The boundary conditions serve to select, from the trial solutions in Table 3.1, those which are applicable to the physical situation of interest. In addition, the use of the boundary conditions allows determination of some of the unknown constants (e.g., A, B, C, and D).

Lateral periodicity condition.   All solutions in Table 3.1 satisfy the Laplace equation; however, some of them are not periodic in x; in fact, the solution is spatially periodic only if k is real[3] and nonzero. Therefore, we have as a solution to the Laplace equation the following velocity potential:

$$\phi(x, z, t) = (A\cos kx + B\sin kx)(Ce^{kz} + De^{-kz})\sin \sigma t \qquad (3.25)$$

To satisfy the periodicity requirement (3.18a) explicitly,

$$A\cos kx + B\sin kx = A\cos k(x + L) + B\sin k(x + L)$$

$$= A(\cos kx \cos kL - \sin kx \sin kL)$$

$$+ B(\sin kx \cos kL + \cos kx \sin kL)$$

which is satisfied for $\cos kL = 1$ and $\sin kL = 0$; which means that $kL = 2\pi$ or k (called the wave number) $= 2\pi/L$.

Using the superposition principle, we can divide ϕ into several parts. Let us keep, for present purposes, only $\phi = A\cos kx(Ce^{kz} + De^{-kz})\sin \sigma t$. Lest

---

[3]For $k = 0$, A is zero. This ultimately yields $\phi = B\sin \sigma t$.

this be thought of as sleight of hand, the $B \sin kx$ term will be added back in later by superposition.

Bottom boundary condition for horizontal bottom. Substituting in the bottom boundary condition yields

$$w = -\frac{\partial \phi}{\partial z} = -A \cos kx (kCe^{kz} - kDe^{-kz}) \sin \sigma t = 0 \qquad \text{on } z = -h \quad (3.26)$$

or

$$-Ak \cos kx (Ce^{-kh} - De^{kh}) \sin \sigma t = 0$$

For this equation to be true for any x and t, the terms within the parentheses must be identically zero, which yields

$$C = De^{2kh}$$

The velocity potential now reads

$$\phi = A \cos kx (De^{2kh} e^{kz} + De^{-kz}) \sin \sigma t$$

or, factoring out De^{kh},

$$\phi = ADe^{kh} \cos kx (e^{k(h+z)} + e^{-k(h+z)}) \sin \sigma t$$

or

$$\phi = G \cos kx \cosh k(h + z) \sin \sigma t \qquad (3.27)$$

where $G = 2ADe^{kh}$, a new constant.

Dynamic free surface boundary condition. As stated previously, the Bernoulli equation can be used to specify a constant pressure on the surface of the water. Yet the Bernoulli equation must be satisfied on $z = \eta(x, t)$, which is a priori unknown. A convenient method used to evaluate the condition, then, is to evaluate it on $z = \eta(x, t)$ by expanding the value of the condition at $z = 0$ (a known location) by the truncated Taylor series.

$$\text{(Bernoulli equation)}_{z=\eta} = \text{(Bernoulli equation)}_{z=0}$$

$$\qquad (3.28)$$

$$+ \eta \frac{\partial}{\partial z} \text{(Bernoulli equation)}_{z=0} + \cdots$$

or

$$\left(gz - \frac{\partial \phi}{\partial t} + \frac{u^2 + w^2}{2} \right)_{z=\eta} = \left(gz - \frac{\partial \phi}{\partial t} + \frac{u^2 + w^2}{2} \right)_{z=0}$$

$$+ \eta \left[g - \frac{\partial^2 \phi}{\partial z \, \partial t} + \frac{1}{2} \frac{\partial}{\partial z} (u^2 + w^2) \right]_{z=0} + \cdots = C(t)$$

where $p = 0$ on $z = \eta$.

Now for infinitesimally small waves, η is small, and therefore it is assumed that velocities and pressures are small; thus any products of these variables are very small: $\eta \ll 1$, but $\eta^2 \ll \eta$, or $u\eta \ll \eta$. If we neglect these small terms, the Bernoulli equation is written as

$$\left(-\frac{\partial \phi}{\partial t} + g\eta \right)_{z=0} = C(t)$$

This process is called linearization. We have retained only the terms that are linear in our variables.[4] The resulting linear dynamic free surface boundary condition relates the instantaneous displacement of the free surface to the time rate of change of the velocity potential,

$$\eta = \frac{1}{g}\frac{\partial \phi}{\partial t}\bigg|_{z=0} + \frac{C(t)}{g} \tag{3.29}$$

If we substitute the velocity potential, as given by Eq. (3.27),

$$\eta = \frac{G\sigma}{g} \cos kx \cosh k(h+z)\cos \sigma t \,|_{z=0} + \frac{C(t)}{g}$$

$$= \left[\frac{G\sigma \cosh kh}{g} \right] \cos kx \cos \sigma t + \frac{C(t)}{g} \tag{3.30}$$

Since by our definition η will have a zero spatial and temporal mean, $C(t) \equiv 0$.[5] The terms within the brackets are constant; therefore, η is given as a constant times periodic terms in space and time plus a function of time. We can rewrite η as

$$\eta = \frac{H}{2}\cos kx \cos \sigma t \tag{3.31}$$

The last substitution came about by comparing the analytical representation of η to the physical model, as shown in Figure 3.7. G can now be obtained from

$$G = \frac{Hg}{2\sigma \cosh kh}$$

The velocity potential is now

$$\phi = \frac{Hg \cosh k(h+z)}{2\sigma \cosh kh} \cos kx \sin \sigma t \tag{3.32}$$

The velocity potential is now prescribed in terms of H, σ, h, and k. The

[4]Linear in the sense that variables are only raised to the first power.
[5]Had we not used $p(\eta) = 0$, how would $C(t)$ be changed?

first three of these would be available from the data or alternatively the wave length might be known and σ unknown.

Kinematic free surface boundary condition.   The remaining free surface boundary condition will be utilized to establish the relationship between σ and k. Using the Taylor series expansion to relate the boundary condition at the unknown elevation, $z = \eta(x, t)$ to $z = 0$, we have

$$\left(w - \frac{\partial \eta}{\partial t} - u \frac{\partial \eta}{\partial x} \right)_{z=\eta} = \left(w - \frac{\partial \eta}{\partial t} - u \frac{\partial \eta}{\partial x} \right)_{z=0}$$

$$+ \eta \frac{\partial}{\partial z} \left(w - \frac{\partial \eta}{\partial t} - u \frac{\partial \eta}{\partial x} \right)_{z=0} + \cdots = 0$$

Again retaining only the terms that are linear in our small parameters, η, u, and w, and recalling that η is not a function of z, the linearized kinematic free surface boundary condition results:

$$w = \frac{\partial \eta}{\partial t} \bigg|_{z=0} \tag{3.33a}$$

or

$$-\frac{\partial \phi}{\partial z} \bigg|_{z=0} = \frac{\partial \eta}{\partial t} \tag{3.33b}$$

Substituting for ϕ and η gives us

$$-\frac{H}{2} \frac{gk}{\sigma} \frac{\sinh k(h + z)}{\cosh kh} \cos kx \sin \sigma t \,|_{z=0}$$

$$= -\frac{H}{2} \sigma \cos kx \sin \sigma t$$

or

$$\boxed{\sigma^2 = gk \tanh kh} \tag{3.34}$$

Rewriting this equation as $\sigma^2 h/gkh = \tanh kh$ and plotting each term versus kh for a particular value of $\sigma^2 h/g$ yields Figure 3.8. The solution is determined by the intersection of the two curves. Therefore, the equation has only one solution or equivalently one value of k for given values of σ and h.

Noting that by definition a propagating wave will travel a distance of one wave length L, in one wave period T, and recalling that $\sigma = 2\pi/T$ and $k = 2\pi/L$, it is clear that the speed of wave propagation C can be expressed from Eq. (3.34) as

$$\left(\frac{2\pi}{T} \right)^2 = g \frac{2\pi}{L} \tanh kh$$

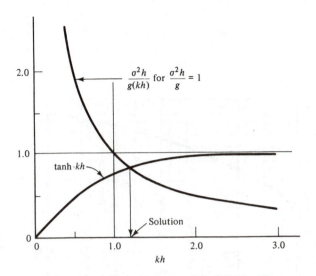

Figure 3.8   Illustrating single root to dispersion equation.

or

$$C^2 = \frac{L^2}{T^2} = \frac{g}{k} \tanh kh \qquad (3.35)$$

A similar algebraic manipulation of Eq. (3.34) will yield a relationship for the wave length,

$$L = \frac{g}{2\pi} T^2 \tanh \frac{2\pi h}{L} \qquad (3.36)$$

In deep water, kh is large and $\tanh 2\pi h/L = 1.0$; therefore, $L = L_0 = gT^2/2\pi$, where the zero subscript is used to denote deep water values. In general, then,

$$L = L_0 \tanh kh \qquad (3.37)$$

Thus the wave length continually decreases with decreasing depth for a constant wave period.

Equations (3.34), (3.35), and (3.37), which are really the same equation expressed in slightly different variables, are referred to as the "dispersion" equation, because they describe the manner in which a field of propagating waves consisting of many frequencies would separate or "disperse" due to the different celerities of the various frequency components.

The wave speed, or celerity, C, has been defined as $C = L/T$. Therefore,

$$C = \frac{L_0}{T} \tanh kh \qquad (3.38a)$$

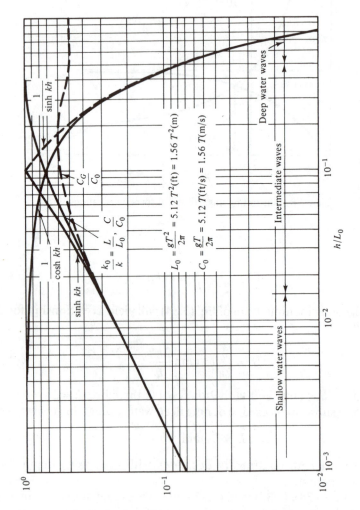

Figure 3.9 Variation of various wave parameters with h/L_0.

or

$$C = C_0 \tanh kh \qquad (3.38b)$$

since, as will be shown later, the wave period does not change with depth. Waves of constant period slow down as they enter shallow water. Figure 3.9 presents, as a function of h/L_0, the ratio $C/C_0 (= L/L_0 = k_0/k)$ and a number of other variables commonly occuring in water wave calculations. This figure provides a convenient graphical means to determine intermediate and shallow water values of these variables.

3.4.3 Summary of Standing Waves

One solution of the boundary value problem for small-amplitude waves has been found to be

$$\phi = \frac{H}{2} \frac{g}{\sigma} \frac{\cosh k(h + z)}{\cosh kh} \cos kx \sin \sigma t$$

$$\eta(x, t) = \frac{1}{g} \frac{\partial \phi}{\partial t}\bigg|_{z=0} = \frac{H}{2} \cos kx \cos \sigma t \qquad (3.39)$$

where $\sigma^2 = gk \tanh kh$.

The wave form is shown in Figure 3.10. At $\sigma t = \pi/2$, the wave form is zero for all x, at $\sigma t = 0$, it has a cosine shape and at other times, the same cosine shape with different magnitudes. This wave form is obviously a "standing wave," as it does not propagate in any direction. At positions $kx = \pi/2$, and $3\pi/2$, and so on, nodes exist; that is, there is no motion of the free surface at these points. Standing waves often occur when incoming waves are completely reflected by vertical walls. At which phase position would the wall be located? See Figure 4.6 for a hint.

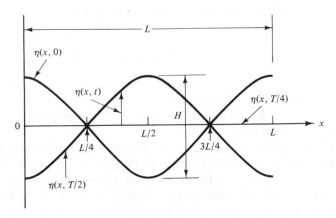

Figure 3.10   Water surface displacement associated with a standing water wave.

3.4.4 Progressive Waves

Consider another standing wave,

$$\phi(x, z, t) = \frac{H}{2} \frac{g}{\sigma} \frac{\cosh k(h + z)}{\cosh kh} \sin kx \cos \sigma t \qquad (3.40)$$

This velocity potential is also a solution to the Laplace equation and all the boundary conditions, as may be verified readily. It is, in fact, one of the solutions that we discarded. It differs from the previous solution in that the x and t terms are 90° out of phase. The associated water surface displacement is

$$\eta(x, t) = -\frac{1}{g} \frac{\partial \phi}{\partial t} \bigg|_{z=0} = -\frac{H}{2} \sin kx \sin \sigma t \qquad (3.41)$$

as determined from the linearized DFSBC. Remembering that the Laplace equation is linear and superposition is valid, we can add or subtract solutions to the linearized boundary value problem to generate new solutions. If we subtract the present velocity potential in Eq. (3.40) from the previous solution we had, Eq. (3.32), we obtain

$$\phi = \frac{H}{2} \frac{g}{\sigma} \frac{\cosh k(h + z)}{\cosh kh} (\cos kx \sin \sigma t - \sin kx \cos \sigma t)$$

$$= -\frac{H}{2} \frac{g}{\sigma} \frac{\cosh k(h + z)}{\cosh kh} \sin (kx - \sigma t) \qquad (3.42)$$

This new velocity potential has a water surface elevation, given as

$$\eta(x, t) = -\frac{1}{g} \frac{\partial \phi}{\partial t} \bigg|_{z=0} = \frac{H}{2} \cos (kx - \sigma t) \qquad (3.43)$$

Had we just subtracted the two $\eta(x, t)$ corresponding to the two velocity potentials, we would have had

$$\eta(x, t) = \frac{H}{2} \cos kx \cos \sigma t + \frac{H}{2} \sin kx \sin \sigma t = \frac{H}{2} \cos (kx - \sigma t)$$

which is the same result. This should not have been a surprise, as the total boundary value problem has been linearized and superposition is valid for all variables in the problem.

Examining the equation for the water surface profile, it is clear that this wave form moves with time. To determine the direction of movement, let us examine the same point on the wave form at two different time values, t_1 and t_2. The x location of the point also changes with time. In Figure 3.11, the locations of the point at time t_1 and t_2 are shown. The speed at which the

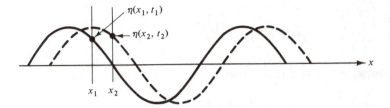

Figure 3.11   Characteristics of a propagating wave form.

wave propagated from one point to the other is C, given as

$$C = \frac{x_2 - x_1}{t_2 - t_1}$$

We further point out that the same point on the wave crest implies that we are examining the wave at the same phase, that is, at constant values of the argument of the trigonometric function of x and t. Therefore, we expect that

$$\eta(x_1, t_1) = \eta(x_2, t_2)$$

or, in fact,

$$kx_1 - \sigma t_1 = kx_2 - \sigma t_2$$
$$k(x_1 - x_2) = \sigma(t_1 - t_2)$$

or

$$\frac{\sigma}{k} = \frac{2\pi/T}{2\pi/L} = C = \frac{x_1 - x_2}{t_1 - t_2} = \frac{x_2 - x_1}{t_2 - t_1}$$

as before. Therefore, if $t_2 > t_1$, $x_2 > x_1$, the wave form propagates from left to right. Had the argument of the trigonometric function been $(kx + \sigma t)$, the waves would propagate from right to left (i.e., in the negative x direction).

Simplifications for shallow and deep water.   The hyperbolic functions have convenient shallow and deep water asymptotes, and often it is helpful to use them to obtain simplified forms of the equations describing wave motion. For example, the function $\cosh kh$, which appears in the denominator for the velocity potential, is defined as

$$\cosh kh = \frac{e^{kh} + e^{-kh}}{2}$$

For a small argument, the exponential function e^z can be expanded to $z = kh$ in a Taylor series about zero as

$$e^{(0+kh)} = e^0 + \frac{de^z}{dz}\bigg|_{z=0} kh + \frac{d^2 e^z}{dz^2}\bigg|_{z=0} \frac{(kh)^2}{2!} + \cdots$$

or

$$e^{kh} = 1 + kh + \frac{(kh)^2}{2} + \cdots$$

Of course, e^{-kh} would then equal

$$e^{-kh} = 1 - kh + \frac{(kh)^2}{2} + \cdots$$

Therefore, for small kh,

$$\cosh kh = \frac{1}{2}\left[(1 + kh + \frac{(kh)^2}{2}\cdots) + (1 - kh + \frac{(kh)^2}{2}\cdots)\right]$$

$$\cong 1 + \frac{(kh)^2}{2}$$

For *large kh*, $\cosh kh = e^{kh}/2$ as e^{-kh} becomes quite small. Table 3.2 presents the asymptotes.

TABLE 3.2   **Asymptotic Forms of Hyperbolic Functions**

Function	Large kh	Small kh
$\cosh kh$	$e^{kh}/2$	1
$\sinh kh$	$e^{kh}/2$	kh
$\tanh kh$	1	kh

It is worthwhile to distinguish the regions within which these asymptotic approximations become valid. Figure 3.12 is a plot of hyperbolic functions together with the asymptotes, $f_1 = kh, f_2 = 1.0, f_3 = e^{kh}/2$. The percentage values presented in Figure 3.12 represent, for particular ranges of kh, the errors incurred by using the asymptotes rather than the actual value of the function. The largest error is 5%. The lower scale on the figure is the relative depth. Note that due to this dimensionless representation a 200-m-long wave in 1000 m of water has the same relative depth as a 0.2-m wave in 1 m of water. Limits for three regions are denoted in the figure: $kh < \pi/10$, $\pi/10 < kh < \pi$, and $kh > \pi$. These regions are defined as the shallow water, intermediate depth, and the deep water regions, respectively. It may be justified to modify the limits of these regions for particular applications.

The dispersion relationship in shallow and deep water.   The dispersion relationship for *shallow water* reduces in the following manner:

$$\sigma^2 = gk \tanh kh = gk^2h$$

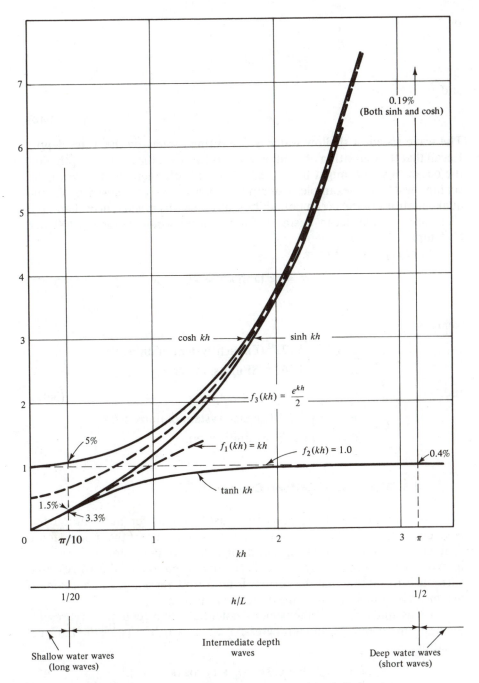

Figure 3.12 Relative depth and asymptotes to hyperbolic functions.

or

$$\frac{\sigma^2}{k^2} = C^2 = gh$$

and

$$C = \sqrt{gh} \tag{3.44}$$

The wave speed in shallow water is determined solely by the water depth. Recall that the definition of shallow water is based on the relative depth. For the ocean, where h might be $\approx$ 1 km, a wave with a length of 20 km is in shallow water. For example, tsunamis, which are waves caused by earthquake motions of the ocean boundaries, have lengths much longer than this. The speed in the ocean basins for long waves would be about 100 m/s (225 mph).

For deep water, $kh > \pi$,

$$\sigma^2 = gk \tanh kh \approx gk$$

$$L = L_0$$

where

$$L_0 = \frac{g}{2\pi} T^2 = \begin{cases} 5.12T^2 \text{ (English system of units, ft)} \\ 1.56T^2 \text{ (SI units, meters)} \end{cases}$$

and   (3.45)

$$C_0 = \frac{g}{2\pi} T = \begin{cases} 5.12T \text{ (English system of units, ft/s)} \\ 1.56T \text{ (SI units, m/s)} \end{cases}$$

3.4.5 Waves with Uniform Current U_0

As an example of the procedure just followed for the solution for progressive and standing waves, it is instructive to repeat the process for a different case: water waves propagating on a current. For example, for waves in rivers or on ocean currents, a first approximation to the waves and currents is to assume that the current is uniform over depth and horizontal distance and flowing in the same direction as the waves.

An assumed form of the velocity potential will be chosen to represent the uniform current U_0 and a progressive wave, which satisfies the Laplace equation.

$$\phi = -U_0 x + A \cosh k(h + z) \cos (kx - \sigma t) \tag{3.46}$$

The form of this solution guarantees periodicity of the wave in space and time and satisfies the no-flow bottom boundary condition. It remains necessary to satisfy the linearized form of the KFSBC and the DFSBC. Yet we cannot just apply the forms that we arrived at earlier, as errors would be incurred because the velocity U_0 is no longer necessarily small; we must rederive the linear boundary conditions.

The dynamic free surface boundary condition.   Again, we will expand the Bernoulli equation about the free surface on which a zero gage pressure is prescribed.

$$\left[\frac{1}{2}(u^2 + w^2) + gz - \frac{\partial\phi}{\partial t}\right]_{z=\eta(x,t)} \simeq \left[\frac{1}{2}(u^2 + w^2) + gz - \frac{\partial\phi}{\partial t}\right]_{z=0}$$

(3.47)

$$+ \eta\frac{\partial}{\partial z}\left[\frac{1}{2}(u^2 + w^2) + gz - \frac{\partial\phi}{\partial t}\right]_{z=0} + \cdots = C(t)$$

Now the horizontal velocity is

$$u = -\frac{\partial\phi}{\partial x} = U_0 + Ak \cosh k(h + z) \sin(kx - \sigma t)$$

Therefore, the u^2 term is

$$u^2 = U_0^2 + 2AkU_0 \cosh k(h + z) \sin(kx - \sigma t)$$
$$+ A^2 k^2 \cosh^2 k(h + z) \sin^2(kx - \sigma t)$$

For infinitesimal waves, it is expected that the wave-induced horizontal velocity component would be small (i.e., Ak small), and therefore $(Ak)^2$ would be much smaller. We will then neglect the last term in the equation above.

The linearized Bernoulli equation [i.e., dropping all terms of order $(Ak)^2$], evaluated on $z = 0$, is now

$$\tfrac{1}{2}[U_0^2 + 2AkU_0 \cosh kh \sin(kx - \sigma t)]$$
$$- A\sigma \cosh kh \sin(kx - \sigma t) + g\eta = C(t)$$

or

$$\eta(x, t) = -\frac{U_0^2}{2g} + \frac{A\sigma}{g}\left(1 - \frac{U_0 k}{\sigma}\right)\cosh kh \sin(kx - \sigma t) + C(t) \qquad (3.48)$$

To determine the Bernoulli term $C(t)$, we average both sides of Eq. (3.48) over space. Since the space average of $\eta(x, t)$ is taken to be zero, it is clear that $C(t) = \text{constant} = U_0^2/2g$. Also, if we define a water surface displacement, $\eta(x, t) = H/2 \sin(kx - \sigma t)$, then

$$A = \frac{gH}{2\sigma(1 - U_0/C)\cosh kh} \qquad (3.49)$$

The kinematic free surface boundary condition. The remaining boundary condition to be satisfied is the linearized form of the KFSBC.

$$\frac{\partial \eta}{\partial t} - \frac{\partial \phi}{\partial x}\frac{\partial \eta}{\partial x} = -\frac{\partial \phi}{\partial z}, \qquad z = \eta$$

Expanding about the still water level, we have

$$\left(\frac{\partial \eta}{\partial t} - \frac{\partial \phi}{\partial x}\frac{\partial \eta}{\partial x}\right) + \eta\frac{\partial}{\partial z}\left(\frac{\partial \eta}{\partial t} - \frac{\partial \phi}{\partial x}\frac{\partial \eta}{\partial x}\right) + \cdots$$

$$= -\frac{\partial \phi}{\partial z} - \eta\frac{\partial}{\partial z}\left(\frac{\partial \phi}{\partial z}\right) + \cdots, \qquad z = 0$$

or, retaining only the linear terms,

$$\frac{\partial \eta}{\partial t} + U_0\frac{\partial \eta}{\partial x} = -\frac{\partial \phi}{\partial z}, \qquad z = 0 \tag{3.50}$$

Substituting for η and ϕ yields the following dispersion equation for the case of a uniform current U_0:

$$\sigma^2 = \frac{gk \tanh kh}{(1 - U_0/C)^2} \tag{3.51}$$

or, another form can be developed by using the relationship $\sigma = kC$:

$$\sigma^2\left(1 - \frac{U_0 k}{\sigma}\right)^2 = gk \tanh kh$$

or

$$\sigma = U_0 k + \sqrt{gk \tanh kh} \tag{3.52}$$

The second term on the right-hand side is the angular frequency formula obtained without a current.

In terms of the celerity, the dispersion relationship can be written as

$$(C - U_0)^2 = \frac{g}{k} \tanh kh \tag{3.53}$$

It is worthwhile noting that it is possible to solve the preceding problem of a uniform current simply by adopting a reference frame which moves with the current U_0. With reference to our new coordinate system, there is no current and the methods, equations, and solutions obtained are therefore identical to those obtained originally for the case of no current.

When relating this moving frame solution for a stationary reference system, it is simply necessary to recognize that (1) the wavelength is the same in both systems; (2) the period T relative to a stationary reference system is related to the period T' relative to the reference system moving with the current U_0 by

$$T = \frac{T'}{1 + U_0/C'} = T'\left(1 - \frac{U_0}{C}\right) \tag{3.54}$$

where C' is the speed relative to the moving observer; and (3) the total water particle velocity is $U_0 + u_w$, where u_w is the wave-induced component. It is noted that in the case of arbitrary depth, when T and h are given, it is necessary to solve for the wave length from Eq. (3.54) by iteration.[6]

For shallow water, we have, from Eq. (3.53),

$$C = \frac{L}{T} = U_0 + \sqrt{gh} \tag{3.55}$$

That is, since the celerity of the wave is independent of wave length, it is simply increased by the advecting current U_0. For deep water, the corresponding result is determined by solving Eq. (3.53) for C using the quadratic solution and replacing k with σ/C, that is,

$$C = \left(U_0 + \frac{g}{2\sigma}\right) + \sqrt{\frac{U_0 g}{\sigma} + \frac{1}{4}\left(\frac{g}{\sigma}\right)^2} \tag{3.56}$$

For small currents with respect to C (i.e., $U_0 < g/\sigma$),

$$C \simeq \frac{g}{\sigma} + 2U_0$$

Capillary waves.   As indicated in Eq. (3.16), the surface tension at the water surface causes a modification to the dynamic free surface boundary condition. To explore the effects of surface tension, we proceed as before by choosing a velocity potential of the form

$$\phi = A \cosh k(h + z) \sin (kx - \sigma t) \tag{3.57}$$

which is appropriate for a progressive water wave, satisfies the Laplace equation, and all boundary conditions except those at the upper surface. The surface displacement associated with Eq. (3.57) will be of the form

$$\eta = \frac{H}{2} \cos (kx - \sigma t) \tag{3.58}$$

Substituting Eqs. (3.57) and (3.58) into the linearized form of Eq. (3.16), and employing the linearized form of the kinematic free surface boundary condition, Eq. (3.33a), the dispersion equation is found to be

$$C^2 = \frac{g}{k}\left(1 + \sigma\frac{k^2}{\rho g}\right) \tanh kh \tag{3.59}$$

---

[6]This technique has been applied to nearly breaking waves by Dalrymple and Dean (1975).

and it can be seen that the effect of surface tension is to increase the celerity for all wave frequencies. The effect of surface tension can be examined most readily by considering the case of deep water waves.

$$C^2 = \frac{g}{k} + \frac{\sigma' k}{\rho} \tag{3.60}$$

That is, the contributions due to the speed of short waves (large wave numbers) is small due to the effect of gravity and large due to the effect of surface tension. There is a minimum speed C_m at which waves can propagate, found in the usual way:

$$\frac{\partial C}{\partial k} = 0 \tag{3.61}$$

which leads to

$$k_m = \sqrt{\frac{g\rho}{\sigma'}} \tag{3.62}$$

$$C_m^2 = \sqrt{\frac{\sigma' g}{\rho}} + \sqrt{\frac{\sigma' g}{\rho}} = 2\sqrt{\frac{\sigma' g}{\rho}} \tag{3.63}$$

That is, the contributions from gravity and surface tension to C_m^2 are equal. For a reasonable value of surface tension, $\sigma' = 7.4 \times 10^{-2}\,\text{N/m}$, $C_m \simeq 23.2\,\text{cm/s}$, which occurs at a wave period of approximately 0.074 s. Figure 3.13 presents

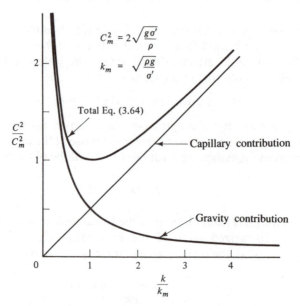

Figure 3.13 Capillary and gravitational components of the square of wave celerity in deep water.

the relationship

$$\frac{C^2}{C_m^2} = \frac{1}{2}\left(\frac{1}{k/k_m} + \frac{k}{k_m}\right) \tag{3.64}$$

3.4.6 The Stream Function for Small-Amplitude Waves

For convenience, the velocity potential has been used to develop the small-amplitude wave theory, yet often it is convenient to use the stream function representation. Therefore, we can use the Cauchy–Riemann equations, Eqs. (2.82), to develop them from the velocity potentials.

Progressive waves.

$$\phi(x, z, t) = -\frac{H}{2}\frac{g}{\sigma}\frac{\cosh k(h + z)}{\cosh kh} \sin (kx - \sigma t) \tag{3.65}$$

$$\psi(x, z, t) = -\frac{H}{2}\frac{g}{\sigma}\frac{\sinh k(h + z)}{\cosh kh} \cos (kx - \sigma t) \tag{3.66}$$

It is often convenient for a progressive wave that propagates without change of form to translate the coordinate system horizontally with the speed of the wave, that is, with the celerity C, as this then gives a steady flow condition.

$$\psi = Cz - \frac{H}{2}\frac{g}{\sigma}\frac{\sinh k(h + z)}{\cosh kh} \cos kx \tag{3.67}$$

Standing waves.   From before,

$$\phi = \frac{H}{2}\frac{g}{\sigma}\frac{\cosh k(h + z)}{\cosh kh} \cos kx \sin \sigma t \tag{3.68}$$

$$\psi = -\frac{H}{2}\frac{g}{\sigma}\frac{\sinh k(h + z)}{\cosh kh} \sin kx \sin \sigma t \tag{3.69}$$

The streamlines and velocity potential for both cases are shown in Figure 3.14. The streamlines and potential lines are lines of constant ψ and ϕ.

3.5 APPENDIX: APPROXIMATE SOLUTIONS TO THE DISPERSION EQUATION

The solution to the dispersion relationship, Eq. (3.34), for k is not difficult to obtain for given σ and h. However, since the relationship is a transcendental

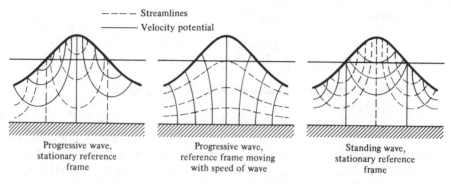

---- Streamlines
———— Velocity potential

Progressive wave,
stationary reference
frame

Progressive wave,
reference frame moving
with speed of wave

Standing wave,
stationary reference
frame

Figure 3.14 Approximate streamlines and lines of constant velocity potential for various types of wave systems and reference frames.

equation, in that it is not algebraic, graphical (see Figure 3.8) and iterative techniques are used (see Problem 3.15).

Eckart (1951) developed an approximate wave theory with a corresponding dispersion relationship,

$$\sigma^2 = gk \sqrt{\tanh\left(\frac{\sigma^2}{g} h\right)}$$

This can be solved directly for k and generally is in error by only a few percent. This equation therefore can be used as a first approximation to k for an iterative technique or can be used to determine k directly if accuracy is not a paramount consideration.

Recently, Hunt (1979) proposed an approximate solution that can be solved directly for kh:

$$(kh)^2 = y^2 + \frac{y}{1 + \sum_{n=1}^{6} d_n y^n}$$

where $y = \sigma^2 h/g = k_0 h$ and $d_1 = 0.666\ldots$, $d_2 = 0.355\ldots$, $d_3 = 0.1608465608$, $d_4 = 0.0632098765$, $d_5 = 0.0217540484$, and $d_6 = 0.0065407983$. The last digits in d_1 and d_2 are repeated seven more times. This formula can be conveniently used on a programmable calculator.

The wave celerity was also obtained

$$\frac{C^2}{gh} = [y + (1 + 0.6522y + 0.4622y^2 + 0.0864y^4 + 0.0675y^5)^{-1}]^{-1}$$

which is accurate to 0.1% for $0 < y < \infty$.

REFERENCES

BLAND, D. R., *Solutions of Laplace's Equation*, Routledge & Kegan Paul, London, 1961.

DALRYMPLE, R. A., and R. G. DEAN, "Waves of Maximum Height on Uniform Currents," *J. Waterways, Harbors Coastal Eng. Div., ASCE*, Vol. 101, No. WW3, pp. 259–268, 1975.

ECKART, C., "Surface Waves on Water of Variable Depth," SIO 51-12, Scripps Institute of Oceanography, Aug. 1951.

HUNT, J. N., "Direct Solution of Wave Dispersion Equation," *J. Waterways, Ports, Coastal Ocean Div., ASCE*, Vol. 105, No. WW4, pp. 457–459, 1979.

PROBLEMS

3.1 The linearization of the kinematic and dynamic free surface boundary conditions involved neglecting nonlinear terms. Show, for both the conditions, that this linearization implies that

$$\frac{kH}{2} \ll 1$$

3.2 Near the bow of a moving submarine, the hull can be represented as a moving parabola,

$$D(z - A)^2 = -(x - Ut)$$

where U is the speed of the submarine, A represents the depth of the centerline of the submarine below the free surface, and D is a constant.

(a) Plot the hull shape at $t = 0$ and $t = 1$ s if the submarine is moving at 2 m/s.

(b) Determine the kinematic boundary condition at the hull.

3.3 The equation for the stationary boundary $\zeta(x)$ of an incompressible fluid is

$$\zeta(x) = Ae^{-Kx}$$

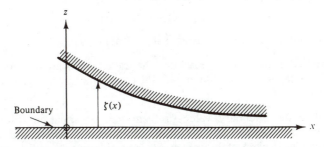

The horizontal velocity component may be regarded to be approximately uniform in the z direction. If $u(x = 0) = 40$ cm/s, $A = 30$ cm, and $K = 0.02$ cm^{-1}, calculate w at the upper boundary for $x = 50$ cm.

3.4 The equation for the upper moving boundary $\zeta_u(x, t)$ of an incompressible fluid is

$$\zeta_u(x, t) = Ae^{-(kx - Mt)}$$

The lower boundary ζ_ℓ is expressed by

$$\zeta_\ell(x, t) = 0$$

$$A = 30 \text{ cm}$$

$$k = 0.02 \text{ cm}^{-1}$$

$$M = 0.1 \text{ s}^{-1}$$

(a) Sketch the boundaries for $t = 0$.
(b) Discuss the motional characteristics of the upper boundary (i.e., speed and direction).
(c) The horizontal velocity component (u) may be regarded to be approximately uniform in the z direction. If

$$u(x = 0, t = 10 \text{ s}) = 40 \text{ cm/s}$$

calculate w at the upper boundary for $x = 50$ cm and $t = 10$ s.

3.5 Using separation of variables, solve in cylindrical coordinates the problem of steady flow past a cylinder. Given Laplace's equation

$$\phi_{rr} + \frac{1}{r} \phi_r + \frac{1}{r^2} \phi_{\theta\theta} = 0 \qquad \text{in two dimensions}$$

in which the subscripts denote partial differentiation with respect to the subscripted variable. The boundary conditions are

$$\phi = Ur \cos \Theta \qquad \text{at } r \text{ large}$$

and

$$\phi_r |_{r=a} = 0$$

3.6 A two-dimensional horizontal flow is described by

$$\phi(x, y) = 10(x^2 - y^2)$$

Find the point of maximum pressure if $p = 0$ at $(x, y) = (1, 1)$.

3.7 A wave field is observed by satellite. The wave lengths are determined to be 312 m in deep water and 200 m over the continental shelf. What is the shelf depth?

3.8 Formulate the boundary value problem for the situation below, which represents a model to study the effects of waves on a harbor with a narrow entrance. The stroke S of the wavemaker is considered to be small compared to the depth h.

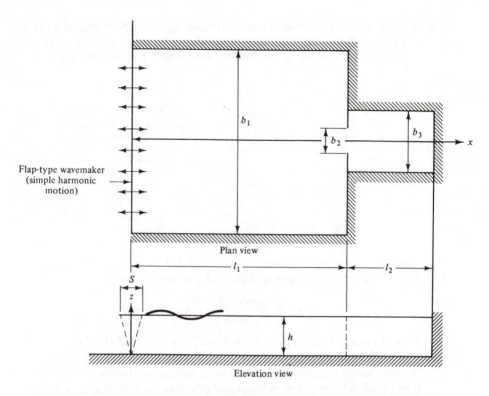

Plan view

Elevation view

3.9 Set up, but do *not* solve, the complete two-dimensional (x, z, t) boundary value problem as illustrated, which was designed to simulate earthquake motions of the continental shelf. The sloping bottom oscillates with a period T and has an amplitude a. State all assumptions.

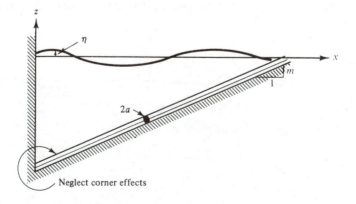

Neglect corner effects

3.10 A horizontal cylindrical wavemaker is oscillating vertically in the free surface. Examining the two-dimensional problem shown below, develop the kinematic boundary condition for the fluid at the cylinder wall. Discuss the results.

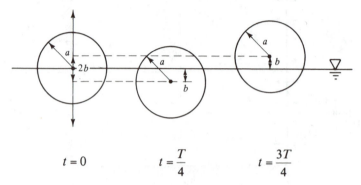

$$t = 0 \qquad\qquad t = \frac{T}{4} \qquad\qquad t = \frac{3T}{4}$$

where T is period of oscillation.

3.11 The stream function for a progressive small-amplitude wave is

$$\psi = -\frac{H}{2}\frac{g}{\sigma}\frac{\sinh k(h+z)}{\cosh kh}\cos(kx - \sigma t)$$

Draw the streamlines for $t = 0$, when $T = 5$ s, $h = 10$ m, and $H = 2.0$ m.

3.12 You are on a ship (100 m in length) on the deep ocean traveling north. The (regular) waves are propagating north also and you note two items of information: (1) when the ship bow is positioned at a crest, the stern is at a trough, and (2) a different crest is positioned at the bow every 20 s.
(a) Do you have enough information to determine the ship speed?
(b) If the answer to part (a) is "no," what additional item(s) of information are required?
(c) If the answer to part (a) is "yes," what is the ship speed?

3.13 A tsunami is detected at 12:00 h on the edge of the continental shelf by a warning system. At what time can the tsunami be expected to reach the shoreline?

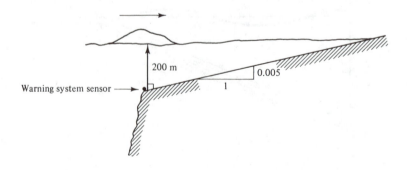

3.14 A rigid sinusoidal form is located as shown in the sketch. The form is forced to move in the $+x$ direction at speed V.

(a) Derive an expression for the velocity potential for the water motion induced by the moving form.

(b) Evaluate $p_c - p_t$ for the following cases:

(1) $V^2 < \dfrac{g}{k} \tanh kh$

(2) $V^2 = \dfrac{g}{k} \tanh kh$

(3) $V^2 > \dfrac{g}{k} \tanh kh$

where p_c and p_t denote the pressure just below the form at the crest and trough, respectively.

(c) Discuss the special significance of b(2).

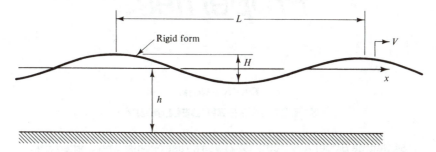

3.15 Develop an iterative technique to solve the dispersion relationship for k given σ and h. *Note*: It is somewhat easier to first solve for kh. (*Hint*: A Newton–Raphson technique could be used.)

3.16 Determine the celerity of a deep water wave on a current equal to 50 cm/s and $T = 5$ s. What is the wave period seen by an observer moving with the current?

3.17 Develop the boundary value problem for small-amplitude waves in terms of the pressure, assuming that Euler's equations are valid and the flow is incompressible.

4

Engineering Wave Properties

Dedication
SIR GEORGE BIDDELL AIRY

Sir George Biddell Airy (1801–1892) was an astronomer who worked in a variety of areas of science, as did his contemporary and personal acquaintance, Laplace. His major work with respect to this book is his development of small-amplitude water wave theory published in an article in the *Encyclopedia Metropolitan*.

Airy was born in Alnwich, Northumberland, England, and attended Trinity College, Cambridge, from 1819 to 1823. In 1826 he was appointed the Lucasian Chair of Mathematics at Cambridge (once held by Isaac Newton). At that time he worked in optics and drew a great deal of attention to the problem of astigmatism, a vision deficiency from which he suffered.

In 1828 he was named the Plumian Professor of Astronomy and Director of the Cambridge Observatory. He became the Astronomer Royal in 1835, a position he held for 46 years. During that time, he and the observatory staff reduced all measurements made by the observatory between 1750 and 1830.

His research (over 377 papers) encompassed magnetism, tides, geography, gravitation, partial differential equations, and sound. In 1867 his paper on suspension bridges received the Telford Medal of the Institution of Civil Engineers.

His *Numerical Theory of Tides* was published in 1886 despite the presence of several inexplicable errors. He attempted (unsuccessfully) to resolve these until 1888. He died in 1892.

4.1 INTRODUCTION

The solutions developed in Chapter 3 for standing and progressive small-amplitude water waves provide the basis for applications to numerous problems of engineering interest. For example, the water particle kinematics and the pressure field within the waves are directly related to the calculation of forces on bodies. The transformation of waves as they propagate toward shore is also important, as in many cases coastal engineering design involves the forecasting of offshore wave climates or the use of offshore data, for example, those obtained from ships. It is obviously necessary to be able to determine any modifications that occur to these waves as they encounter shallower water and approach the shore.

4.2 WATER PARTICLE KINEMATICS FOR PROGRESSIVE WAVES

Consider a progressive wave with water surface displacement given by

$$\eta = \frac{H}{2} \cos (kx - \sigma t)$$

The associated velocity potential is

$$\phi = -\frac{H}{2} \frac{g}{\sigma} \frac{\cosh k(h + z)}{\cosh kh} \sin (kx - \sigma t) \tag{4.1}$$

By introducing the dispersion relationship, $\sigma^2 = gk \tanh kh$, this can be written as

$$\phi = -\frac{H}{2} C \frac{\cosh k(h + z)}{\sinh kh} \sin (kx - \sigma t) \tag{4.2}$$

4.2.1 Particle Velocity Components

The horizontal velocity under the wave is given by definition, Eq. (2.68), as

$$u = -\frac{\partial \phi}{\partial x} = \frac{H}{2} \sigma \frac{\cosh k(h + z)}{\sinh kh} \cos (kx - \sigma t) \tag{4.3a}$$

or

$$u = \frac{gHk}{2\sigma} \frac{\cosh k(h + z)}{\cosh kh} \cos (kx - \sigma t) \tag{4.3b}$$

The local horizontal acceleration is then

$$\frac{\partial u}{\partial t} = \frac{H}{2}\sigma^2 \frac{\cosh k(h+z)}{\sinh kh} \sin (kx - \sigma t) \tag{4.4}$$

and the vertical velocity and local acceleration are

$$w = -\frac{\partial \phi}{\partial z} = \frac{H}{2}\sigma \frac{\sinh k(h+z)}{\sinh kh} \sin (kx - \sigma t) \tag{4.5}$$

$$\frac{\partial w}{\partial t} = -\frac{H}{2}\sigma^2 \frac{\sinh k(h+z)}{\sinh kh} \cos (kx - \sigma t) \tag{4.6}$$

Examining the horizontal and vertical velocity components as a function of position, it is clear that they are 90° out of phase; the extreme values of the horizontal velocity appear at the phase positions $(kx - \sigma t) = 0$, $\pi, \ldots$ (under the crest and trough positions), while the extreme vertical velocities appear at $\pi/2$, $3\pi/2, \ldots$ (where the water surface displacement is zero).

The vertical variation of the velocity components is best viewed by starting at the bottom where $k(h + z) = 0$. Here the hyperbolic terms involving z in both the u and w velocities are at their minima, 1 and 0, respectively. As we progress upward in the fluid, the magnitudes of the velocity components increase. In Figure 4.1, the velocity components are plotted for four phase positions. The accelerations are such that the maximum vertical accelerations occur as the horizontal velocities are extremes and the same is true for the vertical velocities and the horizontal accelerations.

4.2.2 Particle Displacements

A water particle with a mean position of, say, (x_1, z_1) will be displaced by the wave-induced pressures and the instantaneous water particle position will be denoted as $(x_1 + \zeta, z_1 + \xi)$, as shown in Figure 4.2. The displacement components (ζ, ξ) of the water particle can be found by integrating the velocity with respect to time.

$$\zeta(x_1, z_1, t) = \int u(x_1 + \zeta, z_1 + \xi) \, dt \tag{4.7}$$

$$\xi(x_1, z_1, t) = \int w(x_1 + \zeta, z_1 + \xi) \, dt \tag{4.8}$$

In keeping with our small-amplitude wave considerations, ζ and ξ will be small quantities and therefore we can replace $u(x_1 + \zeta, z_1 + \xi)$ with $u(x_1, z_1)$.[1]

---

[1] This involves neglecting terms such as $\frac{\partial u}{\partial x}\zeta$, as can be seen from a Taylor series expansion.

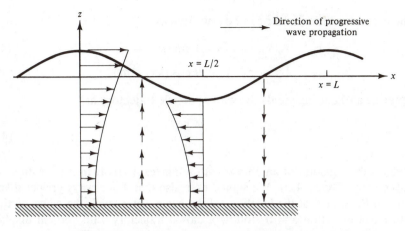

Figure 4.1   Water particle velocities in a progressive wave.

Integrating the equations above then yields

$$\zeta = -\frac{H}{2}\frac{gk}{\sigma^2}\frac{\cosh k(h + z_1)}{\cosh kh}\sin(kx_1 - \sigma t) \tag{4.9}$$

or

$$\zeta = -\frac{H}{2}\frac{\cosh k(h + z_1)}{\sinh kh}\sin(kx_1 - \sigma t)$$

using the dispersion relationship. The vertical displacement is determined similarly:

$$\xi = \frac{H}{2}\frac{\sinh k(h + z_1)}{\sinh kh}\cos(kx_1 - \sigma t) \tag{4.10}$$

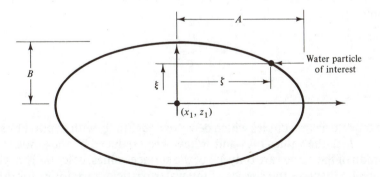

Figure 4.2   Elliptical form of water particle trajectory.

The displacements ζ and ξ can be rewritten as

$$\zeta(x_1, z_1, t) = -A \sin(kx_1 - \sigma t) \tag{4.11}$$

$$\xi(x_1, z_1, t) = B \cos(kx_1 - \sigma t) \tag{4.12}$$

Squaring and adding yields the water particle trajectory as

$$\left(\frac{\zeta}{A}\right)^2 + \left(\frac{\xi}{B}\right)^2 = 1 \tag{4.13}$$

which is the equation of an ellipse with semiaxes A and B in the x-z direction, respectively (Figure 4.2). We should note also that A is always greater than or equal to B. In fact, at the locations of the mean water level, the water particles with mean elevation $z = 0$, follow a closed trajectory with vertical displacement $H/2$; that is, these particles comprise the surface. There are no water particles with mean locations higher than $z = 0$.

In *shallow water* ($h/L < 1/20$), using the shallow water approximations, the major semiaxis reduces to

$$A = \frac{H}{2} \frac{\cosh k(h + z_1)}{\sinh kh} = \frac{H}{2} \frac{1}{kh} = \frac{HL}{4\pi h} = \frac{HT}{4\pi} \sqrt{\frac{g}{h}} \tag{4.14}$$

where the equality for shallow water, $L = CT = \sqrt{gh}\, T$, has been introduced. The minor semiaxis B can be determined similarly.

$$B = \frac{H}{2} \frac{\sinh k(h + z_1)}{\sinh kh} = \frac{H}{2}\left(1 + \frac{z_1}{h}\right) \tag{4.15}$$

Note that A is not a function of elevation. The horizontal excursion of a water particle is a constant distance for all particles under the wave. The total vertical excursion increases linearly with elevation, being zero, of course, at the bottom and being H at the mean water surface, $z = 0$.

For *deep water* waves ($h/L > \frac{1}{2}$) it can be shown that the semiaxes simplify to

$$A = \frac{H}{2} \frac{e^{kh}e^{kz_1}}{e^{kh}} = \frac{H}{2} e^{kz_1} \tag{4.16}$$

$$B = \frac{H}{2} e^{kz_1} = A \tag{4.17}$$

The trajectories are circles which decay exponentially with depth. For a depth of $z = -L/2$, the values of A and B have been reduced by the amount $e^{-\pi}$, or the radii of the circles are only 4% of the surface values, essentially negligible. Figure 4.3 displays the shapes of the water particle trajectories for different relative depths.

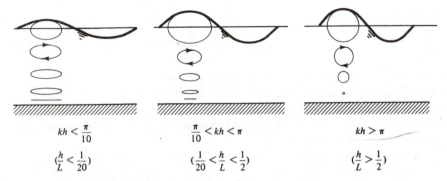

Figure 4.3 Water particle trajectories in progressive water waves of different relative depths.

4.3 PRESSURE FIELD UNDER A PROGRESSIVE WAVE

The pressure field associated with a progressive wave is determined from the unsteady Bernoulli equation developed for an ideal fluid and the velocity potential appropriate to this case, Eq. (2.92):

$$\frac{p}{\rho} + gz + \tfrac{1}{2}\left(u^2 + w^2\right) - \frac{\partial \phi}{\partial t} = C(t) \tag{4.18}$$

Equating the relationship above at any depth z, and at the free surface η, where the pressure is taken as zero, and linearizing yields

$$\left(\frac{p}{\rho} + gz - \frac{\partial \phi}{\partial t}\right)_z = g\eta - \left.\frac{\partial \phi}{\partial t}\right|_{\eta \approx 0} \tag{4.19}$$

Recalling from Chapter 3 that the linearized DFSBC reduces to

$$\eta = \frac{1}{g}\left.\frac{\partial \phi}{\partial t}\right|_{z=0} \tag{4.20}$$

it is seen that the pressure can be expressed as

$$\frac{p}{\rho} = -gz + \frac{\partial \phi}{\partial t} \tag{4.21}$$

where the small velocity squared terms have been neglected.

For a progressive wave described by the velocity potential in Eq. (4.1), we have

$$p = -\rho gz + \rho g\,\frac{H}{2}\,\frac{\cosh k(h+z)}{\cosh kh}\,\cos(kx - \sigma t) \tag{4.22}$$

or

$$p = -\rho gz + \rho g\eta K_p(z) \tag{4.23}$$

where

$$K_p(z) = \frac{\cosh k(h + z)}{\cosh kh} \qquad (4.24)$$

The first term on the right-hand side of the pressure equation (4.23) is, of course, the hydrostatic term, which would exist without the presence of the waves. The second term is called the dynamic pressure. The term $K_p(z)$ is referred to as the "pressure response factor" and below the mean water surface is always less than unity.

The dynamic pressure is a result of two contributions; the first and most obvious contributor is the surcharge of pressure due to the presence of the free surface displacement. If the pressure response factor were unity, the pressure contribution from the free surface displacement would be purely hydrostatic. However, associated with the wave motion is the vertical acceleration, which is 180° out of phase with the free surface displacement. This contribution modifies the pressure from the purely hydrostatic case. The reader may wish to verify that Eq. (4.22) can be obtained by integrating the linearized vertical equation of motion, Eq. (2.38c), from any depth z up to the free surface η. In Figure 4.4, the effect of the dynamic pressure in modifying the hydrostatic pressure is shown.

The pressure response factor has a maximum of unity at $z = 0$, and a minimum of $1/\cosh kh$ at the bottom. To determine the pressure above the mean water level we again must use the Taylor series for a small positive distance z_1 $(0 < z_1 < \eta)$:

$$p(z_1) = (-\rho g z + \rho g \eta K_p)_{z=0} + z_1 \frac{\partial}{\partial z} (-\rho g z + \rho g \eta K_p)_{z=0} + \cdots \qquad (4.25)$$

$$\doteq \rho g \eta - \rho g z_1 \qquad \text{to the first order}$$

$$= \rho g (\eta - z_1) \qquad (4.26)$$

Thus to this approximation the pressure is hydrostatic under the wave crest

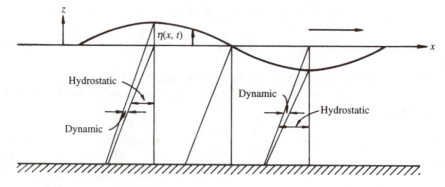

Figure 4.4 Hydrostatic and dynamic pressure components at various phase positions in a progressive water wave.

down to $z = 0$. Below that depth, however, it deviates from the hydrostatic law. Note also that Eq. (4.26) predicts a zero pressure at the instantaneous free surface, $z_1 = \eta$. Figure 4.5 shows the isolines of pressure under a wave for $h/L = 0.2$.

One method of measuring waves in either the laboratory or field is by sensing the pressure fluctuations and then calculating the associated water surface displacements by Eq. (4.23). From Eq. (4.23), a bottom-mounted pressure gage would record a steady hydrostatic pressure plus the oscillating dynamic pressure, which for a particular wave period is proportional to the free surface displacement η, the variable of interest. If the dynamic pressure p_D is isolated by subtracting out the mean hydrostatic pressure, then η is

$$\eta = \frac{p_D}{\rho g K_p(-h)} \quad \text{and} \quad K_p(-h) = \frac{1}{\cosh kh} \tag{4.27}$$

where $K_p(-h)$ is a function of the angular frequency of the waves. Thus the dispersion relationship must be used to determine kh from the frequency of the observed waves. If a mean current is present, the wave number must be computed via Eq. (3.52); otherwise, significant errors can occur.

Even though we have derived the pressure response factor for only one frequency component, it is interesting to note that for cases in which the linear assumption is reasonably valid, Eq. (4.27) can be used to determine the composite wave system containing many (or an infinite) number of components from a measured pressure time series.

Because of the dependency of the pressure response factor on the wave frequency, short-period waves have a very small K_p (at the bottom), while for long-period waves K_p approaches unity. In other words, very short period waves may not even be recorded by the pressure gage. The reader may wish to

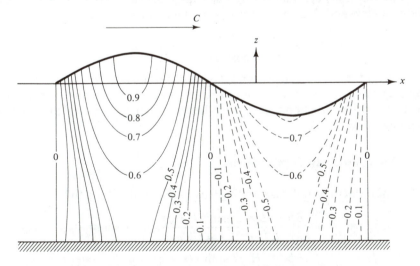

Figure 4.5   Isolines of $p_D/[\gamma(H/2)]$ for progressive wave of $h/L = 0.20$.

show that the shallow and deep water asymptotes for the pressure response factor are unity and e^{kz}, respectively.

4.4 WATER PARTICLE KINEMATICS FOR STANDING WAVES

The original velocity potential we derived represented a pure standing wave,

$$\phi = \frac{H_s g}{2\sigma} \frac{\cosh k(h + z)}{\cosh kh} \cos kx \sin \sigma t \qquad (4.28)$$

with

$$\eta = \frac{H_s}{2} \cos kx \cos \sigma t \qquad (4.29)$$

$$\sigma^2 = gk \tanh kh \qquad (4.30)$$

where H_s denotes the height of the standing wave and is twice the height of each of the two progressive waves forming the standing wave.

The velocity potential for a standing wave can be rederived by subtracting the velocity potential for two progressive waves of the same period with heights H_p propagating in opposite directions.

$$\phi = -\frac{H_p}{2} \frac{g}{\sigma} \frac{\cosh k(h + z)}{\cosh kh} \sin (kx - \sigma t)$$

$$+ \frac{H_p}{2} \frac{g}{\sigma} \frac{\cosh k(h + z)}{\cosh kh} \sin (kx + \sigma t) \qquad (4.31)$$

Sin $(kx \pm \sigma t)$ can be rewritten as $\sin kx \cos \sigma t \pm \cos kx \sin \sigma t$, (from trigonometry) and thus the velocity potential is rewritten as

$$\phi = \frac{H_p g}{\sigma} \frac{\cosh k(h + z)}{\cosh kh} \cos kx \sin \sigma t \qquad (4.32)$$

Comparing the two velocity potentials, it is clear that $H_p = H_s/2$. Therefore, a standing wave of height H_s is composed of two progressive waves propagating in opposite directions, each with height equal to one-half that of the standing wave.

4.4.1 Velocity Components

The velocities under a standing wave are readily found to be

$$u = -\frac{\partial \phi}{\partial x} = \frac{H}{2} \frac{gk}{\sigma} \frac{\cosh k(h + z)}{\cosh kh} \sin kx \sin \sigma t \qquad (4.33)$$

$$w = -\frac{\partial \phi}{\partial z} = -\frac{H}{2} \frac{gk}{\sigma} \frac{\sinh k(h + z)}{\cosh kh} \cos kx \sin \sigma t \qquad (4.34)$$

where for convenience the subscript s has been dropped. Using the dispersion relationship,

$$u = \frac{H}{2}\sigma\frac{\cosh k(h + z)}{\sinh kh}\sin kx \sin \sigma t \qquad (4.35a)$$

$$w = -\frac{H}{2}\sigma\frac{\sinh k(h + z)}{\sinh kh}\cos kx \sin \sigma t \qquad (4.35b)$$

As with the velocities under a progressive wave, these velocities increase with elevation above the bottom. The extreme values of u and w in space occur under the nodes and antinodes of the water surface profile as shown in Figure 4.6, where u and w are zero under the antinodes and nodes, respectively. It is of interest that the horizontal and vertical components of velocity under a standing wave are in phase; that is, the time-varying term "$\sin \sigma t$" modifies both velocity components and, at certain times, the velocity is zero everywhere in the standing wave system. It is therefore evident that at some times all the energy is potential and, by reference to Eqs. (4.35), at other times all the energy is kinetic.

If a progressive wave were normally incident on a vertical wall, it would be reflected backward without a change in height, thus giving a standing wave in front of the wall. The lateral boundary condition at the vertical wall would be one of no flow through the wall, or $u = -\partial\phi/\partial x = 0$ at $x = x_{wall}$, where x_{wall} is the location of the wall. Inspection of the equation for the horizontal velocity, Eq. (4.33), shows that at locations $kx = n\pi$ (where n is an integer), the no-flow boundary condition is satisfied. Therefore, a standing wave could exist within a basin with two walls situated at two antinodes of a standing wave. This is, in fact, the simplest model of uniform depth lakes, estuaries, and harbors where standing waves, called seiches, can be generated by winds, earthquakes, or other phenomena. We examine these waves further in Chapter 5.

The local accelerations under a standing wave are

$$\frac{\partial u}{\partial t} = \frac{H}{2}\sigma^2\frac{\cosh k(h + z)}{\sinh kh}\sin kx \cos \sigma t \qquad (4.36)$$

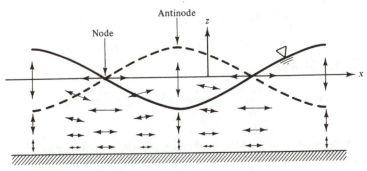

Figure 4.6 Distribution of water particle velocities in a standing water wave.

$$\frac{\partial w}{\partial t} = -\frac{H}{2}\sigma^2 \frac{\sinh k(h+z)}{\sinh kh} \cos kx \cos \sigma t \tag{4.37}$$

Under the wave antinodes, the vertical accelerations are maxima, while the horizontal accelerations are zero, and under the nodes, the opposite is true.

4.4.2 Particle Displacements

The displacements of a water particle (ζ, ξ) from its mean position (x_1, z_1) under a standing wave are defined in a linearized fashion as before.

$$\zeta = \int u(x_1 + \zeta, z_1 + \xi)\, dt \approx \int u(x_1, z_1)\, dt \tag{4.38}$$

$$\xi = \int w(x_1 + \zeta, z_1 + \xi)\, dt \approx \int w(x_1, z_1)\, dt \tag{4.39}$$

or

$$\zeta = -\frac{H}{2}\frac{\cosh k(h+z_1)}{\sinh kh}\sin kx_1 \cos \sigma t = -A \cos \sigma t \tag{4.40}$$

$$\xi = \frac{H}{2}\frac{\sinh k(h+z_1)}{\sinh kh}\cos kx_1 \cos \sigma t = B \cos \sigma t \tag{4.41}$$

The displacement vector is $\mathbf{r} = \zeta\mathbf{i} + \xi\mathbf{k}$; its magnitude $|\mathbf{r}|$ is

$$|\mathbf{r}| = \sqrt{A^2 + B^2}\, \cos \sigma t \tag{4.42}$$

or

$$|\mathbf{r}(t)| = \frac{H}{2}\frac{\cos \sigma t}{\sinh kh}\sqrt{\cosh^2 k(h+z_1)\sin^2 kx_1 + \sinh^2 k(h+z_1)\cos^2 kx_1} \tag{4.43}$$

For infinitesimally small motions, the displacement vector is a straight line,[2] the amplitude and inclination being dependent on position (x_1, z_1). The water particle under the standing wave moves back and forth along the line with time. Substituting the trigonometric identities,

$$\cosh^2 k(h+z_1) = \tfrac{1}{2}[\cosh 2k(h+z_1) + 1]$$

$$\sin^2 kx_1 = \tfrac{1}{2}(1 - \cos 2kx_1)$$

$$\sinh^2 k(h+z_1) = \tfrac{1}{2}[\cosh 2k(h+z_1) - 1]$$

$$\cos^2 kx_1 = \tfrac{1}{2}(1 + \cos 2kx_1)$$

yields from Eq. (4.43),

$$|\mathbf{r}(t)| = \frac{H}{4}\frac{\cos \sigma t}{\sinh kh}\sqrt{2[\cosh 2k(h+z_1) - \cos 2kx_1]}$$

---

[2]From Equations (4.40) and (4.41), we obtain $\xi = -(B/A)\zeta$ which may be compared with Eq. (4.13), the equation for the trajectories of a progressive wave.

Note that at the bottom under the antinodes $|\mathbf{r}|$ is zero. The maximum value of $|\mathbf{r}|$ occurs under the nodes, where $\cos 2kx_1 = -1$.

The motion of the water particles under a standing wave can thus be described as a simple harmonic motion along a straight line. The slope of the displacement vector θ is given by

$$\tan \theta = \frac{\xi}{\zeta} = -\frac{\tanh k(h + z_1)}{\tan kx_1} \tag{4.44}$$

which is not a function of time. Clearly, at the bottom, the trajectories are horizontal $(\theta = 0)$, as is to be expected by the bottom boundary condition. Figure 4.6 portrays the water particle trajectories at several phase positions under a standing wave.

4.5 PRESSURE FIELD UNDER A STANDING WAVE

To find the pressure at any depth under a standing wave, the unsteady Bernoulli equation is used as in the case for progressive waves.

$$\frac{p}{\rho} + \frac{u^2 + w^2}{2} - \frac{\partial \phi}{\partial t} + gz = C(t) \tag{4.45}$$

Linearizing and evaluating as before between the free surface and at some depth (z) in the fluid, the gage pressure is

$$p = -\rho gz + \rho \frac{\partial \phi}{\partial t}$$

or

$$p = -\rho gz + \rho g \frac{H \cosh k(h + z)}{2 \cosh kh} \cos kx \cos \sigma t$$

$$= -\rho gz + \rho g K_p(z)\eta \tag{4.46}$$

where the pressure response factor $K_p(z)$ is the same as determined for progressive waves. Note that under the nodes, the pressure is solely hydrostatic. Again, the dynamic pressure is in phase with the water surface elevation, and as before it is a combined result of the local water surface displacement and the vertical accelerations of the overlying water particles.

The force exerted on a wall at an antinode can be calculated by integrating the pressure over depth per unit width of wall

$$F = \int_{-h}^{\eta_w} p(z)\, dz = \int_{-h}^{0} \left[-\rho gz + \rho g\eta_w \frac{\cosh k(h + z)}{\cosh kh} \right] dz + \int_{0}^{\eta_w} \rho g(\eta_w - z)\, dz$$

from Eqs. (4.26) and (4.46) and where $\eta_w = (H/2) \cos \sigma t$, the water surface displacement at the wall. It should be stressed that this formulation is not entirely consistent, as the second integral on the right-hand side representing the force contribution of the wave crest region is of second order; yet second-order terms in the form of the square of the velocity components have already been dropped from the first term of the right-hand side. Integrating, we get

$$F = \rho g \left(\frac{h^2 + \eta_w^2}{2} \right) + \rho g h \frac{\tanh kh}{kh} \eta_w \qquad (4.47)$$

To first order,

$$F = \rho \frac{gh^2}{2} + \rho g h \frac{\tanh kh}{kh} \eta_w \qquad (4.48)$$

The force on the wall consists of the hydrostatic contribution, plus an oscillatory term due to the dynamic pressure. The maximum force occurs when $\eta_w = H/2$,

$$F_{max} = \rho g \frac{(4h^2 + (H)^2)}{8} + \rho g h \frac{H}{2} \frac{\tanh kh}{kh} \qquad (4.49)$$

4.6 PARTIAL STANDING WAVES

For the case just considered of pure standing waves, two waves of the same period and height, but propagating in opposite directions, were superimposed, as one expects from the perfect reflection of an incident wave from a vertical wall. Quite often in nature, however, when waves are reflected from obstacles, not all of the wave energy is reflected; some is absorbed by the obstacle and some is transmitted past the obstacle. For example, waves are reflected from breakwaters and beaches; in each case wave energy is not perfectly reflected. To examine this case, let us assume that the incident wave has a height H_i, but that the reflected wave has a smaller height H_r and different phase than the incident wave. The wave periods of the incident and reflected waves will be the same. The total wave profile seaward of the obstacle is then

$$\eta_t = \frac{H_i}{2} \cos (kx - \sigma t) + \frac{H_r}{2} \cos (kx + \sigma t + \epsilon) \qquad (4.50a)$$

where ϵ is the phase lag induced by the reflection process. If the water surface displacements are plotted, they appear as in Figure 4.7. Due to the imperfect reflection, there are no true nodes in the wave profile.

Quite often in measuring wave heights in a wave tank, reflections occur and it is necessary to be able to separate out the incident and reflected wave heights. To do this, we rewrite η_t, using trigonometric identities.

$$\eta_t = \frac{H_i}{2}(\cos kx \cos \sigma t + \sin kx \sin \sigma t)$$

$$+ \frac{H_r}{2}(\cos (kx + \epsilon) \cos \sigma t - \sin (kx + \epsilon) \sin \sigma t)$$

Grouping similar time terms,

$$\eta_t = \left[\frac{H_i}{2}\cos kx + \frac{H_r}{2}\cos (kx + \epsilon)\right]\cos \sigma t$$

$$+ \left[\frac{H_i}{2}\sin kx - \frac{H_r}{2}\sin (kx + \epsilon)\right]\sin \sigma t$$

or, for convenience, denoting the bracketed terms by $I(x)$ and $F(x)$,

$$\eta_t = I(x) \cos \sigma t + F(x) \sin \sigma t \tag{4.50b}$$

Thus η_t is a sum of standing waves. To find the extreme values of η_t for any x, that is, the envelope of the wave heights, denoted by the dotted lines in the figure, it is necessary to find the maximas and minimas of η_t with respect to time. Proceeding as usual by taking the first derivative and setting it equal to zero to find the extremes yields

$$\frac{\partial \eta_t}{\partial t} = -I(x)\sigma \sin \sigma t + F(x)\sigma \cos \sigma t = 0 \tag{4.51}$$

or

$$\tan (\sigma t)_m = \frac{F(x)}{I(x)}$$

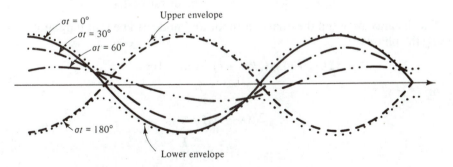

Figure 4.7 Instantaneous water surface displacements and envelope in a partial standing wave system.

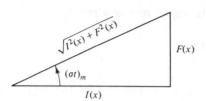

Figure 4.8 Relationships among $(\sigma t)_m$, $F(x)$, and $I(x)$.

Therefore, to find the maxima and minima of η_t, $(\sigma t)_m$ is substituted into Eq. (4.50a). Examining Figure 4.8, it is clear that

$$\cos(\sigma t)_m = \frac{I(x)}{\sqrt{I^2(x) + F^2(x)}}$$

$$\sin(\sigma t)_m = \frac{F(x)}{\sqrt{I^2(x) + F^2(x)}}$$

Substituting into Equation (4.50b),[3] we have

$$(\eta_t)_m = \frac{I^2(x) + F^2(x)}{\sqrt{I^2(x) + F^2(x)}} = \pm \sqrt{I^2(x) + F^2(x)} \qquad (4.52)$$

Substituting for $I(x)$ and $F(x)$ from Eq. (4.50b), it is seen readily that the extreme values of η_t for any location x are

$$[\eta_t(x)]_m = \pm \sqrt{\left(\frac{H_i}{2}\right)^2 + \left(\frac{H_r}{2}\right)^2 + \frac{H_i H_r}{2} \cos(2\,kx + \epsilon)} \qquad (4.53)$$

$[\eta_t(x)]_m$ obviously varies periodically with x. At the phase positions $(2kx_1 + \epsilon) = 2n\pi \ (n = 0, 1, \ldots)$, $[\eta_t(x)]_m$ becomes a maximum of the envelope

$$(\eta_t)_{max} = \tfrac{1}{2}(H_i + H_r), \qquad \text{the quasi-antinodes} \qquad (4.54)$$

whereas at the phase positions, $(2kx_2 + \epsilon) = (2n + 1)\pi \ (n = 0, 1, \ldots)$, the value of $[\eta_t(x)]_m$ becomes a minimum of the envelope:

$$(\eta_t)_{min} = \tfrac{1}{2}(H_i - H_r), \qquad \text{the quasi-nodes} \qquad (4.55)$$

The distance between the quasi-antinode and node can be found by subtracting the phases

$$(2kx_2 + \epsilon) - (2kx_1 + \epsilon) = (2n + 1)\pi - 2n\pi$$

or

$$2k(x_2 - x_1) = \pi$$

$$x_2 - x_1 = \frac{L}{4}$$

[3]This exercise shows simply that the maximum and minimum of $(A \sin \sigma t + B \cos \sigma t)$ are $\pm \sqrt{A^2 + B^2}$.

For a laboratory experiment, where reflection from a beach or an obstacle is present, if the amplitude of the quasi-antinodes and nodes are measured by slowly moving a wave gage along the wave tank, the incident and reflected wave heights are found simply from Eqs. (4.54) and (4.55) as

$$H_i = (\eta_t)_{max} + (\eta_t)_{min} \qquad (4.56)$$

$$H_r = (\eta_t)_{max} - (\eta_t)_{min} \qquad (4.57)$$

The reflection coefficient of the obstacle is defined as

$$\kappa_r = \frac{H_r}{H_i} \qquad (4.58)$$

Figure 4.9 presents such data for the case of extremely small waves and nearly perfect reflection. To find the phase ϵ, it is necessary to find the distance from origin to the nearest maximum or minimum x_1, and to solve one of the following equations:

$$2kx_1 + \epsilon = \begin{cases} 2n\pi, & n = 0, 1, 2,\ldots & \text{for the maximum} \\ (2n + 1)\pi, & n = 0, 1, 2,\ldots & \text{for the minimum} \end{cases}$$

The reader should verify that the dynamic and hydrostatic pressure under a partial standing wave system can be expressed as

$$p(x, z, t) = -\rho g z + \rho g K_p(z)\eta$$

where $\eta(x, t)$ and $K_p(z)$ are given by Eqs. (4.50a) and (4.24), respectively.

4.7 ENERGY AND ENERGY PROPAGATION IN PROGRESSIVE WAVES

The total energy contained in a wave consists of two kinds: the potential energy, resulting from the displacement of the free surface and the kinetic energy, due to the fact that the water particles throughout the fluid are moving. This total energy and its transmission are of importance in determining how waves change in propagating toward shore, the power required to generate waves, and the available power for wave energy extraction devices, for example.

$$\eta_1 + \eta_2, x < 0$$

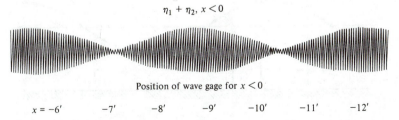

Position of wave gage for $x < 0$

$x = -6'$ $-7'$ $-8'$ $-9'$ $-10'$ $-11'$ $-12'$

Figure 4.9 Water surface displacement as measured from a slowly moving carriage for the case of nearly perfect reflection. (From Dean and Ursell, 1959.)

4.7.1 Potential Energy

Potential energy as it occurs in water waves is the result of displacing a mass from a position of equilibrium against a gravitational field. When water is at rest with a uniform free surface elevation, it can be shown readily that the potential energy is a minimum. However, a displacement of an assemblage of particles resulting in the displacement of the free surface will require that work be done on the system and results in an increase in potential energy.

We will derive the potential energy associated with a sinusoidal wave by two different methods. First consider the wave shown in Figure 4.10; we will determine the average potential energy per unit surface area associated with the wave as the difference between the potential energy with and without the wave present. The potential energy of a small column of fluid shown in Figure 4.10 with mass dm relative to the bottom is

$$d(\text{PE}) = dmg\bar{z} \tag{4.59}$$

in which $\bar{z}$ is the height to the center of gravity of the mass, and can be written as

$$\bar{z} = \frac{h + \eta}{2} \tag{4.60}$$

and the differential mass per unit width is

$$dm = \rho\,(h + \eta)\,dx$$

The potential energy averaged over one wave length for a progressive wave of height H is then

$$(\overline{\text{PE}})_T \equiv \frac{1}{L} \int_x^{x+L} d(\text{PE}) = \frac{1}{L} \int_x^{x+L} \rho g \frac{(h + \eta)^2}{2}\, dx \tag{4.61}$$

$$= \frac{\rho g}{L} \int_x^{x+L} \left[\frac{1}{2}(h^2 + 2\eta h + \eta^2) \right] dx \tag{4.62}$$

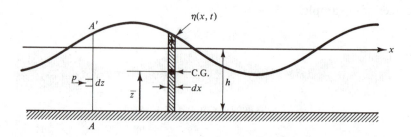

Figure 4.10   Definition sketch for determination of potential energy.

The subscript T signifies that the potential energy of the total water volume is being considered. For $\eta = (H/2) \cos (kx - \sigma t)$, the average potential energy is

$$(\overline{PE})_T = \frac{\rho g}{L} \left(\frac{1}{2} h^2 L + h \int_x^{x+L} \eta \, dx + \frac{1}{2} \int_x^{x+L} \eta^2 \, dx \right) \qquad (4.63)$$

The integration is straightforward. For the last integral, we recognize that the sum of the squares of the sine and cosine functions is identically equal to unity, and it can be seen readily that the average of the square of each function over an integral number of one-half wave lengths is one-half. Since this integration is used often in water wave mechanics, the reader should verify this result. The potential energy is now

$$(\overline{PE})_T = \rho g \frac{h^2}{2} + \rho g \frac{H^2}{16} \qquad (4.64)$$

The potential energy due to the waves is the difference between the potential energy with waves present and with no waves present, that is,

$$(\overline{PE})_{\text{waves}} = (\overline{PE})_T - (\overline{PE})_{\text{w/o}} \qquad (4.65)$$

or

$$\overline{PE} \equiv (\overline{PE})_{\text{waves}} = \frac{\rho g H^2}{16} \qquad (4.66)$$

The potential energy of the waves per unit area depends solely on the wave height. Also, although the development was presented for progressive waves, examination of the details of the derivation will show that the results are equally applicable to the case of standing waves which are sinusoidal in form.

Anticipating the application of this result to more realistic cases, we represent the sea surface η_T by a number N of components, each given by

$$\eta_n = \frac{H_n}{2} \cos (k_n x - \sigma_n t - \epsilon_n) \qquad (4.67)$$

and

$$\eta_T = \sum_{n=1}^{\infty} \eta_n \qquad (4.68)$$

Although it is beyond the scope of the presentation here, the total average potential energy in this case is

$$\overline{PE} = \frac{\rho g}{16} \sum_{n=1}^{N} H_n^2 \qquad (4.69)$$

To return to the case of potential energy due to a sinusoid, it may be worthwhile for the reader to derive the results by a different approach by simply calculating the increase in potential energy required to elevate the water formerly in the trough to the crest location through a vertical distance $2 \cdot z_{cg}$, where z_{cg} is shown in Figure 4.11. Note that this area is $HL/2\pi$ and the vertical distance from the mean waterline to the centers of gravity is $\pi H/16$.

4.7.2 Kinetic Energy

The kinetic energy is due to the moving water particles; the kinetic energy associated with a small parcel of fluid with mass dm is

$$d(KE) = dm \, \frac{u^2 + w^2}{2} = \rho \, dx \, dz \, \frac{u^2 + w^2}{2} \tag{4.70}$$

To find the average kinetic energy per unit surface area, $d(KE)$ must be integrated over depth and averaged over a wave length.

$$\overline{KE} = \frac{1}{L} \int_x^{x+L} \int_{-h}^{\eta} \rho \, \frac{u^2 + w^2}{2} \, dz \, dx \tag{4.71}$$

From the known solution for the velocities under a progressive wave, Eqs. (4.3a) and (4.5), the integral can be written as

$$\overline{KE} = \frac{\rho}{2L} \left(\frac{gHk}{2\sigma} \frac{1}{\cosh kh} \right)^2 \int_x^{x+L} \int_{-h}^{\eta} [\cosh^2 k(h+z) \cos^2 (kx - \sigma t)$$

$$\tag{4.72}$$

$$+ \sinh^2 k(h+z) \sin^2 (kx - \sigma t)] \, dz \, dx$$

Using trigonometric identities (just as was done for the trajectories under a

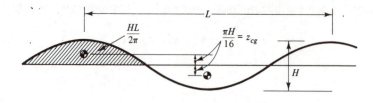

Figure 4.11 Potential energy determined as the result of raising water mass in trough area to crest area.

standing wave), this can be recast as

$$\overline{KE} = \frac{\rho}{2L} \left(\frac{gHk}{2\sigma} \frac{1}{\cosh kh} \right)^2$$

$$\int_x^{x+L} \int_{-h}^{\eta \approx 0} \frac{1}{2} \left[\cosh 2k(h+z) + \cos 2(kx - \sigma t) \right] dz\, dx \tag{4.73}$$

Carrying out the integration and simplifying yields

$$\overline{KE} = \tfrac{1}{16} \rho g H^2 \tag{4.74}$$

This is equal to the magnitude of the potential energy, which is characteristic of conservative (nondissipative) systems in general. The *total* average energy per unit surface area of the wave is then the sum of the potential and kinetic energy. Denoting E as the total average energy per unit surface area

$$E = \overline{KE} + \overline{PE} = \tfrac{1}{8} \rho g H^2 \tag{4.75}$$

The total energy per wave per unit width is then simply

$$E_L = \tfrac{1}{8} \rho g H^2 L \tag{4.76}$$

It is worthwhile emphasizing that neither the average (over a wave length) potential nor kinetic energy per unit area depends on water depth or wave length, but each is simply proportional to the square of the wave height.

4.7.3 Energy Flux

Small-amplitude water waves do not transmit mass as they propagate across a fluid, as the trajectories of the water particles are closed.[4] However, water waves do transmit energy. For example, consider the waves generated by a stone impacting on an initially quiescent water surface. A portion of the kinetic energy of the stone is transformed into wave energy. As these waves travel to and perhaps break on the shoreline, it is clear that there has been a transfer of energy away from the generation area. The rate at which the energy is transferred is called the *energy flux* $\mathcal{F}$, and for linear theory it is the rate at which work is being done by the fluid on one side of a vertical section on the fluid on the other side. For the vertical section AA', shown in Figure 4.10, the instantaneous rate at which work is being done by the dynamic pressure $[p_D = (p + \rho g z)]$ per unit width in the direction of wave propagation is

$$\mathcal{F} = \int_{-h}^{\eta} p_D \cdot u\, dz \tag{4.77}$$

---

[4]For finite-amplitude waves, there is a mass flux; see Chapter 10.

The average energy flux is obtained as before by averaging over a wave period

$$\overline{\mathscr{F}} = \frac{1}{T} \int_{t}^{t+T} \int_{-h}^{\eta} p_D \cdot u \, dz \, dt \tag{4.78}$$

$$= \frac{1}{T} \int_{t}^{t+T} \int_{-h}^{\eta} \left[\rho g \eta \frac{\cosh k(h+z)}{\cosh kh} \right] \left[\frac{gHk}{2\sigma} \frac{\cosh k(h+z)}{\cosh kh} \cos(kx - \sigma t) \right] dz \, dt$$

from Eqs. (4.22) and (4.3b) for p and u, or

$$\overline{\mathscr{F}} = \frac{1}{T} \int_{t}^{t+T} \int_{-h}^{\eta} \left[\rho g \eta \frac{\cosh k(h+z)}{\cosh kh} \right] \left[\sigma \eta \frac{\cosh h(h+z)}{\sinh kh} \right] dz \, dt \tag{4.79}$$

using the dispersion relationship.

To retain terms to the second order in wave height, it is only necessary to integrate up to the mean free surface.

$$\overline{\mathscr{F}} = \frac{1}{T} \int_{t}^{t+T} \int_{-h}^{0} \rho g \sigma \eta^2 \frac{\cosh^2 k(h+z)}{\cosh kh \sinh kh} \, dz \, dt \tag{4.80}$$

$$\overline{\mathscr{F}} = \frac{\rho g \sigma}{4k} \left(\frac{H}{2} \right)^2 \frac{(2kh + \sinh 2kh)}{\sinh 2kh}$$

$$\overline{\mathscr{F}} = \left(\frac{1}{8} \rho g H^2 \right) \frac{\sigma}{k} \left[\frac{1}{2} \left(1 + \frac{2kh}{\sinh 2kh} \right) \right]$$

$$\overline{\mathscr{F}} = ECn \tag{4.81}$$

where Cn is the speed at which the energy is transmitted; this velocity is called the group velocity C_g, for reasons to be explained shortly.

$$C_g = nC \tag{4.82a}$$

or

$$n = \frac{C_g}{C} = \frac{1}{2} \left(1 + \frac{2kh}{\sinh 2kh} \right) \tag{4.82b}$$

The factor n has as deep and shallow water asymptotes the values of $\frac{1}{2}$ and 1, respectively. Therefore, in deep water, the energy is transmitted at only half the speed of the wave profile, and in shallow water, the profile and energy travel at the same speed.

Origin of the term "group velocity."   We have just derived the group velocity in terms of the rate at which energy is being transferred by a train of propagating waves. A more descriptive explanation of the term group velocity results from examining the propagation of a group of waves.

If there are two trains of waves of the same height propagating in the same direction with slightly different frequencies and wave numbers, they

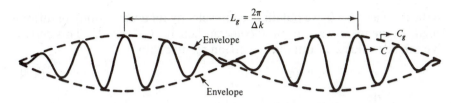

Figure 4.12   Characteristics of a "group" of waves.

are superimposed as

$$\eta = \eta_1 + \eta_2 \tag{4.83}$$

$$= \frac{H}{2} \cos{(k_1 x - \sigma_1 t)} + \frac{H}{2} \cos{(k_2 x - \sigma_2 t)} \tag{4.84}$$

where[5]

$$\sigma_1 = \sigma - \frac{\Delta\sigma}{2}, \qquad k_1 = k - \frac{\Delta k}{2}$$

$$\sigma_2 = \sigma + \frac{\Delta\sigma}{2}, \qquad k_2 = k + \frac{\Delta k}{2} \tag{4.85}$$

Using trigonometric identities, the profiles can be combined in the following manner:

$$\eta = H \cos{\left[\frac{1}{2}[(k_1 + k_2)x - (\sigma_1 + \sigma_2)t]\right]} \cos{\left[\frac{1}{2}[(k_1 - k_2)x - (\sigma_1 - \sigma_2)t]\right]}$$

$$= H \cos{(kx - \sigma t)} \cos{\left[\frac{1}{2}\Delta k \left(x - \frac{\Delta\sigma}{\Delta k}t\right)\right]} \tag{4.86}$$

The resulting profile, consisting of wave forms moving with velocity $C = \sigma/k$, is modulated by an "envelope" that propagates with speed $\Delta\sigma/\Delta k$, which is referred to as the group velocity C_g. The superimposed profile is shown in Figure 4.12. If we recall that the wave energy is proportional to the wave height, it is clear that no energy can propagate past a node as the wave height (and therefore dynamic pressure) is zero there. Therefore, the energy must travel with the speed of the *group* of waves. This velocity is seen to be, from Eq. (4.86),

$$C_g = \frac{\Delta\sigma}{\Delta k} \tag{4.87}$$

---

[5]This derivation is strictly true for small Δk and $\Delta\sigma$, in order that the relationships given in Eq. (4.85) satisfy the dispersion relation.

In the limit as $\Delta k \to 0$, we obtain a group velocity for a wave group of infinite length L_g (hence, a wave train of constant height), $C_g = d\sigma/dk$. This derivative can be evaluated from the dispersion relationship

$$\sigma^2 = gk \tanh kh \tag{4.88}$$

$$2\sigma \frac{d\sigma}{dk} = g \tanh kh + gkh \, \text{sech}^2 kh$$

$$C_g = \frac{d\sigma}{dk} = \frac{(g \tanh kh + gkh \, \text{sech}^2 kh)\sigma}{2 gk \tanh kh}$$

$$= \frac{C}{2}\left(1 + \frac{2kh}{\sinh 2kh}\right) \tag{4.89}$$

Therefore, $C_g = nC$, where again

$$n = \frac{1}{2}\left(1 + \frac{2kh}{\sinh 2kh}\right) \tag{4.90}$$

4.8 TRANSFORMATION OF WAVES ENTERING SHALLOW WATER

Several changes occur as a train of waves propagates into shallow water. One of the most obvious is the change in height as the wave shoals. If energy losses (or additions) are negligible, from observation, it is evident that the waves near the point of breaking at a beach are somewhat higher than those farther offshore. Other changes, such as the previously discussed decrease in wave length with shallower depths and the changes in wave direction (Figure 4.13), are not readily apparent from the beach, but often are clearly observable from the air.

4.8.1 The Conservation of Waves Equation

In all previous derivations it has been assumed that the waves are propagating in the x direction; yet if we are discussing a coastline, it is often convenient to locate the coordinate system such that the x direction is in the onshore direction and the y direction is in the longshore direction. It is rare that waves propagate solely in the x direction once the coordinate system is prescribed.

In general, a wave crest corresponds to a line of constant wave phase. For example, if a wave train is represented as $\eta = H/2 \cos \Omega$, where Ω corresponds to the scalar phase function [recall that for waves propagating in the x direction, $\Omega = (kx - \sigma t)$]. Therefore, crests occur for $\Omega = 2n\pi$, where n is defined here as an integer. From vector analysis, the normal *unit* vector **n**

Figure 4.13 Refraction of waves around a small Caribbean island. (Photo courtesy of the L.S.U. Coastal Studies Institute.)

to a scalar function is related to the normal vector $\mathbf{N}$, which is found by taking the gradient of the function, Eq. (2.55),

$$\mathbf{N} = \nabla\Omega \qquad (4.91)$$

where

$$\mathbf{N} = \mathbf{n}|\nabla\Omega| \qquad (4.92)$$

and where, for purposes here, the gradient operator is only the horizontal operator

$$\nabla \equiv \nabla_h = \frac{\partial}{\partial x}\,\mathbf{i} + \frac{\partial}{\partial y}\,\mathbf{j} \qquad (4.93)$$

as Ω is not a function of elevation z. The vector $\mathbf{N}$ points in the direction of the greatest change of Ω, which is the wave propagation direction.[6]

We will *define* the wave number $\mathbf{k}$ as

$$\mathbf{k} = \mathbf{n}|\nabla\Omega| = \nabla\Omega \qquad (4.94)$$

---

[6]$\nabla\eta = (H/2)\sin\psi\nabla\psi$; thus $\nabla\eta$ is in the same direction as $\nabla\psi$. $\nabla\eta$ is the wave direction.

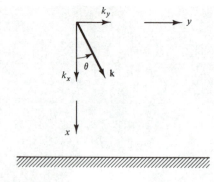

Figure 4.14 Resolution of wave number **k** into orthogonal components.

Note that for waves in the x direction that

$$\mathbf{k} = k\mathbf{i} + 0\mathbf{j} \qquad (4.95a)$$

and

$$|\mathbf{k}| = k \qquad (4.95b)$$

where k is the previously defined wave number. It becomes clear now that the wave number vector is nothing more than the wave number oriented in the wave direction. For waves propagating in an arbitrary direction in x-y space, we have

$$\mathbf{k} = k_x\mathbf{i} + k_y\mathbf{j} \qquad (4.96)$$

and

$$|\mathbf{k}| = k = \sqrt{k_x^2 + k_y^2} \qquad (4.97)$$

If an angle of incidence θ is defined as the angle made between the beach normal (the x direction) and the wave direction, then

$$k_x = |\mathbf{k}| \cos \theta$$
$$k_y = |\mathbf{k}| \sin \theta \qquad (4.98)$$

The phase function[7] is, therefore, $\Omega(x, y, t) = kx \cos \theta + ky \sin \theta - \sigma t = \mathbf{k} \cdot \mathbf{x} - \sigma t$. If the angle of incidence is zero, it is obvious that Ω reverts back to the simple form [Eq. (4.95a)].

The horizontal line along which waves travel is called a wave ray. It is defined (in a manner similar to a streamline) as a line along which the wave number vector is always tangent. As energy travels in the direction of the

[7]This form of the phase function can be obtained in an alternative manner. For waves of length L propagating at an angle to the x axis, the projection of the wave on the x axis has a wave length of L_x. From geometry, $L_x = L/\cos \theta$ and therefore $k_x x = (2\pi/L_x)x = k \cos \theta x$. The y contribution follows similarly.

wave, the wave energy associated with the wave travels along the wave ray also.

The angle made by the wave ray to the x axis can be obtained in the same manner as the local wave direction [see Figure 4.14]:

$$\theta = \tan^{-1} \frac{k_y}{k_x}$$

The wave frequency can be determined from the phase function as

$$\sigma = -\frac{\partial \Omega}{\partial t} \qquad (4.99)$$

It is readily seen that the following expression is identically zero:

$$\frac{\partial}{\partial t}(\nabla \Omega) + \nabla \left(-\frac{\partial \Omega}{\partial t} \right) = 0 \qquad (4.100)$$

which using Eqs. (4.94) and (4.99) can be written as

$$\frac{\partial \mathbf{k}}{\partial t} + \nabla \sigma = 0 \qquad (4.101)$$

This equation states that any temporal variation in the wave number vector must be balanced by spatial changes in the wave angular frequency. If the wave field is constant in time, then $\nabla \sigma = 0$, or the wave period does not change with space. It is constant even as the water depth changes. If the waves encounter a steady current, it was shown in Chapter 3 that $\sigma = \mathbf{k} \cdot \mathbf{U} + \sqrt{gk \tanh kh}$, where $\mathbf{U} =$ mean current vector. Even for this case $\sigma \neq f(x, y)$, that is, only changes in $\mathbf{k}$ occur to compensate for the variable current.

To examine the conservation of waves relationship further, it is best to rederive it in a more intuitive manner. For a small length dx in the direction of wave travel, shown in Figure 4.15, we will relate the number of waves entering and leaving the block of fluid to the accumulation of waves within it. The rate at which waves enter the column is $1/T$ or $\sigma/2\pi$. The rate at which waves are leaving the column a distance dx away is found by using the first-order Taylor series. The difference in inflow and efflux of waves must be equal

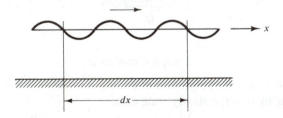

Figure 4.15   Consideration of conservation of waves.

to the accumulation of waves within the region with time, that is, the time rate of change of the number N of waves within the column,

$$\frac{\partial N}{\partial t} = \frac{\partial}{\partial t}\frac{dx}{L} = \frac{\partial}{\partial t}\left(k\frac{dx}{2\pi}\right) = \frac{dx}{2\pi}\frac{\partial k}{\partial t} \qquad (4.102)$$

Equating, we have

$$\frac{\sigma}{2\pi} - \left(\frac{\sigma}{2\pi} + \frac{1}{2\pi}\frac{\partial\sigma}{\partial x}dx\right) = +\frac{dx}{2\pi}\frac{\partial k}{\partial t}$$

or

$$\frac{\partial k}{\partial t} + \frac{\partial\sigma}{\partial x} = 0 \qquad (4.103)$$

which agrees with Eq. (4.101) when applied in the direction of the waves.

4.8.2 Refraction

Referring back to Eq. (4.94), the wave number vector is the gradient of a scalar. If we take the curl of $\mathbf{k}$, we find that

$$\nabla \times \mathbf{k} = 0 \qquad (4.104)$$

by the identity that the curl of a gradient is zero. This irrotationality condition on $\mathbf{k}$ indicates that the line integral $\int\mathbf{k}\cdot d\mathbf{l}$ is independent of path (Chapter 2). Rewriting the integral, we have $\int\nabla\Omega\cdot d\mathbf{l} = \int d\Omega$. Therefore, the irrotationality implies that $\Omega(x, y, t)$ is uniquely determined at each point (for fixed t).

Substituting the components of $\mathbf{k}$ yields

$$\frac{\partial(k\sin\theta)}{\partial x} - \frac{\partial(k\cos\theta)}{\partial y} = 0 \qquad (4.105)$$

For a shoreline where the alongshore variations in the y direction of all variables are zero, that is, there are straight and parallel offshore contours, this equation reduces to

$$\frac{d(k\sin\theta)}{dx} = 0 \qquad (4.106)$$

or

$$k\sin\theta = \text{constant} \qquad (4.107)$$

Therefore, the longshore projection of the wave number is a constant.

Dividing by σ in the steady-state case,

$$\frac{\sin\theta}{C} = \text{constant} \qquad (4.108)$$

The constant is most readily evaluated in deep water, yielding Snell's law:

$$\frac{\sin \theta}{C} = \frac{\sin \theta_0}{C_0} \qquad (4.109)$$

This equation, originally found in geometric optics, relates the change in direction of a wave to the change in wave celerity. Yet from before we know that waves slow down in shallower water; therefore, Snell's law indicates that for coastlines with straight and parallel contours, the wave direction θ decreases as the wave shoals, tending to make the waves approach shore normally.

In general, however, offshore contours are irregular and vary along a coast, so that the full equation must be used.

$$\frac{\partial\, k \sin \theta}{\partial x} - \frac{\partial\, k \cos \theta}{\partial y} = 0 \qquad (4.110)$$

or

$$k \cos \theta \frac{\partial \theta}{\partial x} + k \sin \theta \frac{\partial \theta}{\partial y} = \cos \theta \frac{\partial k}{\partial y} - \sin \theta \frac{\partial k}{\partial x} \qquad (4.111)$$

This first-order nonlinear partial differential equation for θ must be solved by computer techniques for a general coastline (see Noda *et al.*, 1974) to give the wave directions for various locations and water depths.

Historically, ray-tracing techniques were developed to solve this equation following the path of the waves. We can transform Eq. (4.111) into one valid for a coordinate system (s, n) such that s is in the wave direction and n normal to it (see Figure 4.16), defined as

$$x = s \cos \theta - n \sin \theta \qquad (4.112a)$$

$$y = s \sin \theta + n \cos \theta$$

Using the chain rule the derivative operators in the s and n directions can be

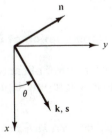

Figure 4.16   Coordinate system (s, n) defined by direction of wave number vector **k**.

established,

$$\frac{\partial}{\partial s} = \frac{dx}{ds}\frac{\partial}{\partial x} + \frac{dy}{ds}\frac{\partial}{\partial y} \tag{4.112b}$$

$$= \cos\theta\,\frac{\partial}{\partial x} + \sin\theta\,\frac{\partial}{\partial y}$$

and correspondingly,

$$\frac{\partial}{\partial n} = -\sin\theta\,\frac{\partial}{\partial x} + \cos\theta\,\frac{\partial}{\partial y} \tag{4.112c}$$

It is clear that the equation governing the wave angle can be rewritten as

$$\frac{\partial\theta}{\partial s} = \frac{1}{k}\frac{\partial k}{\partial n} = -\frac{1}{C}\frac{\partial C}{\partial n} \tag{4.113}$$

with $k = \sigma/C$. This equation relates the curvature of the wave ray to the logarithmic derivative of the wave number normal to the wave direction.

Ray tracing is often done by hand calculation,[8] as well as by computer programs. The procedure involves using Snell's law locally at each contour line of the offshore bathymetry that must be known. First a "smoothing" procedure is used to remove sharp changes of direction of the contour lines. The proper amount of smoothing is unfortunately a matter of judgment. Then the deep water wave period and angle of incidence must be known. Drawing the deep water wave crest on the bathymetry chart offshore of the $(h/L_0 = 0.5)$ contour provides the starting point for each of the rays, which are spaced at equal intervals. These intervals are chosen to give sufficient detail in the nearshore zone. For each of the contours representing a known depth, the wave celerity is determined. A ray is then drawn from the deep water crest location to the first intersection of a contour for which the wave feels bottom. At this point, a locally straight contour line is assumed and constructed by making a line segment tangent to the point of intersection. The normal to this line provides a means to calculate the angle of incidence with respect to the contour. Using Snell's law [Eq. (4.109)], the angle to which the wave is refracted is computed. The ray is then extended to the next contour and the process repeated. This can be tedious and several aids have been constructed to aid in this process (see the *Shore Protection Manual*).

4.8.3 Conservation of Energy

For conservation of energy, in a steady-state case, where there are not any energy losses or inputs, equations are developed readily relating the wave

[8]See, for example, the *Shore Protection Manual* (1977).

heights at two points of interest, especially for the case of straight and parallel bottom contours as in Figure 4.17. Recognizing that there is no energy flux across the wave rays, the energy flux $\mathcal{F}$ across b_0 is the same as across b_1 and b_2. Due to the convergence or divergence of the wave rays, resulting from either refraction or actual physical boundaries, and due to changes in depth, the energy per unit area changes between b_1 and b_2. Assuming no wave reflection, the conservation of energy, Eq. (4.81), requires

$$(EnC)_1 b_1 = (EnC)_2 b_2 \qquad (4.114)$$

or, using our definition for E as

$$E = \tfrac{1}{8} \rho g H^2 \qquad (4.115)$$

we can solve for the wave height H_2:

$$H_2 = H_1 \sqrt{\frac{C_{g_1}}{C_{g_2}}} \sqrt{\frac{b_1}{b_2}} \qquad (4.116)$$

If it is recognized that waves do not change period with depth (i.e., the wave period is a constant), then we have between deep and intermediate or shallow

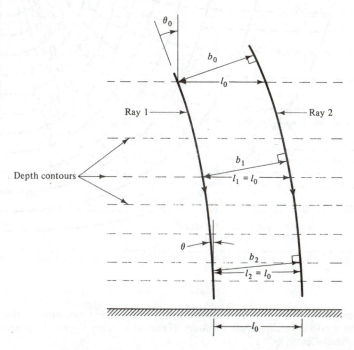

Figure 4.17 Characteristics of wave rays during refraction over idealized bathymetry.

depth water

$$H_2 = H_0 \sqrt{\frac{C_0}{2C_{g_2}}} \sqrt{\frac{b_0}{b_2}}$$

$$= H_0 K_s K_r$$

(4.117)

where K_s is the shoaling coefficient and K_r the refraction coefficient. The shoaling coefficient is plotted in Figure 3.9.

In water with straight and parallel offshore contours, it is possible to determine the refraction coefficient, $(b_0/b_2)^{1/2}$, directly. In Figure 4.17 two rays are shown propagating to shore. Intuitively, since each wave refracts at the same rate along the beach, it should be expected that ray 2 is merely ray 1 displaced a constant distance l_0 in the longshore direction. This is, in fact, the interpretation of the constancy of longshore wave number given by Snell's law, $k_0 \sin \theta_0 = k \sin \theta$. From the diagram it can be seen that $b_0 = l_0 \cos \theta_0$ and

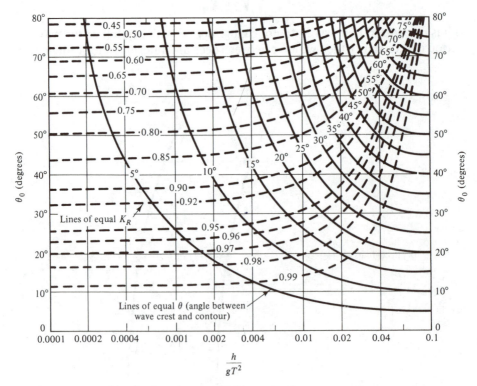

Figure 4.18 Changes in wave direction and height due to refraction on slopes with straight, parallel depth contours. (From U.S. Army Coastal Engineering Research Center, 1977.)

$b_2 = l_0 \cos \theta_2$. Therefore, the refraction coefficient K_r is

$$K_r = \left(\frac{b_0}{b_2}\right)^{1/2} = \left(\frac{\cos \theta_0}{\cos \theta_2}\right)^{1/2} = \left(\frac{1 - \sin^2 \theta_0}{1 - \sin^2 \theta_2}\right)^{1/4} \qquad (4.118)$$

which is always less than unity. The perpendicular spacing between the rays always becomes greater as the wave shoals. Figure 4.18 presents a convenient means to determine K_r and wave directions from deep water characteristics. Since K_r depends on h/gT^2 and θ_0 and K_s depends only on h/gT^2, it is possible to present the product $K_r K_s$ as a function of h/gT^2 and θ_0, as shown in Figure 4.19.

Example 4.1

A wave of 2 m height in deep water approaches shore with straight and parallel contours at a 30° angle and has a wave period of 15 s. In water of 8 m, what is the direction of the wave, and what is its wave height?

Solution.   Using Figure 4.18, $h/gT^2 = 0.0036$ and therefore $\theta \simeq 10.5°$ and $K_r = 0.94$. The value of K_s, using the C_g/C_0 curve of Figure 3.9, is computed to be 1.2. $H = 2(0.94)(1.2) = 2.26$ m. This result can also be obtained directly from Figure 4.19 [i.e., $K_r K_s = 1.13$ and $H = 2(1.13) = 2.26$ m].

In ray-tracing procedures, the separation distance b can be found analytically (Munk and Arthur, 1952). From Figure 4.20 it can be seen, for waves traveling with celerity C in the s direction, that the velocity components are

$$\frac{ds}{dt} = C, \qquad \frac{dx}{dt} = C \cos \theta, \qquad \frac{dy}{dt} = C \sin \theta \qquad (4.119)$$

Given C and θ, these equations serve to provide the locations along the ray path.

At A, $d\theta = (\partial\theta/\partial n)b$ and, also $db = d\theta \, ds$, which is the first-order change in arc length due to the angle increment $d\theta$. Substituting for $d\theta$ in these two equations yields

$$\frac{1}{b}\frac{\partial b}{\partial s} = \frac{\partial \theta}{\partial n} \qquad (4.120a)$$

or, defining $\beta = b/b_0$, where b_0 is an initial reference spacing of the wave ray, we obtain

$$\frac{1}{\beta}\frac{\partial \beta}{\partial s} = \frac{\partial \theta}{\partial n} \qquad (4.120b)$$

This equation, which relates the change in spacing along the ray to the change in θ in the normal direction, is similar in form to Eq. (4.113), which also involves θ.

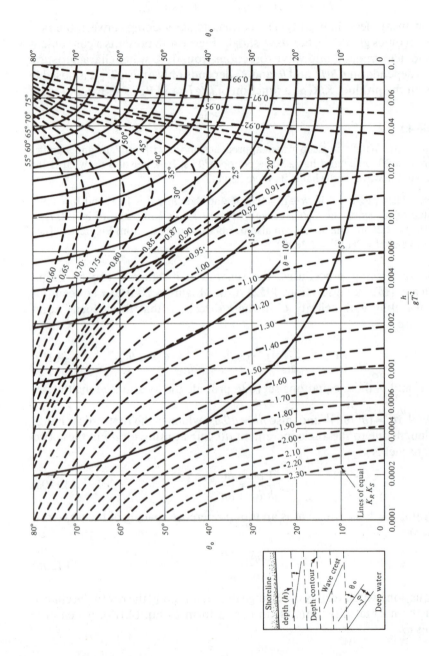

Figure 4.19 Change in wave direction and height due to refraction on slopes with straight, parallel depth contours including shoaling. (From U.S. Army Coastal Engineering Research Center, 1977.)

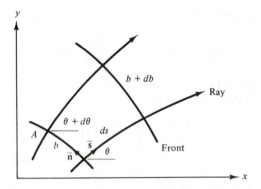

Figure 4.20   Schematic diagram showing adjacent rays.

An ordinary differential equation can be obtained for β by computing the mixed derivatives

$$\frac{\partial}{\partial n}\frac{\partial \theta}{\partial s} - \frac{\partial}{\partial s}\frac{\partial \theta}{\partial n}$$

Using the defnitions for the $\partial/\partial n$, $\partial/\partial s$ operators [Eqs. (4.112b) and (4.112c)], we obtain

$$\frac{\partial}{\partial n}\frac{\partial \theta}{\partial s} - \frac{\partial}{\partial s}\frac{\partial \theta}{\partial n} = \left(\frac{\partial \theta}{\partial x}\right)^2 + \left(\frac{\partial \theta}{\partial y}\right)^2 = \left(\frac{\partial \theta}{\partial s}\right)^2 + \left(\frac{\partial \theta}{\partial n}\right)^2$$

$$= \frac{1}{C^2}\left(\frac{\partial C}{\partial n}\right)^2 + \frac{1}{\beta^2}\left(\frac{\partial \beta}{\partial s}\right)^2$$

after substituting from Eqs. (4.113) and (4.120b). Note that the right-hand side is nonzero; this is due to the fact that the derivative operators are functions of θ.

If we cross-differentiate Eqs. (4.113) and (4.120b) directly for the mixed derivative expressions, the following results:

$$\frac{\partial}{\partial n}\frac{\partial \theta}{\partial s} - \frac{\partial}{\partial s}\frac{\partial \theta}{\partial n} = -\frac{1}{C}\frac{\partial^2 C}{\partial n^2} + \frac{1}{C^2}\left(\frac{\partial C}{\partial n}\right)^2 - \frac{1}{\beta}\frac{\partial^2 \beta}{\partial s^2} + \frac{1}{\beta^2}\left(\frac{\partial \beta}{\partial s}\right)^2$$

again, a nonzero right-hand side. If we now equate the two right-hand sides, we have

$$\frac{\partial^2 \beta}{\partial s^2} + \frac{1}{C}\frac{\partial^2 C}{\partial n^2}\beta = 0 \qquad (4.121a)$$

This equation can be used to obtain β; however, it involves knowledge of the wave fronts in order to determine derivatives in the n direction. If we evaluate the second term, we have

$$\frac{1}{C}\frac{\partial^2 C}{\partial n^2} = \frac{1}{C}\left(\sin^2\theta\,\frac{\partial^2 C}{\partial x^2} - 2\sin\theta\cos\theta\,\frac{\partial^2 C}{\partial x\,\partial y} + \cos^2\theta\,\frac{\partial^2 C}{\partial y^2} - \frac{\partial C}{\partial s}\frac{\partial\theta}{\partial n}\right)$$

but $\partial\theta/\partial n = (1/\beta)\,(\partial\beta/\partial s)$ from Eq. (4.120b).

Therefore, finally β is given by

$$\frac{d^2\beta}{ds^2} + p\,\frac{d\beta}{ds} + q\beta = 0 \qquad\qquad (4.121\mathrm{b})$$

where

$$p(s) = -\frac{\cos\theta}{C}\frac{dC}{dx} - \frac{\sin\theta}{C}\frac{\partial C}{\partial y}$$

and

$$q(s) = \frac{\sin^2\theta}{C}\frac{\partial^2 C}{\partial x^2} - 2\frac{\sin\theta\cos\theta}{C}\frac{\partial^2 C}{\partial x\,\partial y} + \frac{\cos^2\theta}{C}\frac{\partial^2 C}{\partial y^2}$$

Equations (4.121b) and (4.119) provide four ordinary differential equations which can be solved simultaneously to provide locations along the ray and the spacing between the rays over a given bathymetry for which $C(x, y)$ is available (through the dispersion relationship). Numerous ray-tracing programs have been written (see, e.g., Wilson, 1966) and a recent example from Noda (1974) is presented in Figure 4.21.

Wave heights along a ray are related to β, as shown in the preceding section. Similarly to Eq. (4.117), we have

$$H = H_0\sqrt{\frac{C_0}{2C_g}}\sqrt{\frac{1}{\beta}}$$

4.8.4 Waves Breaking in Shallow Water

The shoaling coefficient indicates that the wave height will approach infinity in very shallow water, which clearly is unrealistic. At some depth, a wave of given characteristics will become unstable and break, dissipating energy in the form of turbulence and work against bottom friction. When designing a structure which at times may be inside the surf zone it becomes necessary to be able to predict the location of the breaker line.

The means by which waves break depends on the nature of the bottom and the characteristics of the wave. See Figure 4.22. For very mildly sloping beaches, typically the waves are *spilling* breakers and numerous waves occur within the surf zone (defined as that region where the waves are breaking, extending from the dry beach to the seaward limit of the breaking). *Plunging* breakers occur on steeper beaches and are characterized by the crest of the

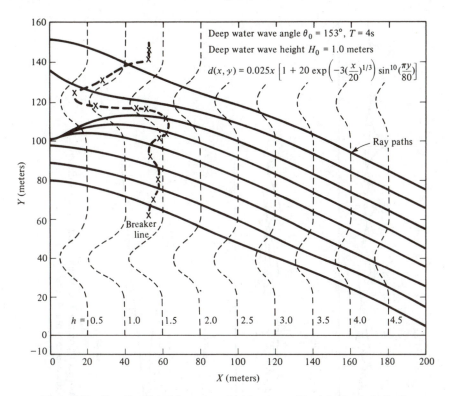

Figure 4.21 Ray lines for oblique wave incidence on a beach in the periodic rip channels. (From Noda, 1974.)

wave curling over forward and impinging onto part of the wave trough. These waves can be spectacular when air, trapped inside the "tube" formed by the wave crest, escapes by bursting through the back of the wave or by blowing out at a nonbreaking section of wave crest. *Surging* breakers occur on very steep beaches and are characterized by narrow or nonexistent surf zones and high reflection. Galvin (1968) has identified *collapsing* as a fourth classification, which is a combination of plunging and surging.

The earliest breaker criterion was that of McCowan (1894), who determined that waves break when their height becomes equal to a fraction of the water depth

$$H_b = \kappa h_b \tag{4.122a}$$

where $\kappa = 0.78$ and the subscript b denotes the value at breaking. Weggel (1972) reinterpreted many laboratory results, showing a dependency of breaker height on beach slope m. His results were

$$\kappa = b(m) - a(m) \frac{H_b}{gT^2} \tag{4.122b}$$

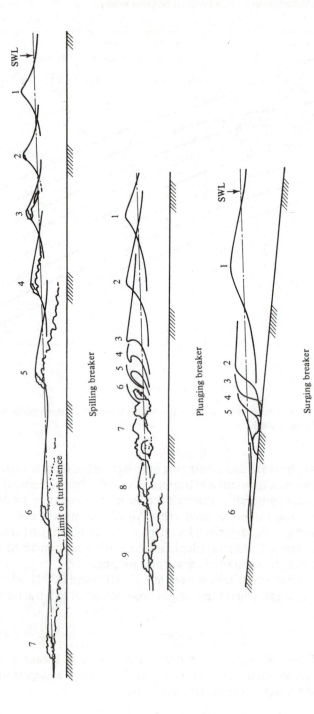

Figure 4.22 Three types of wave breaking on beaches. Small figures denote different stages of the breaking process. (Figure courtesy of I. A. Svendsen.)

where

$$a(m) = 43.8(1.0 - e^{-19m})$$

$$b(m) = 1.56(1.0 + e^{-19.5m})^{-1}$$

which approaches $\kappa = 0.78$ as the beach slope m approaches zero.[9] See Figure 12.7.

As a first approximation, the depth of wave breaking can be determined by the shoaling and refraction formulas for straight and parallel contours if the offshore wave characteristics are known.

$$H = H_0 \left(\frac{C_0}{2nC} \right)^{1/2} \left(\frac{\cos \theta_0}{\cos \theta} \right)^{1/2} \tag{4.123}$$

For shallow water, this is approximately equal to

$$H = H_0 \left(\frac{C_0}{2\sqrt{gh}} \right)^{1/2} \left(\frac{\cos \theta_0}{1} \right)^{1/2} \tag{4.124}$$

if it is assumed that the breaking angle is small. Using McCowan's breaking criterion, we have

$$\kappa h_b = H_0 \left[\frac{C_0}{2\sqrt{gh_b}} (\cos \theta_0) \right]^{1/2} \tag{4.125}$$

and solving for h_b yields

$$h_b = \frac{1}{g^{1/5} \kappa^{4/5}} \left(\frac{H_0^2 C_0 \cos \theta_0}{2} \right)^{2/5} \tag{4.126}$$

or for a plane beach where $h = mx$ and $m = \tan \beta$, the beach slope, the distance to the breaker line from shore is

$$x_b = \frac{h_b}{m} = \frac{1}{mg^{1/5} \kappa^{4/5}} \left(\frac{H_0^2 C_0 \cos \theta_0}{2} \right)^{2/5} \tag{4.127}$$

Finally, the breaking wave height is estimated to be

$$H_b = \kappa m x_b = \left(\frac{\kappa}{g} \right)^{1/5} \left(\frac{H_0^2 C_0 \cos \theta_0}{2} \right)^{2/5} \tag{4.128}$$

Komar and Gaughan (1972), using the conservation of wave energy flux in the manner of Munk (1949) for solitary waves, developed an equation similar to Eq. (4.128) for normally incident waves ($\theta = 0°$). Dalrymple et al. (1977) included the deep water wave angle as developed above. By comparing to a number of laboratory data sets, it appears that Eq. (4.128) underpredicts the breaking wave height by approximately 12% (with $\kappa = 0.8$). See Figure

[9]The $a(m)$ parameter originally defined by Weggel was dimensional and required use of the English system of units. The parameters $a(m)$ and $b(m)$ presented here are dimensionless.

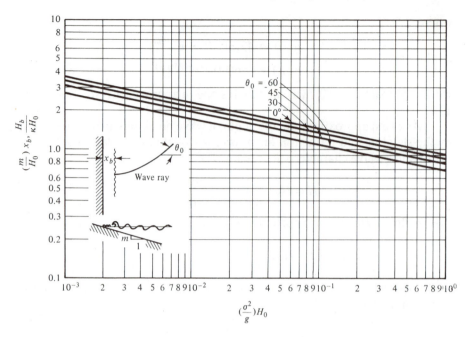

Figure 4.23 Surf zone width x_b and breaking wave height H_b versus deep water wave height H_0 in dimensionless form and as a function of θ_0, the deep water incident angle $\kappa = 0.8$.

4.23 for a dimensionless representation of Eq. (4.128). Wave breaking, with its complexities of turbulence and wave nonlinearities, is still an area of active research. The reader who must deal with design in the surf zone is referred to the literature for the most accurate prediction of surf zone width, breaking wave height, and other surf zone parameters. As an example, see Svendsen and Buhr Hansen (1976).

4.9 WAVE DIFFRACTION

Wave diffraction is the process by which energy spreads laterally perpendicular to the dominant direction of wave propagation. A simple illustration is presented in Figure 4.24, in which a wave propagates normal to a breakwater of finite length and diffraction occurs on the sheltered side of the breakwater such that a wave disturbance is transmitted into the "geometric shadow zone." It is clear that a quantitative understanding of the effects of wave diffraction is relevant to the planning and evaluation of various harbor layouts, including the extent and location of various wave-absorbing features on the perimeter. Diffraction is also important in the case of wave propagation across long distances, in which classical wave refraction effects considered alone would indicate zones of wave convergences and extremely high concentrations of wave energy. As the energy tends to be concentrated

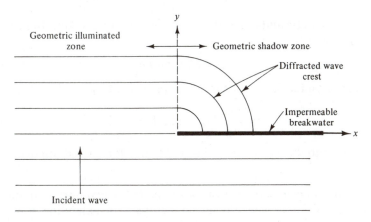

Figure 4.24 Diffraction of wave energy into geometric shadow zone behind a structure.

between a pair of converging wave orthogonals, some of this energy will "leak" across the rays toward regions of less wave energy density. Most present methodologies for computing wave energy distribution along a shoreline due to wave propagation across a shelf do not account for diffraction and may result in greatly exaggerated distributions of wave energy. In the following sections, the main contributions contained in the classical paper by Penney and Price (1952) which relate to diffraction around breakwater-like structures will be reviewed.

4.9.1 Diffraction Due to Wave–Structure Interaction

The three-dimensional linearized boundary value problem formulation for this situation is similar to that presented before [Eqs. (3.19), (3.20), (3.29), and (3.30)] for two dimensions with the exception of the no-flow condition on the structure boundary and will not be presented here. Considering water of uniform depth, the vertical dependency $Z(z)$ satisfying the no-flow bottom boundary condition is

$$Z(z) = \cosh k(h + z) \tag{4.129}$$

and the velocity potential ϕ is represented by

$$\phi(x, y, z, t) = Z(z)F(x, y)\, e^{i\sigma t} \tag{4.130}$$

where $F(x, y)$ is a complex function and $i = \sqrt{-1}$. Substituting Eq. (4.130) into the Laplace equation yields the Helmholtz equation in $F(x, y)$:

$$\frac{\partial^2 F}{\partial x^2} + \frac{\partial^2 F}{\partial y^2} + k^2 F(x, y) = 0 \tag{4.131}$$

The kinematic and dynamic free surface boundary conditions yield the usual dispersion equation

$$\sigma^2 = gk \tanh kh$$

and an equation for the water surface displacement η given by

$$\eta = \frac{i\sigma}{g} F(x, y) \cosh kh \, e^{i\sigma t} \tag{4.132}$$

The solutions to this equation will be examined for several important cases.

Normal wave incidence on a semi-infinite breakwater.   An ideal (perfectly reflecting) breakwater aligned on the x axis and extending from $x = 0$ to $x = +\infty$ will require the boundary condition

$$\frac{\partial F}{\partial y} = 0, \qquad 0 < x < +\infty, \quad y = 0 \tag{4.133}$$

For the boundary condition for $x < 0$, we require that the waves be purely progressive in the positive y direction, that is,

$$F(x, y) = Ae^{-iky}, \qquad x \to -\infty, \quad \text{all } y \tag{4.134}$$

which, when combined with Eq. (4.132), yields the desired result.

The solution of the governing equations was developed by Sommerfeld (1896) and is expressed as

$$F(x, y) = \frac{1 + i}{2} \left[e^{-iky} \int_{-\infty}^{\beta} e^{-i(\pi/2)u^2} \, du + e^{iky} \int_{-\infty}^{\beta'} e^{-i(\pi/2)u^2} \, du \right] \tag{4.135}$$

where β, β', and r are defined by

$$\beta^2 \equiv \frac{4}{L}(r - y), \qquad \beta'^2 \equiv \frac{4}{L}(r + y), \qquad r = \sqrt{x^2 + y^2} \tag{4.136}$$

and the signs of β and β' to be taken depend on the quadrant in which the solution is being applied (see Figure 4.25). With considerable algebra, it can be verified that $F(x, y)$ as given by Eq. (4.135) satisfies both the Helmholtz equation and the boundary condition given by Eq. (4.133). The solution for $F(x, y)$ may be evaluated in terms of Fresnel integrals

$$\int_0^u \cos \tfrac{1}{2} \pi u^2 \, du \qquad \text{and} \qquad \int_0^u \sin \tfrac{1}{2} \pi u^2 \, du \tag{4.137}$$

which are tabluated in Abramowitz and Stegen (1965).

As $F(x, y)$ is complex, it contains both wave amplitude and phase information. As expected, at large x and $y < 0$ a standing wave is formed, at large x and y the waves approach zero, and for $x \to -\infty$, and all y, the wave is unaffected by the presence of the breakwater. Figure 4.26 represents wave

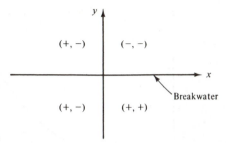

Figure 4.25 Sign criterion for (β, β').

fronts and isolines of relative wave height for $y > 0$; the horizontal scales are rendered dimensionless in terms of wave lengths.

Although the solution for $F(x, y)$ is algebraically complicated, there are several simple features that are of engineering relevance. First for large y, the relative wave height approaches one-half on a line separating the geometric shadow and illuminated regions ($x = 0$) (see Figure 4.27). Second, for $y/L > 2$, isolines of wave height behind a breakwater may be determined in

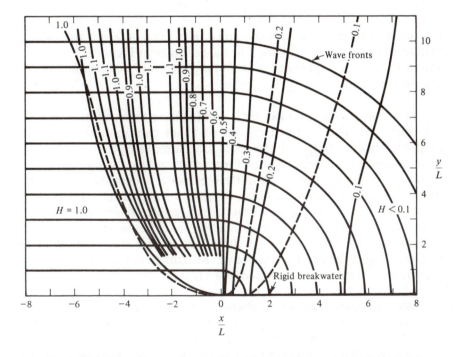

Figure 4.26 Wave fronts and contour lines of maximum wave heights in the lee of a rigid breakwater, and waves being incident normally. (————) exact solution, (- - - - -) approximate solution based on Eq. (4.138) and Figure 4.27. (After Penney and Price, 1952.)

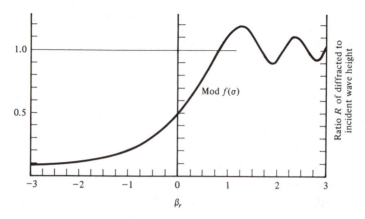

Figure 4.27 Relative diffracted wave height R versus distance parameter β_r. (From Penney and Price, 1952.)

accordance with the following parabolic equation:

$$\frac{x}{L} = \sqrt{\frac{\beta_R^4}{16} + \frac{\beta_R^2}{2}\frac{y}{L}} \qquad (4.138)$$

in which β_R is the abscissa value obtained from Figure 4.27 for any value of relative wave height, $R = H/H_I$. The dashed lines in Figure 4.26 compare several isolines obtained from Eq. (4.138) and Figure 4.27 with those from the complete solution.

Obliquely incident waves on a semi-infinite breakwater. For this case, there will also be three regions or zones corresponding to (1) the geometric shadow zone, (2) the geometric illuminated zone outside the region of direct reflection from the breakwater, and (3) the up-wave region within which direct reflection from the breakwater occurs. An example of a diffraction diagram showing isolines of relative wave height is presented in Figure 4.28 for $\theta_0 = 30°$. Plots for other directions are presented in the *Shore Protection Manual* (1977). The diffracted wave fronts in the geometric shadow zone are approximated well by circles with their centers at the breakwater tip. As before, the relative wave height along a line separating the geometric sheltered and illuminated zones is approximately one-half.

Wave diffraction behind an offshore breakwater of finite length. For an offshore breakwater of finite length, an approximate diffraction diagram can be developed by considering the maximum wave height to be the sum of the two waves diffracting around each of the two ends of the breakwater. The resulting diffraction coefficients would therefore represent an upper limit, since only in very special locations would the waves reinforce completely. Figure 4.29 presents approximate isolines of diffraction coefficients for an offshore breakwater which is 10 wave lengths long.

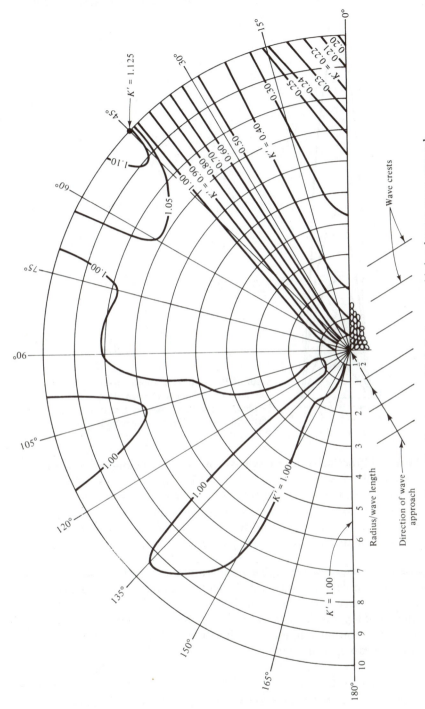

Figure 4.28 Wave diffraction by a semi-infinite impermeable breakwater, wave approach direction = 30°. (After Wiegel, 1962.)

121

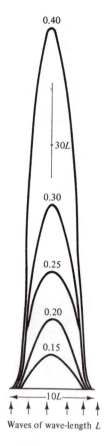

Figure 4.29 Isolines of approximate diffraction coefficients for normal wave incidence behind a breakwater that is 10 wavelengths long. (From Penney and Price, 1952.)

Wave diffraction due to waves of normal incidence propagating through a breakwater gap. For a gap width that is in excess of one wave length, it can be shown that the diffracted wave solution is very nearly given by the superposition of terms in the diffraction solution selected to approximately satisfy the boundary conditions on the two breakwater segments. Figure 4.30 presents an example for a gap that is 2.5 wave lengths long.

Waves propagating through a breakwater gap narrower than one wave length. For this case, the waves in the lee of the breakwater propagate as if from a point source and in accordance with energy conservation relationships; the wave heights decrease as $r^{-1/2}$ with distance from the center of the gap. The expression for relative wave height as a function of r for locations not too near a gap of width b is

$$\frac{H(r)}{H_0} = \frac{\pi\sqrt{b/\pi r}}{2\sqrt{kb[(\ell n\ kb/8 + \gamma)^2 + \pi^2/4]}} \tag{4.139}$$

in which γ is the Euler constant ($= 0.577\ldots$).

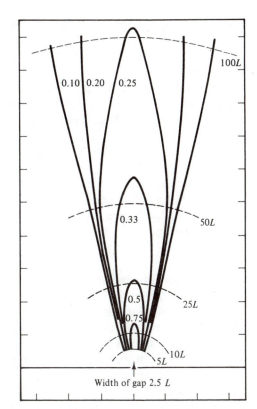

Figure 4.30 Isolines of approximate diffraction coefficients for normal wave incidence and a breakwater gap width of 2.5 wavelengths. (From Penney and Price, 1952.)

4.10 COMBINED REFRACTION–DIFFRACTION

Refraction, which involves wave direction and height changes due to depth variations, and diffraction, caused by discontinuities in the wave field resulting from the wave's interaction with structures, often occur simultaneously. For example, at the tip of a breakwater, diffraction is of utmost importance, yet if a large scour hole exists there or if a beach is nearby, refraction is important as well. It therefore is necessary to be able to treat both phenomena simultaneously.

Theoretically, the problem is difficult, demanding the solution of the Laplace equation in an irregularly varying domain. Therefore, approximations must be made to simplify the problem. The crudest approach, most often used in practice, is to assume that diffraction predominates within several wave lengths of the structure and farther away, only refraction. In the last decade, however, a newer approach has evolved through the use of a model equation. Berkhoff (1972), seeking an equation governing the propagating wave mode [which has a $\cosh k(h + z)$ dependency over the depth], multiplied the Laplace equation by $\cosh k(h + z)$ and integrated over the depth. This reduces the equation to the two horizontal dimensions and yields

$$\nabla_h \cdot (CC_g \, \nabla_h F) + \sigma^2 \left(\frac{C_g}{C} \right) F = 0 \tag{4.140}$$

where ∇_h is the horizontal gradient operator and C and C_g are the wave and group velocity, respectively. The F is a complex function which represents the wave amplitude and phase. The total velocity potential then is

$$\phi(x, y, z) = F \cdot \frac{\cosh k(h + z)}{\cosh kh} \tag{4.141}$$

In deriving this equation it was assumed that the bottom slopes are mild. This model equation, while approximate in intermediate depth, is exact in both deep and shallow water. In deep water it reduces to Eq. (4.131), while in shallow water it is

$$g\nabla_h \cdot (h\nabla_h F) + \sigma^2 F = 0 \tag{4.142}$$

which is a two-dimensional equivalent of Eq. (5.37), valid for long waves, as discussed in Chapter 5.

Analytical solutions to the model equation are few; Jonsson and Brink-Kjaer (1973) and Smith and Sprinks (1975) present the case of waves encountering a circular island, and for Smith and Sprinks, the case for edge waves and waves propagating over a step are also treated. Kirby et al. (1981) used the model equation to study edge waves on irregular beach profiles. Numerical finite element techniques have been used by Berkhoff to treat arbitrary boundary problems such as harbors and islands.

A second approach, developed by Radder (1979) and Lozano and Liu (1980), utilizes a parabolic approximation to the elliptic Laplace equation, which makes the solution more easily obtainable as only initial condition must be specified as opposed to all the lateral boundary conditions. These methods are computationally quicker than Berkhoff's.

REFERENCES

ABRAMOWITZ, M., and I. A. STEGUN, *Handbook of Mathematical Functions*, Dover, New York, 1965.

BERKHOFF, J. C. W., "Computation of Combined Refraction–Diffraction," *Proc. 13th Conf. Coastal Eng., ASCE*, Vancouver, 1972.

DALRYMPLE, R. A., R. A. EUBANKS, and W. A. BIRKEMEIER, "Wave-Induced Circulation in Shallow Basins," *J. Waterways, Ports, Coastal Ocean Div., ASCE*, Vol. 103, Feb. 1977.

DEAN, R. G. and F. URSELL, "Interation of a Fixed Circular Cylinder with a Train of Surface Waves," MIT Hydrodynamics Laboratory Rept. T.R. No. 37, Sept. 1959.

GALVIN, C. J., "Breaker type classifications of three laboratory beaches," *J. Geophys. Res.*, Vol. 73, No. 12, 1968.

JONSSON, I. G., and O. BRINK-KJAER, "A Comparison between Two Reduced Wave Equations for Gradually Varying Depth," ISVA Prog. Rep. 31, Tech. Univ., Denmark, 1973.

KIRBY, J. T., R. A. DALRYMPLE, and P. L.-F. LIU, "Modification of Edge Waves by Barred Beach Topography," *Coastal Eng.*, Vol. 5, 1981.

KOMAR, P. D., and M. K. GAUGHAN, "Airy Wave Theory and Breaker Wave Height Prediction," *Proc. 13th Conf. Coastal Eng., ASCE*, Vancouver, 1972.

LOZANO, C. J., and P. L.-F. LIU, "Refraction–Diffraction Model for Linear Surface Water Waves," *J. Fluid Mech.*, Vol. 101, 1980.

McCOWAN, J., "On the Highest Wave of Permanent Type," *Philos. Mag. J. Sci.*, Vol. 38, 1894.

MUNK, W. H., "The Solitary Wave Theory and Its Applications to Surf Problems," *Ann. N. Y. Acad. Sci.*, Vol. 51, 1949.

MUNK, W. H., and R. S. ARTHUR, "Wave Intensity along a Refracted Ray in Gravity Waves," Natl. Bur. Stand. Circ. 521, Washington, D. C., 1952.

NODA, E. K., "Wave-Induced Nearshore Circulation," *J. Geophys. Res.*, Vol. 79, No. 27, 1974, pp. 4097–4106.

NODA, E. K., C. J. SONU, V. C. RUPERT, and J. I. COLLINS, "Nearshore Circulation under Sea Breeze Conditions and Wave–Current Interaction in the Surf Zone," Rep. TETRA T-P-72-149-4, Tetra Tech, Inc., Pasadena, Calif., 1974.

PENNEY, W. G., and A. T. PRICE, "The Diffraction Theory of Sea Waves and the Shelter Afforded by Breakwaters," *Philos. Trans. Roy. Soc. A*, Vol. 244 (882), pp. 236–253, 1952.

RADDER, A. C., "On the Parabolic Equation Method for Water-Wave Propagation," *J. Fluid Mech.*, Vol. 95, 1979.

SMITH, R., and T. SPRINKS, "Scattering of Surface Waves by a Conical Island," *J. Fluid Mech.*, Vol. 72, 1975.

SOMMERFELD, A., "Mathematische Theorie der Diffraction," *Math. Ann.*, Vol. 47, pp. 317–374, 1896.

SVENDSEN, I. A. and J. BUHR HANSON, "Deformation Up to Breaking of Periodic Waves on a Beach," *Proc. 15th Conf. Coastal Eng., ASCE*, Honolulu, 1976.

U.S. ARMY, Coastal Engineering Research Center, *Shore Protection Manual*, Vol. I, U.S. Government Printing Office, Washington, D.C., 1977.

WEGGEL, J. R., "Maximum Breaker Height," *J. Waterways, Harbors Coastal Eng. Div., ASCE*, Vol. 98, No. WW4, Nov. 1972.

WIEGEL, R. L., "Transmission of Waves Past a Rigid Vertical Thin Barrier," *J. Waterways Harbors Div., ASCE*, Vol. 86, No. WW1, Mar. 1960.

WIEGEL, R. L., "Diffraction of Waves by a Semi-infinite Breakwater," *J. Hydraulics Div., ASCE*, Vol. 88, No. HY1, pp. 27–44, Jan. 1962.

WILSON, W. S., "A Method for Calculating and Plotting Surface Wave Rays," Tech. Memo 17, U.S. Army, Coastal Engineering Research Center, 1966.

PROBLEMS

4.1 **(a)** A wave train is propagating normally toward the coastline over bottom topography with straight and parallel contours. The deep water wave length and height are 300 m and 2 m, respectively. What are the wave length, height, and group velocity at a depth of 30 m?
(b) What is the average energy per unit surface area at the site of interest?
(c) Work part (a) for the case of the same deep water characteristics, but with deep water crests oriented at 60° to the bottom contours.

4.2 Derive the relationship for the average potential energy per unit interface area associated with the interface displacement: (*Note*: Neglect capillary effects.)

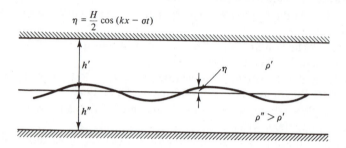

$$\eta = \frac{H}{2} \cos (kx - \sigma t)$$

4.3 The harbor entrance shown below is designed for the following deep water wave conditions:

$$H_0 = 5 \text{ m}$$

$$T = 18 \text{ s}$$

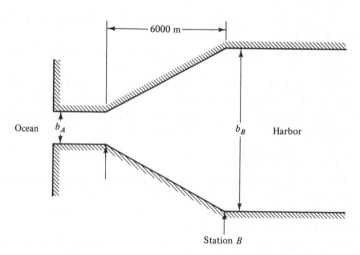

Station *B*

It is desired to design the width at station *B* such that the wave height at station *B* resulting from the design wave is 2 m. What must be the slope of the side walls between A and B for this criterion to be satisfied? Use the following information:

$$b_A = 100 \text{ m}$$

$$h_A = 15 \text{ m}$$

$$h_B = 10 \text{ m}$$

and assume that the wave height is uniform across the harbor width at station B and that the spacing between orthogonals at station A is one-half that in deep water.

4.4 Observations of the water particle motions in a small-amplitude wave system have resulted in the following data for a total water depth of 1 m.

$$\text{major semiaxis} = 0.1 \text{ m}$$

$$\text{minor semiaxis} = 0.05 \text{ m}$$

These observations apply for a particle whose mean position is at middepth. What are the wave height, period, and wave length?

4.5 As a first approximation, the decrease in wave amplitude due to viscous effects can be considered to occur exponentially. For example, for a progressive wave η,

$$\eta = \frac{H_i}{2} e^{-\mu x} \cos (kx - \sigma t)$$

(a) Develop an expression corresponding to that above for the wave system resulting from a wave of height H generated at the wave maker, propagating (and suffering a loss in wave height due to viscosity) to the barrier which is at $x = \ell$, reflecting back (reflection coefficient $= 1.0$) and propagating back to the wavemaker. Do not consider secondary reflections from the wavemaker.

(b) Outline a laboratory procedure for determining the wave system amplitude envelope $|\eta|$.

(c) Show that

$$|\eta| = \frac{H}{2} e^{-\mu \ell} \sqrt{2 \left[\cos 2k(x - \ell) + \cosh 2\mu(x - \ell) \right]}$$

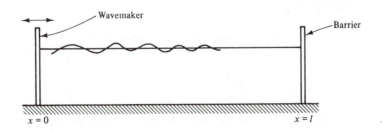

4.6 A wave of 10 s period is propagating toward the rubble mound breakwater. The recording determined by the traversing pressure sensor is shown below. Calculate the rate (per meter of width) of energy dissipation by the breakwater. At what separation distance do the pressure maxima occur?

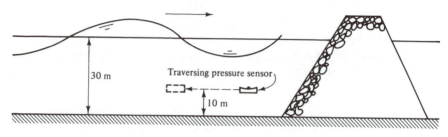

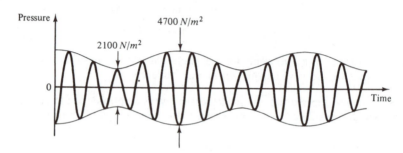

Pressure record from traversing sensor

4.7 An important problem in beach erosion control is the scour in front of vertical walls due to reflected waves. Assuming perfect reflection from a wall and shallow water conditions, determine the resulting water depth under the node nearest the wall if the wave height and period are known at the wall. Assume that the equilibrium scour depth h is one for which the maximum horizontal velocity at the bottom is less than or equal to 3 m/s.

4.8 For a group of waves in deep water, determine the time for each individual wave to pass through the group and the distance traveled by the group during that time if the spacing between the nodes of the group is L_1 and the wave period of the constituent wave is T. There are n waves in the group.

4.9 Two pressure sensors are located as shown in the sketch. For an 8-s progressive wave, the dynamic pressure amplitudes at sensors 1 and 2 are 2.07×10^4 N/m^2 and 2.56×10^4 N/m^2, respectively. What are the water depth, wave height, and wave length?

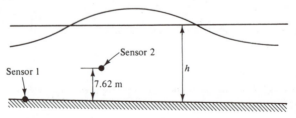

The z axis is oriented vertically upward, that is, in a direction opposed to the gravity vector. The following values may be used:

$$g = 9.81 \text{ m/s}^2$$

$$\rho = 992 \text{ kg/m}^3$$

4.10 An experiment is being conducted on the wave reflection–transmission by the step–barrier combination shown in the drawing that follows. The characteristics of the two wave envelopes are shown.
 (a) What is the height Δ of the step?
 (b) Is enough information given to determine whether or not energy is conserved at the step–barrier?
 (c) If the answer to part (b) is "no," what additional information is required? If the answer to part (b) is "yes," determine whether energy losses occur at the step–barrier.

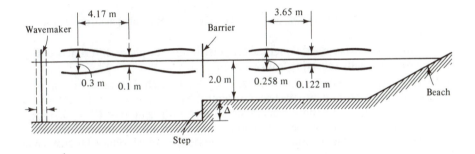

4.11 An axially symmetric wavemaker is oscillating vertically in the free surface, generating circular waves propagating radially outward. At some distance (say R_0) from the wavemaker, the crests are nearly straight over a short distance and the results derived for plane waves may be regarded as valid for the wave kinematics and dynamics at any point. The wave height at R_0 is $H(R_0)$. Derive an expression for $H(r)$, where $r > R_0$. (The depth is uniform.)

4.12 A wave with the following deep water characteristics is propagating toward the coast:

$$H_0 = 1 \text{ m}$$

$$T = 15 \text{ s}$$

At a particular nearshore site (depth = 5 m) a refraction diagram indicates that the spacing between orthogonals is one-half the deep water spacing.
 (a) Find the wave height and wave length at the nearshore site.
 (b) Assuming no wave refraction, but the same deep water information as in part (a), and that the wave will break when the ratio H/h reaches 0.8, in what depth does the wave break?

4.13 A wave with the following deep water characteristics is propagating toward the shore in an area where the bottom contours are all straight and parallel to the coastline:

$$H_0 = 3 \text{ m}$$

$$T = 10 \text{ s}$$

The bottom is composed of a sand of 0.1 mm diameter. If a water particle velocity of 30 cm/s is required to initiate sediment motion, what is the greatest depth in which sediment motion can occur?

4.14 For the wave system formed by the two progressive wave components

$$\eta_i = \frac{H_i}{2} \cos (kx - \sigma t + \epsilon_i)$$

$$\eta_r = \frac{H_r}{2} \cos (kx + \sigma t - \epsilon_r)$$

derive the expression for the average rate of energy propagation in the $+x$ direction.

4.15 Develop an experimental method for determining the phase shift ϵ incurred by a wave partially reflecting from a barrier.

4.16 Develop an equation for the transmitted wave height behind a vertical wall extending a depth d into the water of depth h based on the concept that the wall allows all the wave power below depth d to propagate past (Wiegel, 1960). Qualitatively, do you believe that your equation for the transmitted wave height would underestimate or overestimate the actual value? Discuss your reasons.

4.17 What is the *physical* reason that the pressure is hydrostatic under the nodes of a standing wave (to first order in wave height)?

4.18 Consider an intuitive treatment for the sum of an incident wave of height H_i and reflected wave of height H_r and show that the same envelope results are determined as obtained in the text. Represent the incident wave as two components: one of height H_r and the second as $H_i - H_r$. Now the combination of the first incident component with the reflected yields a pure standing wave and the second incident component is a pure progressive wave. Simply add the envelopes for the pure standing and progressive wave systems.

4.19 Develop the pressure response factor by integrating the linearized equation of motion from some arbitrary elevation z up to the free surface $z = \eta$.

4.20 Using as a breaking criterion that the horizontal water particle at the wave crest exceeds the wave celerity, determine breaking criteria for deep and shallow water. Why does the latter one differ from that of McCowan?

5

Long Waves

Dedication
LORD KELVIN

Sir William Thompson (Lord Kelvin) (1824–1907), born in Belfast, contributed significantly to the field of hydrodynamics, from its theoretical basis to the solution of numerous wave problems. Here he is cited for his work in long waves with Coriolis and gravitational forcing, but he addressed a variety of problems, as is evidenced by his 661 papers and 56 patents. (See *Mathematical and Physical Papers*, Cambridge, 1882.)

When he was 11 years old, he entered the University of Glasgow, leaving in 1841 to enter Peterhouse, Cambridge University, to further his education. During this time he made a trip to Paris University to meet Biot, Liouville, Sturm, and Foucault. In 1846 he became Professor of Natural Philosophy at Glasgow, a post he held for 53 years.

A contemporary of Joule (whom he had met at Oxford) as well as Carnot, Rankine, and Helmholtz, Kelvin pursued a variety of research areas, including heat and heat conduction. Between 1851 and 1854, he fully elucidated the first two laws of thermodynamics, and suggested the concept of refrigeration by the expansion of compressed cold air.

Kelvin contributed actively to the early development of submarine cables. He interacted with cable companies and developed means of testing the purity of copper in the cables after he showed that the purity affected its conductivity. He was knighted in 1866 for his cable work. Before the Institution of Civil Engineers in 1883, Kelvin remarked, "There cannot be a greater mistake than that of looking superciliously upon practical applications of sciences." This philosophy led him to invent numerous electrical devices such as a galvanometer and an ampere gauge, and to set up an electrical company, Kelvin and White, Limited.

He became the first Baron Kelvin of Largs in 1892. He died in 1907 and was buried in Westminster Abbey.

5.1 INTRODUCTION

Waves propagating in shallow water, $kh < \pi/10$, are often called long waves or shallow water waves. Tidal waves, tsunamis (erroneously called tidal waves), and other waves with extremely long periods and wave lengths are shallow water waves, even in the deep ocean.

   The study of long waves is of importance to the engineer in the design of harbors and in studying estuaries and lagoons. Because long wave energy is effectively reflected by structures or even by beaches of mild slope, harbors, which have waves propagating into them, can be excited into resonance by long waves of the proper period, obviously not a desirable state. Tidal propagation in estuaries is affected greatly by the geometry of the estuary; resonance, as in a harbor, can also occur, yielding large tides (50+ ft at the Bay of Fundy).

   In this chapter selected long wave topics are presented, after the equations governing them are derived.

5.2 ASYMPTOTIC LONG WAVES

Previously, the velocity potential and the corresponding velocities and free surface profile for small amplitude waves were derived. The velocities and the surface profile for a progressive wave are described by these equations:

$$u = \frac{H}{2} \frac{gk}{\sigma} \frac{\cosh k(h + z)}{\cosh kh} \cos (kx - \sigma t)$$

$$w = \frac{H}{2} \frac{gk}{\sigma} \frac{\sinh k(h + z)}{\cosh kh} \sin (kx - \sigma t)$$

$$\eta = \frac{H}{2} \cos (kx - \sigma t) \tag{5.1}$$

Using the shallow water asymptotic forms of the hyperbolic functions, we can arrive at equations for the water particle velocities of long waves, $kh << \pi/10$,

$$u_s = \frac{gHk}{2\sigma} \cos (kx - \sigma t) = \frac{\eta C}{h} \tag{5.2}$$

where the shallow water wave celerity $C = \sqrt{gh}$ was introduced and the subscript s denotes shallow water. Interestingly, u_s is not a function of elevation; the horizontal velocity is uniform over depth. For the vertical water particle velocity,

$$w_s = \frac{gHk}{2\sigma}[k(h + z)] \sin (kx - \sigma t)$$

$$= \frac{H}{2} \frac{C}{h}(kh) \left(1 + \frac{z}{h} \right) \sin (kx - \sigma t) = -C \left(1 + \frac{z}{h} \right) \frac{\partial \eta}{\partial x} \tag{5.3}$$

The vertical velocity varies linearly with depth from zero at the bottom to a maximum at the surface and is much smaller in magnitude than u_s. The ratio of their maximum values is

$$\frac{(u_s)_{max}}{(w_s)_{max}} = \frac{1}{kh} \tag{5.4}$$

where kh is small. The pressure under these long waves is found by Eq. (4.22):

$$p = -\rho g z + \rho g \eta \frac{\cosh k(h+z)}{\cosh kh}$$

or

$$p_s = -\rho g z + \rho g \eta$$

$$= \rho g(\eta - z) \tag{5.5}$$

The pressure under these long waves is thus hydrostatic, as might be expected since the vertical accelerations can be shown to be small.

5.3 LONG WAVE THEORY

In Chapter 3 the equations and boundary conditions necessary to solve for two-dimensional water waves were presented. If we assume that the pressure under long waves is hydrostatic at the outset, we can integrate the governing equations over the water depth to get the long wave equations directly rather than asymptotically. Integrating over depth should not be a surprising technique here, particularly when we know that the horizontal velocity is not a function of depth. As a further generalization of the results, the flow will be allowed to be three-dimensional.

5.3.1 Continuity Equation

The three-dimensional conservation of mass equation for an incompressible fluid is

$$\frac{\partial u}{\partial x} + \frac{\partial v}{\partial y} + \frac{\partial w}{\partial z} = 0 \tag{5.6}$$

This is true everywhere in the fluid. Integrating over depth, we have

$$\int_{-h}^{\eta} \left(\frac{\partial u}{\partial x} + \frac{\partial v}{\partial y} + \frac{\partial w}{\partial z} \right) dz \tag{5.7}$$

$$= \int_{-h}^{\eta} \frac{\partial u}{\partial x} \, dz + \int_{-h}^{\eta} \frac{\partial v}{\partial y} \, dz + w(x, y, \eta) - w(x, y, -h) = 0$$

The Leibniz rule of integration is used to integrate terms such as the first two on the right-hand side of this expression. In general, it is stated as

$$\frac{\partial}{\partial x} \int_{\alpha(x)}^{\beta(x)} Q(x, y) \, dy \tag{5.8}$$

$$= \int_{\alpha(x)}^{\beta(x)} \frac{\partial}{\partial x} Q(x, y) \, dy + Q(x, \beta(x)) \frac{\partial \beta(x)}{\partial x} - Q(x, \alpha(x)) \frac{\partial \alpha(x)}{\partial x}$$

Note that if the limits of the integral are constants relative to the variable of integration, the differential operator can be moved into or out of the integral without generating additional terms.

Therefore, the integrated continuity equation is rewritten as

$$\frac{\partial}{\partial x} \int_{-h}^{\eta} u \, dz - u(x, y, \eta) \frac{\partial \eta}{\partial x} - u(x, y, -h) \frac{\partial h}{\partial x} + w(x, y, \eta) - w(x, y, -h)$$

$$+ \frac{\partial}{\partial y} \int_{-h}^{\eta} v \, dz - v(x, y, \eta) \frac{\partial \eta}{\partial y} - v(x, y, -h) \frac{\partial h}{\partial y} = 0 \tag{5.9}$$

If we define

$$U = \frac{1}{h + \eta} \int_{-h}^{\eta} u \, dz \qquad \text{and} \qquad V = \frac{1}{h + \eta} \int_{-h}^{\eta} v \, dz$$

through the use of the mathematical definition of an average (thereby incorporating any possible vertical variation in horizontal velocity), or if we just assume that u and v are constants over the depth, U and V, the continuity equation can be written as

$$\frac{\partial}{\partial x}[U(h + \eta)] + \frac{\partial}{\partial y}[V(h + \eta)] - v(x, y, \eta) \frac{\partial \eta}{\partial y} - v(x, y, -h) \frac{\partial h}{\partial y} \tag{5.10}$$

$$- u(x, y, \eta) \frac{\partial \eta}{\partial x} - u(x, y, -h) \frac{\partial h}{\partial x} + w(x, y, \eta) - w(x, y, -h) = 0$$

Further simplification will result through the use of boundary conditions. The kinematic free surface boundary condition is, in three dimensions,

$$\frac{\partial \eta}{\partial t} + u(x, y, \eta) \frac{\partial \eta}{\partial x} + v(x, y, \eta) \frac{\partial \eta}{\partial y} = w(x, y, \eta) \tag{5.11}$$

The bottom boundary condition for a fixed (with time) surface is

$$w(x, y, -h) = -u(x, y, -h) \frac{\partial h}{\partial x} - v(x, y, -h) \frac{\partial h}{\partial y} \tag{5.12}$$

Substituting these conditions into the vertically integrated continuity equation yields the final form of the continuity equation

$$\frac{\partial[U(h + \eta)]}{\partial x} + \frac{\partial[V(h + \eta)]}{\partial y} = -\frac{\partial \eta}{\partial t} \tag{5.13}$$

This equation can also be derived by considering a column of water of area $dx\,dy$ and height $(h + \eta)$. The continuity equation states that the sum of all the net fluid flows into the column must be balanced by an increase of fluid in the column, which, since it is an incompressible fluid, is manifested by a change in height (volume) of the column (see Figure 5.1). This exercise is recommended to the reader.

5.3.2 Equations of Motion

The equation of motion in the x direction for a fluid is [Eq. (2.35)]

$$\frac{\partial u}{\partial t} + u\frac{\partial u}{\partial x} + v\frac{\partial u}{\partial y} + w\frac{\partial u}{\partial z} = -\frac{1}{\rho}\frac{\partial p}{\partial x} + \frac{1}{\rho}\left(\frac{\partial \tau_{xx}}{\partial x} + \frac{\partial \tau_{yx}}{\partial y} + \frac{\partial \tau_{zx}}{\partial z}\right) \qquad (5.14)$$

Using the equation for pressure under a long wave [Eq. (5.5)], $p = \rho g(\eta - z)$, which states that the pressure is hydrostatic, the first term on the right-hand side becomes

$$-\frac{1}{\rho}\frac{\partial p}{\partial x} = -g\frac{\partial \eta}{\partial x} \qquad (5.15)$$

which is constant over depth. After adding the continuity equation, and vertically integrating using Leibniz's rule, as well as using the kinematic boundary conditions at the surface and the bottom, the horizontal momentum equation becomes

$$\frac{\partial U(h + \eta)}{\partial t} + \frac{\partial}{\partial x}[\beta_{xx}U^2(h + \eta)] + \frac{\partial}{\partial y}[\beta_{yx}UV(h + \eta)] \qquad (5.16)$$

$$= -g(h + \eta)\frac{\partial \eta}{\partial x} + \frac{1}{\rho}(h + \eta)\left(\frac{\partial \tau_{xx}}{\partial x} + \frac{\partial \tau_{yx}}{\partial y}\right) + \frac{\tau_{zx}(\eta) - \tau_{zx}(-h)}{\rho}$$

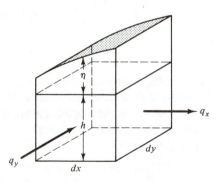

Figure 5.1   Control volume for conservation of mass. The q_x, q_y denote $U(h + \eta)$ and $V(h + \eta)$, respectively.

where

$$\beta_{xx} = \frac{1}{(h+\eta)U^2} \int_{-h}^{\eta} u^2 \, dz$$

$$\beta_{yx} = \frac{1}{(h+\eta)UV} \int_{-h}^{\eta} uv \, dz$$

$$U = \frac{1}{h+\eta} \int_{-h}^{\eta} u \, dz$$

Equation (5.16) is based on the assumption that τ_{xx} and τ_{yx} do not depend on z. The β parameters are momentum correction factors, β_{xx} is slightly greater than unity, and they are used in hydraulics in order to permit the substitution of the squared mean velocity for the mean of the velocity squared.

The y equation becomes

$$\frac{\partial V(h+\eta)}{\partial t} + \frac{\partial}{\partial x}[\beta_{yx}UV(h+\eta)] + \frac{\partial}{\partial y}[\beta_{yy}V^2(h+\eta)] \tag{5.17}$$

$$= -g(h+\eta)\frac{\partial \eta}{\partial y} + \frac{1}{\rho}(h+\eta)\left(\frac{\partial \tau_{xy}}{\partial x} + \frac{\partial \tau_{yy}}{\partial y}\right) + \frac{\tau_{zy}(\eta) - \tau_{zy}(-h)}{\rho}$$

Quite often in practice the momentum correction factor is considered to be unity, and, employing the continuity equation, the equations may be simplified to

$$\frac{\partial U}{\partial t} + U\frac{\partial U}{\partial x} + V\frac{\partial U}{\partial y} = -g\frac{\partial \eta}{\partial x} + \frac{1}{\rho}\left(\frac{\partial \tau_{xx}}{\partial x} + \frac{\partial \tau_{yx}}{\partial y}\right) + \frac{1}{\rho(h+\eta)}[\tau_{zx}(\eta) - \tau_{zx}(-h)]$$

$$\tag{5.18}$$

$$\frac{\partial V}{\partial t} + U\frac{\partial V}{\partial x} + V\frac{\partial V}{\partial y} = -g\frac{\partial \eta}{\partial y} + \frac{1}{\rho}\left(\frac{\partial \tau_{xy}}{\partial x} + \frac{\partial \tau_{yy}}{\partial y}\right) + \frac{1}{\rho(h+\eta)}[\tau_{zy}(\eta) - \tau_{zy}(-h)]$$

$$\tag{5.19}$$

The governing equations, continuity and the equations of motion, are nonlinear. To linearize them to facilitate analytical solutions, we again argue that U, V, and η are small; therefore, their products are also small. The linear equations become, in the absence of shear stresses:

Linearized continuity equation—

$$\frac{\partial(Uh)}{\partial x} + \frac{\partial(Vh)}{\partial y} = -\frac{\partial \eta}{\partial t} \tag{5.20}$$

Linearized frictionless long wave equations of motion—

$$\frac{\partial U}{\partial t} = -g \frac{\partial \eta}{\partial x} \tag{5.21}$$

$$\frac{\partial V}{\partial t} = -g \frac{\partial \eta}{\partial y} \tag{5.22}$$

If the bottom is horizontal, the equations can be cross-differentiated to eliminate U and V, yielding

$$C^2 \left(\frac{\partial^2 \eta}{\partial x^2} + \frac{\partial^2 \eta}{\partial y^2} \right) = \frac{\partial^2 \eta}{\partial t^2} \tag{5.23}$$

where $C = \sqrt{gh}$. This is known as the "wave equation," which occurs quite often in other fields; it governs, for example, membrane vibrations and planar sound waves. To compare with the previous asymptotic results, a solution of the wave equation will be sought for only the x direction. The solution to this equation for a progressive long wave is

$$\eta = \frac{H}{2} \cos (kx - \sigma t) \tag{5.24}$$

Substituting into the x equation of motion [Eq. (5.21)] yields

$$\frac{\partial U}{\partial t} = g \frac{H}{2} k \sin (kx - \sigma t) \tag{5.25}$$

or

$$U = g \frac{H}{2\sigma} k \cos (kx - \sigma t) = \frac{\eta C}{h} \tag{5.26}$$

the same as found by asymptotic means before.

Substituting into the continuity equation yields $C^2 = gh$, the long wave form of the dispersion relationship as derived in Chapter 3.

5.3.3 The Energy and Energy Flux in a Long Wave

For a progressive long wave, the total average energy may be obtained as before as the contributions from the kinetic (KE) and potential energy (PE) components. Because the vertical velocity component is much smaller than the horizontal velocity component, it is not necessary to account for the vertical velocity (for the same order of accuracy). The appropriate expressions are

$$KE = \frac{1}{L} \int_x^{x+L} \int_{-h}^{0} \rho \frac{U^2}{2} \, dz \, dx \tag{5.27}$$

and

$$PE = \frac{1}{L} \int_x^{x+L} \int_{-h}^{\eta} \rho g(h + z) \, dz \, dx - \frac{\rho g h^2}{2} \qquad (5.28)$$

Substituting Eqs. (5.24) and (5.26) for U and η, respectively, and integrating, it is found that

$$KE = PE = \tfrac{1}{16} \rho g H^2$$

and, as before, the total energy per unit surface area is

$$E = KE + PE = \tfrac{1}{8} \rho g H^2 \qquad (5.29)$$

The average energy flux can be shown to be

$$\overline{\mathcal{F}} = EnC = E \sqrt{gh}$$

which again shows that the wave energy travels with the phase speed of the shallow water wave. If we examine the change in wave height due to changes in water depth and channel width via conservation of energy flux, we find that

$$H_2 = H_1 \left(\frac{h_1}{h_2}\right)^{1/4} \left(\frac{b_1}{b_2}\right)^{1/2}$$

which is the shallow water approximation to Eq. (4.116). For the special case of $b_1 = b_2$, this relationship is called Green's law.

5.4 ONE-DIMENSIONAL TIDES IN IDEALIZED CHANNELS

5.4.1 Co-oscillating Tide

As a simple example of tidal wave propagation into a channel, consider a long wave propagating from the deep ocean into a channel of constant depth which has a reflecting wall at one end. This configuration is depicted in Figure 5.2. The wall requires that there be an antinode of a standing wave system there.

Adding two long waves (remember, the equations are linear and superposition is still valid), we have

$$\eta = \eta_i + \eta_r = \frac{H}{2} \cos (kx - \sigma t) + \frac{H}{2} \cos (kx + \sigma t)$$

$$= H \cos \sigma t \cos kx \qquad (5.30)$$

a pure standing wave system as before. Note that the total water surface elevation has a range twice that of the incident tidal height and $\sigma = 2\pi/T$, where T is the tidal period. For a semidiurnal tide, two highs and two lows

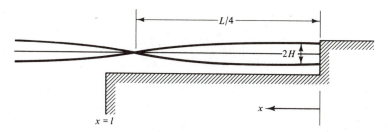

Figure 5.2   Co-oscillating tide in a channel of length l.

during a lunar day, the tidal period is 12.4 h. The distance to the node is found by equating the spatial phase function of η_T to $\pi/2$, that is, finding the phase position for which η equals zero.

$$kx_{\text{node}} = \frac{\pi}{2} \tag{5.31a}$$

or

$$x_{\text{node}} = \frac{L}{4} \tag{5.31b}$$

The *range* of the tide at the entrance to the channel is

$$2|\eta(l)| = 2H|\cos kl| \cdot \tag{5.32}$$

Relating $\eta(l)$ to $\eta(0)$, the amplitude of the tide at the wall, we have

$$\left|\frac{\eta(0)}{\eta(l)}\right| = \frac{1}{|\cos kl|} \tag{5.33}$$

For channels for which l approaches $(2n-1)(L/4)$ and $n = 1, 2, \ldots$, the ratio $|\eta(0)/\eta(l)|$ approaches infinity (i.e., this represents a resonant condition).

5.4.2 Channels with Variable Cross Sections

In deriving the equations of motion and continuity, had we not taken a unit width in the derivation, but considered a channel of width b, the linearized one-dimensional equations valid along the channel centerline would have been

$$\frac{\partial(Uhb)}{\partial x} = -b\frac{\partial \eta}{\partial t} \tag{5.34a}$$

$$b\frac{\partial U}{\partial t} = -gb\frac{\partial \eta}{\partial x} \tag{5.34b}$$

These can be verified by integrating Eqs. (5.6) and (5.14) with respect to y prior to the integration over depth. Differentiating the first equation (5.34a) with respect to time,

$$\frac{\partial}{\partial x}\left(hb \frac{\partial U}{\partial t} \right) = -b \frac{\partial^2 \eta}{\partial t^2} \tag{5.35}$$

Substituting the second equation (5.34b) yields

$$\frac{g}{b} \frac{\partial}{\partial x}\left(bh \frac{\partial \eta}{\partial x} \right) = \frac{\partial^2 \eta}{\partial t^2} \tag{5.36}$$

which reduces to the wave equation if b and h are constant. As in the previous case for the constant depth basin, assume that $\eta(x, t)$ can be written as $\eta(x, t) = \eta(x) \cos \sigma t$. The equation then becomes

$$\frac{g}{b} \frac{d}{dx}\left[bh \frac{d\eta(x)}{dx} \right] + \sigma^2 \eta = 0 \tag{5.37}$$

Several examples of the application of this equation to estuaries with linearly varying widths, depths, or both are provided by Lamb (1945) in Article 186. One case is discussed below. In all these examples, the resulting wave height is different from that predicted by Green's law, as Eq. (5.37) allows for the reflection of waves by the topographic changes, while Green's law assumes that the bathymetric changes are so gradual as to not cause reflection.

Example 5.1

Consider an estuary of uniform depth whose width increases linearly (from zero) with distance toward the mouth at $x = l$. Determine the tidal surface elevations within the estuary, due to the co-oscillating tide.

Solution. Let $b = ax$, where a is equal to b_l/l and b_l is the width of the bay at the mouth. Substituting into Eq. (5.37) the following equation results directly:

$$\frac{d^2\eta(x)}{dx^2} + \frac{1}{x} \frac{d\eta(x)}{dx} + k^2\eta(x) = 0 \tag{5.38}$$

where $k^2 = \sigma^2/C^2 = \sigma^2/gh$. This equation is a Bessel equation of order zero which is solved in terms of Bessel functions. The general solution is

$$\eta(x, t) = [C_1 J_0(kx) + C_2 Y_0(kx)] \cos \sigma t \tag{5.39a}$$

where C_1 and C_2 are constants to be determined. At $x = 0$, the end of the channel, $Y_0(0)$, is infinite, which would be unrealistic for $\eta(0, t)$; therefore, $C_2 = 0$. To evaluate C_1, the tide at $x = l$, the mouth, is taken to be $(H/2) \cos \sigma t$, where, again, H is the local tide range.

$$\eta(l, t) = C_1 J_0(kl) \cos \sigma t = \frac{H}{2} \cos \sigma t$$

or

$$C_1 = \frac{H}{2J_0(kl)}$$

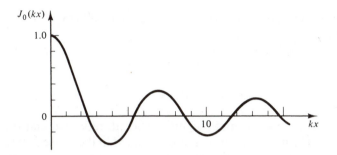

Figure 5.3   Standing waves in a pie-shaped estuary of uniform depth.

Finally, the solution is

$$\eta(x, t) = \frac{H}{2} \frac{J_0(kx)}{J_0(kl)} \cos \sigma t \qquad (5.39b)$$

As shown in Figure 5.3 the zeroth-order Bessel function calls for a large increase in tidal height into the estuary or bay, with a corresponding wave length decrease in the near field (about 25% over the first half wave length). If the estuary length l corresponds to a zero of the Bessel function, then again the possibility for resonance occurs.

5.5 REFLECTION AND TRANSMISSION PAST AN ABRUPT TRANSITION

A more dramatic example of long wave reflection (and transmission) occurs when there is an abrupt change in depth or channel width. Also in this case, Green's law does not apply due to the presence of a reflected wave. Figure 5.4 shows the geometry of the transition region. The fluid domain is divided into regions 1 and 2 as shown. The incoming wave η_i will be assumed to propagate in the positive x direction with height H_i. At the step, it is expected that a portion of the wave will be reflected and some of it transmitted. Therefore, in

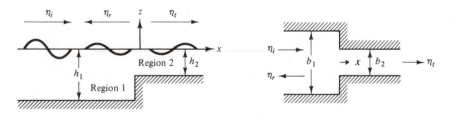

Figure 5.4   Elevation and plan views of an abrupt channel transition.

each region, the total wave forms are assumed as follows:

$$\eta_1 = \eta_i + \eta_r = \frac{H_i}{2} \cos(k_1 x - \sigma t) + \frac{H_r}{2} \cos(k_1 x + \sigma t + \epsilon_r)$$

$$\eta_2 = \eta_t = \frac{H_t}{2} \cos(k_2 x - \sigma t + \epsilon_t)$$

(5.40)

where the subscripts i, r, and t signify incident, reflected, and transmitted, respectively. The difference in sign modifying σt in the phase function for the reflected wave means that this wave is propagating in the negative x direction. In each region the angular frequencies are the same; however, the wave numbers are different due to the change in water depths. The two phase angles, ϵ_r and ϵ_t, are included to allow for the possible phase differences caused by the reflection process.

At the step there are two boundary considerations that must be met by the wave forms η_1 and η_2. First (at $x = \pm\delta$, where δ is infinitesimally small), the water levels on each side of the step should be the same, as, from the long wave equations of motion, any finite water level change over an infinitely small distance 2δ would give rise to infinite accelerations of the fluid particles. Second, from continuity considerations, the mass flow rate from region 1 must equal that into region 2. For a homogeneous fluid, this merely reduces to matching volumetric flow rates between regions. Applying the first condition gives us

$$\eta_i + \eta_r = \eta_t \qquad \text{at } x = 0 \tag{5.41}$$

or, through a trigonometric expansion after substitution,

$$\cos \sigma t \left(\frac{H_i}{2} + \frac{H_r}{2} \cos \epsilon_r - \frac{H_t}{2} \cos \epsilon_t \right) - \sin \sigma t \left(\frac{H_r}{2} \sin \epsilon_r + \frac{H_t}{2} \sin \epsilon_t \right) = 0$$

(5.42)

As this condition must be valid for all time t, two independent conditions result by equating each bracketed term separately to zero:

$$H_i + H_r \cos \epsilon_r = H_t \cos \epsilon_t \tag{5.43a}$$

$$H_r \sin \epsilon_r = -H_t \sin \epsilon_t \tag{5.43b}$$

The continuity of flow condition can be written in terms of the horizontal water particle velocity of the wave multiplied by the cross-sectional area for each region [from the width-integrated continuity equation, Eq. (5.34a)].

$$(Ubh)_1 = (Ubh)_2 \qquad \text{at } x = 0 \tag{5.44}$$

or, recalling that for a long wave,

$$U = \frac{\eta C}{h}$$

in the direction of the wave, we can write

$$b_1 C_1 (\eta_i - \eta_r) = b_2 C_2 \eta_t \tag{5.45}$$

Again we have two conditions, after trigonometric expansion and equating the terms modifying the cosine and sine, respectively:

$$b_1 C_1 H_i - b_1 C_1 H_r \cos \epsilon_r = b_2 C_2 H_t \cos \epsilon_t \tag{5.46}$$

$$b_1 C_1 H_r \sin \epsilon_r = b_2 C_2 H_t \sin \epsilon_t \tag{5.47}$$

Denoting the reflection and transmission coefficients by κ_r $(= H_r/H_i)$ and κ_t $(= H_t/H_i)$, respectively, the four equations (5.43a, 5.46, 5.43b, and 5.47) in terms of the four unknowns $(\kappa_r, \kappa_t, \epsilon_r,$ and $\epsilon_t)$ are

$$1 + \kappa_r \cos \epsilon_r = \kappa_t \cos \epsilon_t \tag{5.48}$$

$$1 - \kappa_r \cos \epsilon_r = \kappa_t \frac{b_2 C_2}{b_1 C_1} \cos \epsilon_t \tag{5.49}$$

$$\kappa_r \sin \epsilon_r = -\kappa_t \sin . \epsilon_t \tag{5.50}$$

$$\kappa_r \sin \epsilon_r = \kappa_t \frac{b_2 C_2}{b_1 C_1} \sin \epsilon_t \tag{5.51}$$

Subtraction of the last two equations yields

$$\kappa_t \left(1 + \frac{b_2 C_2}{b_1 C_1} \right) \sin \epsilon_t = 0 \tag{5.52}$$

which requires that ϵ_t be $\pm n\pi$ for non-trivial values of κ_t. Multiplying Eq. (5.50) by $b_2 C_2/b_1 C_1$ and adding to Eq. (5.51) also indicates that $\epsilon_r = \pm n\pi$ for nontrivial solutions. The four governing equations can therefore be condensed to the following two:

$$1 \pm \kappa_r = \pm \kappa_t \tag{5.53}$$

$$1 \mp \kappa_r = \pm \kappa_t \frac{b_2 C_2}{b_1 C_1} \tag{5.54}$$

in which the plus and minus signs follow from the requirements on ϵ_t and ϵ_r. It is only known that the signs on the right-hand side of each equation are the same and those on the left-hand side are in opposition. The correct signs will be determined later from physical reasoning. Adding Eqs. (5.53) and (5.54), we find that

$$\kappa_t = \pm \frac{2}{1 + b_2 C_2 / b_1 C_1} \tag{5.55}$$

and here it is clear that the + sign is to be taken because for $b_2 C_2 = b_1 C_1$, that is, the case of a uniform channel, the transmission coefficient is obviously unity. Multiplying Eq. (5.53) by $b_2 C_2 / b_1 C_1$ and subtracting from Eq. (5.54) gives us

$$\kappa_r = \mp \left[\frac{(b_2 C_2 / b_1 C_1) - 1}{(b_2 C_2 / b_1 C_1) + 1} \right] \tag{5.56}$$

and here the minus sign is to be taken since for the limiting case of a vanishing channel, $b_2 C_2 / b_1 C_1 = 0$, the reflection coefficient should be +1, that is,

$$\kappa_r = \frac{1 - b_2 C_2 / b_1 C_1}{1 + b_2 C_2 / b_1 C_1} \tag{5.57}$$

Several interesting cases can be examined for $b_1 = b_2$. If the long wave assumptions are still valid, yet $h_1 \gg h_2$, then $\kappa_t \to 2$ and $\kappa_r \to 1$. This case corresponds to a pure standing wave in region 1 and transmitted wave of the same height as the standing wave. But if the situation is reversed, that is, if long waves in very shallow water propagate to a region of greater depth, $h_2 \gg h_1$, then $\kappa_t \to 0$ and $\kappa_r \to -1$. (A negative reflection coefficient means only that the phase of the wave ϵ_r, which we had taken as zero degrees, is shifted to 180°.) It is thus very difficult for waves to propagate from shallow to deeper water. This in fact is true for short waves also. [Hilaly (1969) shows interesting experiments for waves unable to propagate over steps.] Figure 5.5 presents the variations of κ_r and κ_t with the parameter $(b_2/b_1)\sqrt{h_2/h_1}$.

Dean (1964), using this approach and Eq. (5.37), has examined numerous cases of cross-sectional channel changes and obtained the transmission and reflection coefficients.

5.5.1 Seiching

In previous sections, the oscillations of the water in a basin were forced by the tide at a frequency corresponding to the tidal frequency. However, any natural basin, closed or open to a larger body of water, will oscillate at its natural frequency if it is excited in some fashion, such as by earthquake motion, impulsive winds, or other effects.

To predict these oscillations, the equation developed previously can be used. As an example, the seiching in a long rectangular lake with essentially a constant depth will be examined first. A solution to Eq. (5.23) for standing

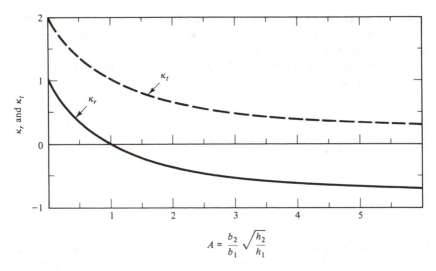

$$A = \frac{b_2}{b_1} \sqrt{\frac{h_2}{h_1}}$$

Figure 5.5   Reflection and transmission coefficients for long waves propagating past an abrupt transition.

waves in this basin is, as before,

$$\eta = \frac{H}{2} \cos kx \cos \sigma t. \tag{5.58}$$

except that σ and k are both unknown. At the ends of the basin, the horizontal velocities must be zero. This requirement can be satisfied using Eq. (5.21) or using the knowledge that the antinodes must be situated at the walls, $x = 0, l$. This requirement yields $\sin kx = 0$ for $x = 0, l$. Therefore, $kl = n\pi$, where n is the number of oscillations of the wave within the basin (equivalently the number of nodes). Substituting for k gives us

$$L = \frac{2l}{n} \tag{5.59}$$

For three values of n, the wave lengths are shown for the basin in Figure 5.6. Each possible type of oscillation is called a mode, and the mode that occurs in

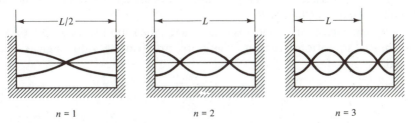

| $n = 1$ | $n = 2$ | $n = 3$ |

Figure 5.6   Standing waves in a simple rectangular basin. The first three modes are shown.

seiching is determined by the cause (forces) that induces seiching. In reality, however, the lower modes are most prevalent since the energy in the higher modes is dissipated more rapidly.

To determine the period of seiching, the dispersion relationship for shallow water waves is used, with Eq. (5.59):

$$C = \frac{L}{T} = \sqrt{gh}$$

or

$$T = \frac{2l}{n\sqrt{gh}} \tag{5.60}$$

This formula is known as the Merian formula. Proudman (1953) gives several examples for actual lakes. For Lake Baikal in Siberia, the length is 664 km and the average depth is 680 m. The Merian formula predicts $T = 4.52$ h, compared to a measured period of 4.64 h.

For more complex one-dimensional basins, a modified Merian formula can be used. Wilson (1966) has summarized the results for a number of geometries and these are presented in Table 5.1. More recently, Wilson (1972) has developed more analytical seiching models and also reviews the literature.

5.6 LONG WAVES WITH BOTTOM FRICTION

The bottom shear stress τ_b retarding the motion of the fluid in unidirectional open channel flow can be expressed in terms of a quadratic friction law:

$$\tau_b = \frac{\rho f U^2}{8} \tag{5.61}$$

where f is the Darcy–Weisbach friction factor and U is the fluid velocity. This equation has been developed through dimensional analysis and experimental data have been used to develop values of f. Further discussion of bottom friction appears in Chapter 9.

For an oscillatory flow, it is clear that as the fluid reverses direction, so also must the bottom friction. Therefore, an absolute value sign is introduced.

$$\tau_b = \frac{\rho f U |U|}{8} \tag{5.62}$$

For wave motions, the bottom friction is a nonlinear function and due to the absolute value sign becomes difficult to work with directly. A common procedure is to linearize the friction term.

Consider U as a periodic function in time, $U = U_m \cos \sigma t$, where U_m is the maximum magnitude of U. If we expand the shear stress term in a Fourier cosine[1] series, we have

$$U|U| = a_0 + \sum_{n=1}^{\infty} a_n \cos n\sigma t \qquad (5.63)$$

where

$$a_0 = \frac{U_m^2}{T} \int_0^T \cos \sigma t \ |\cos \sigma t| \ dt \qquad (5.64)$$

and

$$a_n = \frac{2U_m^2}{T} \int_0^T \cos \sigma t \ |\cos \sigma t| \cos n\sigma t \ dt \qquad (5.65)$$

Evaluating several of these integrals yields

$$a_0 = 0$$

$$a_1 = \frac{8U_m^2}{3\pi}$$

$$a_2 = 0$$

$$a_3 = \frac{8U_m^2}{15\pi}$$

All of the even harmonics are zero while the odd harmonics are nonzero. It is interesting that the quadratic friction law has introduced higher harmonics (which is expected as friction is a nonlinear process). Keeping only the first term in the Fourier expansion (recognizing, however, that the next term in the series expansion is only one-fifth of the leading term),

$$\tau_b = \frac{\rho f U_m^2}{3\pi} \cos \sigma t = \frac{\rho f U_m U}{3\pi} \qquad (5.66)$$

This linearization was first developed by Lorentz (1926) utilizing a dissipation argument and is sometimes referred to as the Lorentz concept. For uniform depth the vertically integrated equation of motion in the x direction can now be written with $\tau_b \equiv \tau_{zx}(-h)$, from Eq. (5.18), as

$$\frac{\partial U}{\partial t} = -g \frac{\partial \eta}{\partial x} - \frac{\tau_{zx}(-h)}{\rho h} = -g \frac{\partial \eta}{\partial x} - AU \qquad (5.67)$$

where $A = f U_m / 3\pi h$, typically a small number, much less than unity. The continuity equation, Eq. (5.13), remains unchanged, of course. Cross-differentiating the two equations assuming A is locally constant and substi-

---

[1] A cosine series is chosen as U and τ_b are even functions of time.

TABLE 5.1 Modes of Free Oscillation in Basins of Simple Geometrical Shape (Constant Width) (From Wilson, 1966)

Basin Type		Profile Equation	Periods of Free Oscillation				
Description	Dimensions		Fundamental T_1	Mode Ratios T_n/T_1			
				$n=1$	2	3	4
Rectangular		$h(x) = h_0$	$\dfrac{2L}{\sqrt{gh_0}}$	1.000	0.500	0.333	0.250
Triangular (isosceles)		$h(x) = h_0\left(1 - \dfrac{2x}{L}\right)$	$1.305\,\dfrac{2L}{\sqrt{gh_0}}$	1.000	0.628	0.436	0.343
Parabolic		$h(x) = h_0\left(1 - \dfrac{4x^2}{L^2}\right)$	$1.110\,\dfrac{2L}{\sqrt{gh_0}}$	1.000	0.577	0.408	0.316

		$h(x)$	Period				
Quartic		$h(x) = h_0\left(1 - \dfrac{4x^2}{L^2}\right)^2$	$1.242\,\dfrac{2L}{\sqrt{gh_0}}$	1.000	0.686	0.500	0.388
Triangular (right-angled)		$h(x) = \dfrac{h_1 x}{L}$	$1.640\,\dfrac{2L}{\sqrt{gh_1}}$	1.000	0.546	0.377	0.288
Coupled, rectangular		$h(x) = h_1 \;(x < 0)$ $h(x) = h_2 \;(x > 0)$ $\left(\dfrac{h_1}{h_2} = \dfrac{1}{4}\right)$	$\dfrac{L_1}{L_2} = \dfrac{1}{2} \quad \dfrac{4L_2}{\sqrt{gh_2}}$	1.000	0.500	0.250	0.125
			$\dfrac{L_1}{L_2} = \dfrac{1}{3} \quad \dfrac{3.13 L_2}{\sqrt{gh_2}}$	1.000	0.559	0.344	0.217
			$\dfrac{L_1}{L_2} = \dfrac{1}{4} \quad \dfrac{2.73 L_2}{\sqrt{gh_2}}$	1.000	0.579	0.367	0.252
			$\dfrac{L_1}{L_2} = \dfrac{1}{8} \quad \dfrac{2.31 L_2}{\sqrt{gh_2}}$	1.000	0.525	0.371	0.279

tuting, the wave equation can be derived, including friction:

$$\frac{\partial^2 \eta}{\partial t^2} + A \frac{\partial \eta}{\partial t} = gh \frac{\partial^2 \eta}{\partial x^2} \tag{5.68}$$

5.6.1 Standing Waves with Frictional Damping

If a solution is assumed of the form

$$\eta = \frac{H_I}{2} f(\sigma_I t) \cos k_I x \tag{5.69}$$

where k remains fixed, such as would occur with a standing wave in a basin with fixed length, and $f(\sigma_I t)$ is some unknown function of time, then the equation is

$$\frac{d^2 f}{dt^2} + A \frac{df}{dt} + ghk_I^2 f = 0 \tag{5.70}$$

The total solution is then found to be

$$\eta = \frac{H_I}{2} e^{(-A/2)t} \cos\left[\sigma_I \sqrt{1 - \frac{1}{4}\left(\frac{A}{\sigma_I}\right)^2}\, t\right] \cos k_I x \tag{5.71}$$

where $\sigma_I = k_I C_I$ (the subscript I refers to undamped conditions), $C_I = \sqrt{gh}$ and H_I is the initial wave height (at $t = 0$), or

$$\eta = \frac{H_I}{2} e^{-\sigma_i t} \cos \sigma_r t \cos kx$$

where

$$\sigma_i \equiv \frac{A}{2} \quad \text{and} \quad \sigma_r = \sigma_I \sqrt{1 - \frac{1}{4}\left(\frac{A}{\sigma_I}\right)^2}$$

The horizontal velocity can be found using the continuity equation

$$U = -\int \frac{1}{h} \frac{\partial \eta}{\partial t}\, dx$$

or

$$U = \frac{H_I}{2k_I h} \sqrt{\sigma_r^2 + \sigma_i^2}\, e^{-\sigma_i t} \sin(\sigma_r t + \epsilon) \sin k_I x \tag{5.72}$$

where

$$\epsilon = \tan^{-1} \frac{\sigma_i}{\sigma_r}$$

The parameters σ_i and σ_r are plotted in Figure 5.7 versus the ratio A/σ_I. As σ_r decreases with friction, the period of oscillation increases; friction slows

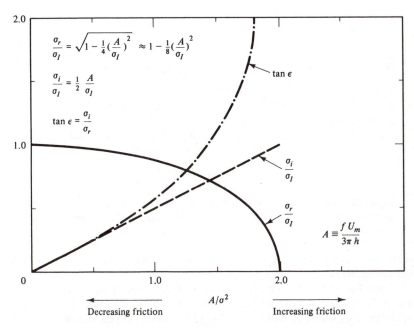

Figure 5.7   Wave number and phase angle for a damped standing wave.

the wave motion. It is clear that the damping ratio A/σ_I in the expression for σ_r must be less than 2; otherwise, excessive damping occurs and there is no wave-like motion (such as might occur with a basin full of molasses).

The relative reduction in amplitude over one wave period is a constant value and is expressed as

$$\frac{\eta(t + T)}{\eta(t)} = e^{-(A/2)T} = e^{-(\pi A/\sigma_I)} = e^{-\sigma_i T} \tag{5.73}$$

which decreases rapidly with increasing σ_i or A. For example, for A/σ_I as small as 0.05, this ratio is 0.85, or a 15% reduction in height within one wave period.

Example 5.2

Shiau and Rumer (1974) carried out a series of experiments to examine the decay of shallow water standing waves (seiches) in a basin. The experiments were conducted in very shallow water ($0.15 < h < 8.5$ cm). Assuming that the motion is laminar, a friction factor can be chosen to compare the above model with their experimental results. Stokes's (1851) second problem, that of an oscillating (with frequency σ) flat plate beneath a still fluid, yields a shear stress on the plate with a magnitude

$$\tau_{b_{max}} = \rho\sqrt{\sigma\nu}\, U_m \tag{5.74}$$

where ν is the kinematic viscosity of the fluid and U_m is the magnitude of the oscillating velocity. This problem is directly analogous to the case under considera-

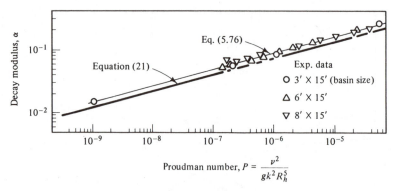

Figure 5.8 Decay modulus versus Proudman number for an assumed laminar friction factor. [From Shiau and Rumer (1974). Equation (21) in figure refers to their solution.]

tion; the only change is that of the reference frame, which is taken as one that is fixed to the oscillating plate.

Since Eq. (5.74) for the shear stress is linear and the preceding treatment represents a linearized form of the shear stress, the laminar flow problem can be treated directly. Comparison of Eqs. (5.74), (5.67), and (5.18) shows that

$$A = \frac{\sqrt{\sigma \nu}}{h} \tag{5.75}$$

The Shiau and Rumer study determined the decay modulus α, which can be obtained from Eq. (5.73) as

$$\alpha = \frac{\pi A}{\sigma_I} \tag{5.76}$$

or from Eq. (5.75) can be expressed as

$$\alpha = \frac{\pi}{h} \sqrt{\frac{\nu}{\sigma}} = \pi P^{1/4} \tag{5.77}$$

where P is the Proudman number, $P = \nu^2/gk^2h^5$. Figure 5.8 shows the theoretical value of α compared with the experimental data. As can be seen, the agreement is excellent for this case with laminar conditions. For deeper relative water depth, when the flow conditions become turbulent, the friction factor becomes more like that for turbulent open channel flow.

5.6.2 Progressive Waves with Frictional Damping

For a periodic progressive wave, the free surface is assumed of a similar form as before, except for a spatial amplitude dependence,

$$\eta = \frac{H_I}{2} e^{-k_i x} \cos (k_r x - \sigma t) \tag{5.78}$$

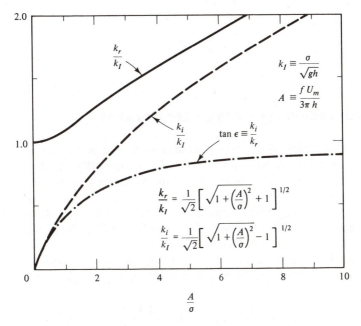

Figure 5.9 Wave number and phase angle for a damped progressive long wave.

The k_r and k_i are determined from the differential equation, Eq. (5.68):

$$k_r = \frac{k_I}{\sqrt{2}}\left[\sqrt{1 + \left(\frac{A}{\sigma}\right)^2} + 1\right]^{1/2} \approx k_I\left[1 + \frac{1}{8}\left(\frac{A}{\sigma}\right)^2\right] \qquad (5.79)$$

where the second expression is valid for small A/σ and $k_I = \sigma/\sqrt{gh}$.

$$k_i = \frac{k_I}{\sqrt{2}}\left[\sqrt{1 + \left(\frac{A}{\sigma}\right)^2} - 1\right]^{1/2} \approx \frac{k_I}{2}\frac{A}{\sigma} \qquad \text{for small } A/\sigma \qquad (5.80)$$

These wave numbers are plotted in Figure 5.9 as a function of A/σ. As can be seen, k_r increases with A/σ; therefore, friction decreases the wave length of the wave, thus slowing it.

The change in wave amplitude over one wave length of travel can be readily found to be

$$\frac{\eta(x+L)}{\eta(x)} = e^{-k_i L} = e^{-2\pi(k_i/k_r)} \approx e^{-\pi(A/\sigma)} \qquad (5.81)$$

which decreases rapidly with increasing A/σ. For example, with $A/\sigma = 0.05$, this ratio is 0.85, or a 15% reduction in wave height. The horizontal velocity is then found by the same means as before.

$$U = \frac{H_I \sigma e^{-k_i x}}{2h\sqrt{k_i^2 + k_r^2}}\cos(k_r x - \sigma t - \epsilon) \qquad (5.82)$$

where

$$\epsilon = \tan^{-1} \frac{k_i}{k_r}$$

5.7 GEOSTROPHIC EFFECTS ON LONG WAVES

The earth's rotation plays an important role in long wave motion when the Coriolis acceleration becomes significant, or equivalently when the wave frequency σ is the same order as f_c, the Coriolis parameter defined as $2\omega \sin \phi$, where ϕ is the earth's latitude measured positive and negative in the northern and southern hemispheres, respectively, and ω is the rotation rate of the earth, $\omega = 7.27 \times 10^{-5}$ rad/s^{-1}. Typically, the Coriolis acceleration can produce significant effects in tidal waves.

The frictionless equations of motion for long waves on a rotating surface are modified by the introduction of two terms as follows:

$$\frac{\partial U}{\partial t} + U \frac{\partial U}{\partial x} + V \frac{\partial U}{\partial y} - f_c V = -g \frac{\partial \eta}{\partial x} \qquad (5.83a)$$

$$\frac{\partial V}{\partial t} + U \frac{\partial V}{\partial x} + V \frac{\partial V}{\partial y} + f_c U = -g \frac{\partial \eta}{\partial y} \qquad (5.83b)$$

where shear stresses have been neglected. The continuity equation is the same as before:

$$\frac{\partial \eta}{\partial t} + \frac{\partial U(h + \eta)}{\partial x} + \frac{\partial V(h + \eta)}{\partial y} = 0 \qquad (5.84)$$

To illustrate the effects of the Coriolis acceleration, consider the propagation of long progressive waves in an infinitely long straight canal in the x direction with a flat bottom. The transverse velocity V is considered negligible. The equation of motion in the x direction, therefore, is not affected by the presence of the Coriolis force. In the y direction the equation reduces to

$$f_c U = -g \frac{\partial \eta}{\partial y} \qquad (5.85)$$

which states that the Coriolis force is balanced by a cross-channel hydrostatic force in the form of a water surface slope, which varies in magnitude and sign with the longitudinal velocities in the channel.

If we linearize the equation of motion in the x direction (5.83a), a solution can be assumed as

$$\eta = \hat{\eta}(y) \cos (kx - \sigma t)$$

$$U = \frac{C}{h} \hat{\eta}(y) \cos (kx - \sigma t)$$

The y equation of motion is now

$$-\frac{f_c C}{g}\frac{\hat{\eta}}{h} = \frac{d\hat{\eta}}{dy} \tag{5.86}$$

or

$$\hat{\eta} = \frac{H}{2} e^{-f_c y/C}$$

where $C = \sqrt{gh}$. The total water surface profile and horizontal water profile motions are now

$$\eta = \frac{H}{2} e^{-f_c y/C} \cos(kx - \sigma t) \tag{5.87}$$

$$U = \frac{H}{2}\frac{C}{h} e^{-f_c y/C} \cos(kx - \sigma t) \tag{5.88}$$

At the wave crest, the wave amplitude and velocity decrease across the channel (y increasing) while at the wave trough (when the velocities are reversed) the amplitude increases. (Recall that we are dealing with a right-handed coordinate system.) The wave is called a Kelvin wave after Lord Kelvin (Sir W. Thomson), who derived an expression for it in 1879. The speed of propagation of the Kelvin wave is found by the continuity equation and it is the same as any other long wave, $C = \sqrt{gh}$.

The deviation in tidal ranges between the French and English coasts of the English channel can be largely explained by a northward-propagating Kelvin wave, which causes the French tides to be roughly twice as large (Proudman, 1953).

5.7.1 Amphidromic Waves in Canals

Consider the superposition of two Kelvin waves, traveling in opposite directions but with the same height:

$$\eta = \frac{H}{2} e^{-f_c y/C} \cos(kx - \sigma t) - \frac{H}{2} e^{f_c y/C} \cos(kx + \sigma t) \tag{5.89}$$

The resulting water surface elevation is always zero at the origin, $(x, y) = 0$; however, the wave amplitudes reinforce across the channel. The wave propagating in the positive x direction has a surface slope increasing in the negative y direction, while the wave propagating in the negative x direction has a positive surface slope in the positive y direction. Lines of maximum water surface elevation may be found by maximizing $\eta(x, t)$ as a function of time,

$$\frac{\partial \eta}{\partial t} = 0$$

or, after some rearranging,

$$\tanh\frac{f_c y}{C} = -\cot \sigma t \, \tan kx \tag{5.90}$$

Near the origin the equation for the tidal maxima is given by

$$\frac{f_c y}{C} = -kx \cot \sigma t$$

or

$$y = -\frac{Ckx}{f_c}\cot \sigma t \tag{5.91}$$

which is a straight line varying with time. A plot of the lines of high tide as a function of time is shown in Figure 5.10. These lines are called cotidal lines. The origin is called an amphidromic point and the tides are seen to apparently rotate around the origin. However, there is no transverse V velocity and the motion is purely in the X direction. Amphidromic tides of this nature are frequently seen in semienclosed bodies of water; Proudman (1953) cites the Adriatic Sea and Taylor (1920) discusses the Irish Sea. The mechanism for opposite traveling Kelvin waves requires a narrow channel in order that the motion be rectilinear and either two connected seas or a reflecting end to the channel. Taylor (1920) discusses the problem of the reflection of Kelvin waves and also seiching in a rectangular basin with the influence of Coriolis forces. For a further discussion of long waves with Coriolis effects, see Platzman (1971).

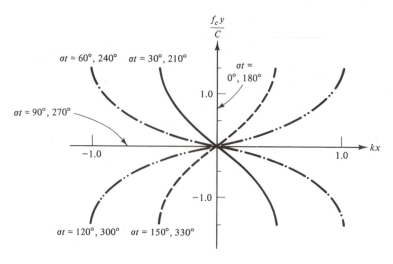

Figure 5.10　Cotidal lines.

5.8 LONG WAVES IN IRREGULAR-SHAPED BASINS OR BAYS

Quite often, a study of long waves or tides in a basin, lagoon, or near the coast requires the use of a computer, due to the complicated bathymetry, basin shape, and forcing due to winds or tide. To study these problems adequately, recourse must be made to computer techniques. Numerous studies have been made of tidal propagation by computer—too numerous to mention, in fact; however, many are referenced in two papers by Hinwood and Wallis (1975a,b).

5.9 STORM SURGE

The long wave equations can be used to describe the change in water level induced by wind blowing over bodies of water such as a continental shelf (Freeman et al., 1957) or a lake. Although the wind shear stress is usually very small, its effect, when integrated over a large body of water, can be catastrophic. Hurricanes, blowing over the shallow continental shelf of the Gulf of Mexico, have caused rises in water levels (storm surges, but not tidal waves!) in excess of 6 m at the coast.

The wind shear stress acting on the water surface τ_w, is represented as

$$\tau_w = \rho k \mathbf{W} |\mathbf{W}| \tag{5.92}$$

where ρ is the mass density of water, $\mathbf{W}$ the wind speed vector at a reference elevation of 10 m, and k a friction factor of order 10^{-6}. Numerous studies have been made for k (see Wu, 1969) and one of the more widely used sets of results is that of Van Dorn (1953),

$$k = \begin{cases} 1.2 \times 10^{-6}, & |\mathbf{W}| \leqslant W_c \\ 1.2 \times 10^{-6} + 2.25 \times 10^{-6}\left(1 - \dfrac{W_c}{|\mathbf{W}|}\right)^2, & |\mathbf{W}| > W_c \end{cases} \tag{5.93}$$

where $W_c = 5.6$ m/s.

If we adopt a coordinate system normal to a coastline, and the wind blows at an angle θ to the coast normal (Figure 5.11), then the onshore wind shear stress is $\tau_{w_x} = |\tau_w| \cos \theta$. The linearized equation of motion in this direction is [from Eq. (5.18), neglecting lateral shear stresses]

$$\frac{\partial U}{\partial t} = -g\frac{\partial \eta}{\partial x} + \frac{1}{\rho(h + \eta)}[\tau_{zx}(\eta) - \tau_{zx}(-h)] \tag{5.94}$$

After a long time, the flow U in the x direction must be zero, due to the presence of the coast, and therefore the steady-state equations show that the wind shear stress is balanced by the bottom shear stress as well as a hydrostatic pressure gradient. As we can no longer define the bottom friction in

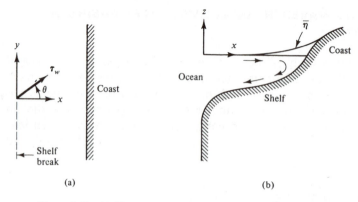

Figure 5.11   (a) Plan and (b) cross-sectional view of the coast.

terms of the mean (zero) flow U, it is convenient to define a factor n such that

$$n\tau_{zx}(\eta) \equiv \tau_{zx}(\eta) - \tau_{zx}(-h)$$

or

$$n = 1 - \frac{\tau_{zx}(-h)}{\tau_{zx}(\eta)} \tag{5.95}$$

This factor, which lumps the effect of the bottom friction in with the wind shear stress, is greater than 1, as the bottom shear stress in our convention (Figure 2.4) is negative. Typical values are $n = 1.15$ to 1.30 (*Shore Protection Manual*, 1977).

The equation is now

$$\boxed{\frac{\partial \eta}{\partial x} = \frac{n\tau_{zx}(\eta)}{\rho g(h + \eta)}} \tag{5.96}$$

Example 5.3

Calculate the wind setup due to a constant and uniform wind (τ_w is not a function of x) blowing over a continental shelf of width l. Assume (a) that the depth is a constant, h_0; and (b) that h is linearly varying, $h = h_0(1 - x/l)$.

Solution.   To begin, the governing equation can be written as

$$(h + \eta)\frac{\partial \eta}{\partial x} = \frac{n\tau_{w_x}}{\rho g}$$

(a) Since h_0 is not a function of x,

$$\frac{1}{2}\frac{d(h_0 + \eta)^2}{dx} = \frac{n\tau_{w_x}}{\rho g}$$

Solving gives us

$$(h_0 + \eta)^2 = \frac{2n\tau_{w_x}x}{\rho g} + C$$

To evaluate the constant of integration, we require the setup to be zero at $x = 0$. This condition arises from the fact that where h is very large, there is no surface gradient (why?) and thus no setup in deep water. After substitution for C, we have

$$(h_0 + \eta)^2 = \frac{2n\tau_{w_x}x}{\rho g} + h_0^2$$

or

$$\eta(x) = + \sqrt{h_0^2 + \frac{2n\tau_{w_x}x}{\rho g}} - h_0 \qquad (5.97a)$$

In dimensionless form, η is

$$\frac{\eta(x)}{h_0} = \sqrt{1 + \frac{2n\tau_{w_x}l}{\rho g h_0^2}\frac{x}{l}} - 1 = \sqrt{1 + \frac{2Ax}{l}} - 1 \qquad (5.97b)$$

where $A = n\tau_{w_x}l/\rho g h_0^2$, a ratio of shear to hydrostatic forces.

(b) For a sloping bottom, the governing equation, Eq. (5.96), can be rewritten as

$$(h + \eta)\frac{d(h + \eta)}{dx} - (h + \eta)\frac{dh}{dx} = \frac{n\tau_{w_x}}{\rho g} \qquad (5.98)$$

where $dh/dx = -h_0/l$, a constant. Separation of variables leads to

$$-\frac{(h + \eta)d(h + \eta)}{\dfrac{h_0^2}{l}\left(\dfrac{h + \eta}{h_0} - A\right)} = dx,$$

with A again defined as $n\tau_{w_x}l/\rho g h_0^2$.
Solving yields

$$x + C = l\left[\left(1 - \frac{h + \eta}{h_0}\right) - A \ln\left(\frac{h + \eta}{h_0} - A\right)\right]$$

Evaluating C as before, we have

$$x = l\left[\left(1 - \frac{h + \eta}{h_0}\right) - A \ln\left(\frac{\dfrac{h + \eta}{h_0} - A}{1 - A}\right)\right] \qquad (5.99a)$$

or, in dimensionless form,

$$\frac{x}{l} = \left(1 - \frac{h + \eta}{h_0}\right) - A \ln\left(\frac{\dfrac{h + \eta}{h_0} - A}{1 - A}\right) \qquad (5.99b)$$

These two solutions [Eqs. (5.99b) and (5.97b)] are plotted in Figure 5.12 to show the effect of the bottom slope on the storm surge. Clearly, the sloping

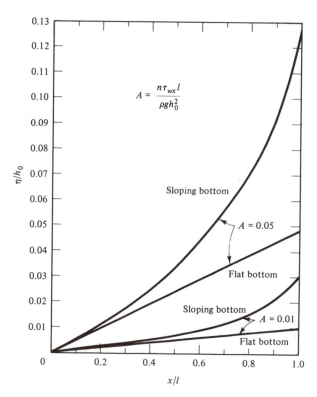

Figure 5.12 Dimensionless storm surge versus dimensionless distance of a continental shelf for two cases of dimensionless wind shear stress.

bottom causes an increase in the storm surge height; this can be explained by referring to Eq. (5.96), which indicates that for a given wind stress, the water surface slope depends on the local water depth in such a way that the shallower the depth, the greater the slope. In Figure 5.13, the storm surge at the coast ($x/l = 1$) is shown for a sloping shelf as a function of the dimensionless onshore shear stress. The solution for x/l is usually obtained for given values of $(h + \eta)/h_0$. However, to obtain $(h + \eta)/h_0$ directly for a given x/l value, then it is usually more convenient to solve the equation iteratively for $(h + \eta)/h_0$. The Newton–Raphson technique works well here.

The solution of Eqs. (5.99) is generally not computed for x shoreward of the shoreline ($x/l = 1$); however, it is often useful to determine backshore inundation (i.e., when h is negative). This can be done with this equation up to the point where

$$(h + \eta)\frac{h_0}{l} = \frac{n\tau_{w_x}}{\rho g}$$

At this point the water surface slope is equal to the bottom slope [from Eq. (5.98)] and a uniform steady surge is reached, analogous to steady open

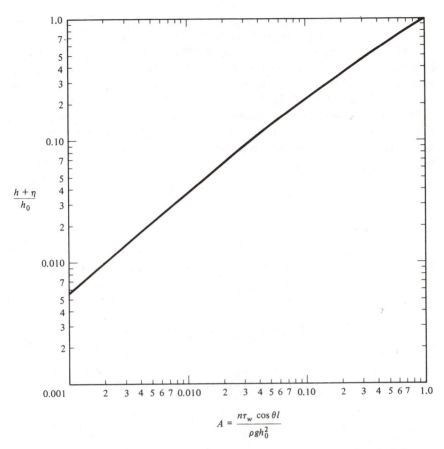

$$A = \frac{n\tau_w \cos\theta l}{\rho g h_0^2}$$

Figure 5.13 Storm tide for $x/l = 1.0$ for a sloping shelf. For the case of no Coriolis force, the ordinate is equal to η/h_0, the storm surge at the coast, as $h = 0$ at $x/l = 1$.

channel flow, in the sense that the downstream component of fluid weight is supported by the surface and bottom shear stresses. In a practical problem, the backshore region terminates in a wall or else significant flooding can occur.

5.9.1 Bathystrophic Storm Tide

For large-scale systems the influence of the Coriolis forces cannot be neglected. If the wind blows at an angle θ to the coast, such that a longshore current is generated, then if the current is moving in such a direction that the coastline is to the right (in the northern hemisphere), the Coriolis force requires a balancing hydrostatic gradient, as in the Kelvin wave. This gradient adds to the surface gradient induced by the wind. If the wind were blowing in the opposite direction, of course, the Coriolis forces would reduce the surge; however, large storms, such as hurricanes (due to their circular wind patterns), will induce longshore flows in *both* directions.

The analytical solution will be developed for a wind that begins abruptly at $t = 0$, with a magnitude W and a direction θ. To simplify the problem, we will assume that (a) the onshore flow and the return flows are continually in balance, so that $U = 0$ for all times, and (b) the wind system is uniform, so that there is no variability in the y direction. Assumption (a) is not always true, as a certain amount of water must flow into the shelf region to generate the surge. For these conditions the equations of motion in the x and y directions are

$$x: \quad (h + \eta)\left[\frac{d(h + \eta)}{dx} - \frac{dh}{dx} - \frac{f_c V}{g}\right] = \frac{n\tau_{w_x}}{\rho g} \tag{5.100}$$

$$y: \quad \frac{\partial V}{\partial t} = \frac{\tau_{w_y} - \tau_{z_y}(-h)}{\rho(h + \eta)} = \frac{\tau_{w_y}}{\rho(h + \eta)} - \frac{fV^2}{8(h + \eta)} \tag{5.101}$$

where a Darcy–Weisbach friction factor f is introduced for the bottom shear stress in the y direction. If we now consider $\eta \ll h$, we can solve the last equation:

$$V = \sqrt{\frac{8k \sin \theta}{f}}\, W \cdot \tanh\left(\sqrt{\frac{kf \sin \theta}{8}}\, \frac{Wt}{h}\right) \tag{5.102}$$

where k is defined in Eqs. (5.92) and (5.93). The longshore velocity increases from $V = 0$ at $t = 0$ to the steady-state value of

$$V_S = \sqrt{\frac{8k \sin \theta}{f}}\, W \tag{5.103}$$

for $t = \infty$. Effectively, the time to steady state is determined by setting the V argument of the hyperbolic tangent to π (tanh $\pi = 0.996$) or

$$\sqrt{\frac{kf \sin \theta}{8}}\, \frac{W}{h}\, t = \pi \tag{5.104}$$

Solving for t, we get

$$t = \frac{\pi h}{W\sqrt{\dfrac{kf \sin \theta}{8}}} \tag{5.105}$$

The time to steady state varies with the depth, with the shallower depths reaching the terminal velocity more rapidly than the offshore regions. As an example, for $h = 10$ m and $W = 20$ m/s, about 8 h is necessary for steady state to be reached. At this time, Eq. (5.103) shows that V_S is about 3% of the wind speed.

If V is now introduced into the x-momentum equation,

$$(h + \eta)\left[\frac{d(h + \eta)}{dx} - \frac{dh}{dx} - \frac{f_c}{g} V_S \tanh t'\right] = \frac{n\tau_{w_x}}{\rho g} \tag{5.106}$$

where

$$t' = \sqrt{\frac{kf \sin \theta}{8}} \frac{W}{h} t$$

Again solving by separation of variables yields

$$\frac{x}{l^*} = \left(1 - \frac{h + \eta}{h_0}\right) - A^*\ell n \left\{\frac{\frac{h + \eta}{h_0} - A^*}{1 - A^*}\right\} \tag{5.107}$$

where

$$A^* = \frac{n\tau_{w_x} l^*}{\rho g h_0^2}$$

and

$$l^* = l \left/ \left(1 - f_c \sqrt{\frac{8k \sin \theta}{f}} \frac{Wl}{gh_0} \tanh t'\right)\right. \tag{5.108a}$$

or

$$l^* = l \left/ \left(1 - f_c \sqrt{\frac{8k \sin \theta}{f}} \frac{Wl}{gh_0}\right)\right. \tag{5.108b}$$

for large t.

This solution in dimensionless form is exactly the same as the solution for a surge over a sloping beach without the Coriolis terms except that l is replaced by l^*, and we see that the Coriolis force simply serves to "modify" the bottom slope.

The effect of wind angle becomes important in this problem as τ_{w_x} is important for the direct wind stress component of the surge, while τ_{w_y} is important for the Coriolis force contribution. Figure 5.14 shows the effect of wind angle for the setup at the shoreline at $x = l$.

5.10 LONG WAVES FORCED BY A MOVING ATMOSPHERIC PRESSURE DISTURBANCE

Consider the case of an atmospheric pressure disturbance p_0 moving with speed U in the positive x direction:

$$p_0 = f(Ut - x) \tag{5.109}$$

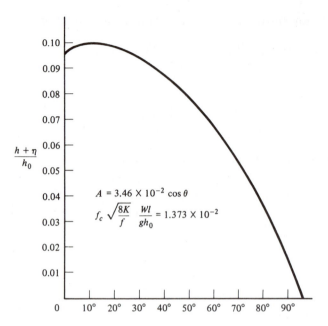

Figure 5.14   Maximum storm surge at $x = l$ from the bathystrophic storm tide.

where the parentheses indicate a functional relationship. The governing equations include the momentum and continuity equations. The linearized momentum equation is

$$h \frac{\partial u}{\partial t} = -gh \frac{\partial \eta}{\partial x} - \frac{h}{\rho} \frac{\partial p_0}{\partial x} \tag{5.110}$$

The continuity equation will be developed by selecting a coordinate system moving with the wave that renders the system stationary with a horizontal velocity component $u - U$. Realizing that the discharge Q past any given point is invariant and that the wave-induced particle velocity is proportional to the water surface displacement,

$$Q = (u - U)(h + \eta) = -Uh$$

or

$$u = U \frac{\eta}{h + \eta} \approx U \frac{\eta}{h} \tag{5.111}$$

which has been linearized. Assuming η of the form

$$\eta = G(Ut - x) \tag{5.112}$$

it is clear that

$$\frac{\partial \eta}{\partial t} = -U \frac{\partial \eta}{\partial x} \tag{5.113}$$

and combining Eqs. (5.109), (5.110), (5.112), and (5.113), we get

$$-\frac{\partial \eta}{\partial x}(U^2 - gh) = -\frac{h}{\rho}\frac{\partial p_0}{\partial x}$$

which is an exact differential and can be integrated from a location from where both η and p_0 are nonexistent to

$$\frac{\eta}{h} = \frac{p_0/\rho}{U^2 - gh} \tag{5.114}$$

From Eq. (5.114), it is seen that for a static condition, $\eta_s = -p_0/\rho g$, whereas for cases in which the speed of translation approaches that of a long free wave $(C = \sqrt{gh})$ there is an amplification which becomes unbounded due to the lack of any damping terms. Moreover, when $U < C$, the pressure and displacement are exactly out of phase, whereas for $U > C$, the two are in phase. For values of $U \gg C$, the response approaches zero as the time interval over which the force is applied is not sufficient for the liquid to respond. The solid line in Figure 5.15 presents the amplification factor $|\eta|/|\eta_s|$ for no damping

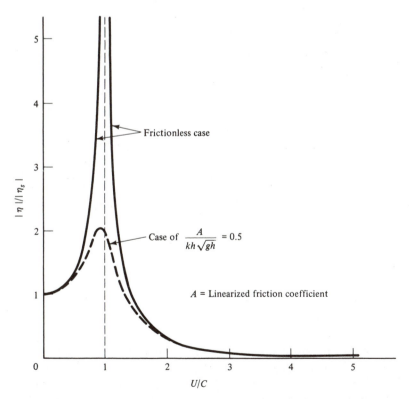

Figure 5.15 Dynamic response of translating pressure disturbance, with and without friction.

in which η_s is the static water displacement for a pressure anomaly,

$$|\eta_s| = \frac{|p_0|}{\rho g} \tag{5.115}$$

It is noted that the effect of friction is to reduce the maximum amplification due to a finite value as shown by the dashed line in Figure 5.14; see also Problem 5.19.

Finally, it is noted that the "forcing function" present in Eq. (5.109) could have been generalized to include the surface shear stress.

5.11 LONG WAVES FORCED BY A TRANSLATING BOTTOM DISPLACEMENT

A displacement of the bottom η_0, which translates at speed U, will cause an associated surface displacement, much as in the case for a moving pressure displacement discussed in the preceding section. In this case, the linearized momentum equation is simply

$$\frac{\partial u}{\partial t} = -g \frac{\partial \eta_1}{\partial x} \tag{5.116}$$

where η_1 and η_0 pertain to the air–water and bottom interface displacements, respectively, given by the forms

$$\eta_0 = f_0 (Ut - x) \tag{5.117a}$$

$$\eta_1 = f_1 (Ut - x) \tag{5.117b}$$

The continuity equation can be determined in the same manner as before:

$$u = \frac{U(\eta_1 - \eta_0)}{h + (\eta_1 - \eta_0)} \approx \frac{U(\eta_1 - \eta_0)}{h} \tag{5.118}$$

Combining Eqs. (5.116), (5.117), and (5.118), the following exact differential results:

$$\frac{\partial \eta_1}{\partial x} = \left(\frac{U^2}{U^2 - gh} \right) \frac{\partial \eta_0}{\partial x} \tag{5.119}$$

or

$$\eta_1 = \eta_0 \frac{U^2}{U^2 - gh} \tag{5.120}$$

which, as in the previous case, increases without bound as U approaches the speed, $C (= \sqrt{gh})$ of a long free wave. For $U = 0$, of course, there is no upper surface displacement and for large U, the upper surface displacement η_1 approaches the lower surface displacement η_0. The latter can be interpreted as due to the bottom motions occurring so rapidly that the upper surface does not have time to respond laterally (i.e., for the liquid to be mobilized in the horizontal direction).

REFERENCES

DEAN, R. G., "Long Wave Modification by Linear Transitions," *J. Waterways Harbors Div., ASCE*, Vol. 90, No. WW1, pp. 1–29, 1964.

FREEMAN, J. C., L. BAER, and G. H. JUNG, "The Bathystrophic Storm Tide," *J. Mar. Res.*, Vol. 16, No. 1, 1957.

HILALY, N., "Water Waves over a Rectangular Channel through a Reef," *J. Waterways Harbors Div., ASCE*, Vol. 95, No. WW1, pp. 77–94, Feb. 1969.

HINWOOD, J. B., and I. G. WALLIS, "Classification of Models of Tidal Waters," *J. Hydraulics Div., ASCE*, Vol. 101, No. HY10, pp. 1315–1331, Oct. 1975a.

HINWOOD, J. B., and I. G. WALLIS, "Review of Models of Tidal Waters," *J. Hydraulics Div., ASCE*, Vol. 101, No. HY11, pp. 1405–1421, Nov. 1975b.

LAMB, H., *Hydrodynamics*, 6th ed., Dover, New York, 1945.

LORENTZ, H. A., "Verslag Staatscommissie Zuiderzee 1918–1926," Staatsdrukkerij, The Hague, The Netherlands, 1926.

PLATZMAN, G. W., "Ocean Tides and Related Waves," in *Lectures in Applied Mathematics*, Vol. 14, Pt. 2, W. H. Reid, ed., American Mathematical Society, Providence, R.I., 1971.

PROUDMAN, J., *Dynamical Oceanography*, Wiley, New York, 1953, p. 239.

SHIAU, J., and R. R. RUMER, Jr., "Decay of Mass Oscillations in Rectangular Basins," *J. Hydraulics Div., ASCE*, Vol. 100, No. HY1, Jan. 1974.

STOKES, G. G., "On the Effects of the Internal Friction of Fluids on the Motion of Pendulums," *Trans. Camb. Philos. Soc.*, Vol. 9, No. 8, 1851.

TAYLOR, Sir G., II, *Proc. Lond. Math. Soc. (2)*, Vol. 20, p. 148 (1920).

THOMSON, W. (Lord Kelvin), "On Gravitational Oscillations of Rotating Water," *Proc. Roy. Soc. Edin.*, Vol. 10, 1879, p. 92.

U.S. Army, Coastal Engineering Research Center, *Shore Protection Manual*, U.S. Government Printing Office, Washington, D.C., 1977.

VAN DORN, W. C., "Wind Stress on an Artificial Pond," *J. Mar. Res.*, Vol. 12, 1953.

WILSON, B. S., in *Encyclopedia of Oceanography*, R. W. Fairbridge, ed., Academic Press, New York, 1966.

WILSON, B. S., "Seiches," in *Advances in Hydroscience*, Ven Te Chow, ed., Vol. 8, Academic Press, New York, 1972.

WU, J., "Wind Stress and Surface Roughness at Sea Interface," *J. Geophys. Res.*, Vol. 74, pp. 444–453, 1969.

PROBLEMS

5.1 Compare the fundamental periods of *seiching* for a long narrow basin with length 1 km and maximum depth of 10 m, if its bottom is flat or sloped. Explain the differences.

5.2 Making reasonable assumptions, calculate the time necessary for the seiching in Problem 5.1 to reduce to 10% of the original value.

5.3 Determine the water surface elevation of a long standing wave in an estuary with linearly increasing depth and constant width. What assumptions have been made? $h = h_0$ at $x = l$, the mouth of the estuary.

5.4 Show that a linearized equation for seiching in two dimensions would be

$$\frac{\partial^2 \eta}{\partial t^2} = g\left[\frac{\partial}{\partial x}\left(h\frac{\partial \eta}{\partial x}\right) + \frac{\partial}{\partial y}\left(h\frac{\partial \eta}{\partial y}\right)\right]$$

With this equation, determine the seiching periods in a rectangular basin of length l and width b with constant depth h.

5.5 Verify that long wave reflection from an abrupt step conserves the flux of wave energy.

5.6 An edge wave is a progressive wave that propagates parallel to a coast. For a sloping beach given by $h = mx$, show that

$$\eta = Ae^{-\lambda_n x} L_n\left(2\lambda_n x\right) \cos\left(\lambda_n y - \sigma t\right)$$

is a solution where $L_n\left(2\lambda_n x\right)$ is the Laguerre polynomial of order n and λ_n and σ are related by $\sigma^2 = g\lambda_n\left(2n + 1\right)m$.

5.7 A large dock extends from above the free surface down to a depth d. Assuming long waves and that the dock is rigid, calculate the reflection and transmission coefficients for the dock, which has a width of l.

5.8 Determine the Kelvin wave in a long narrow canal with bottom friction.

5.9 Develop the condition for the constant of integration C for the case of a storm surge in a closed basin of constant depth h_0. A numerical solution will be necessary.

5.10 Calculate an equation for the "blow-down" on a sloping continental shelf of width l due to a strong directly offshore wind. Determine the location of the mean water line.

5.11 Show from the continuity equation, Eq. (5.6), that the vertical velocity $W(z)$ under a long wave varies linearly with depth and can be expressed as

$$W(z) = \frac{D\eta}{Dt} + (\eta - z)\left(\frac{\partial U}{\partial x} + \frac{\partial V}{\partial y}\right)$$

if U and V are assumed to be independent of depth.

5.12 Determine the seiching period of a circular tank of radius a. Use the wave equation in cylindrical form and find only the first mode, which has a $\cos \theta$ dependency (Lamb, 1945). Compare your results to reality by shaking a coffee cup.

5.13 Compare the transmission coefficient determined in the abrupt step problem to the one calculated by Green's law. Account for the differences.

5.14 Develop an equation for the ratio R of kinetic energy in the horizontal component of water particle velocity to the total kinetic energy. Solve for the shallow and deep water asymptotes. Plot this ratio versus h/L_0.

5.15 For a bay of uniform depth and pie-shaped plan form as discussed in Example 5.1, develop an expression for the ratio R,

$$R(x) = \frac{\eta(x)}{\eta(l)}$$

as determined by Green's law and the complete solution, Eq. (5.39b). Plot and discuss the ratio R for the case of $l/L = 10$ and $l/L = 2$.

5.16 Which continental shelf configuration allows the greatest storm surges at the coast: (a) shelf width l_0; maximum depth h_0; (b) shelf width l_1 ($> l_0$), same maximum depth h_0; or (c) shelf width l_1 ($> l_0$), maximum depth h_1 ($> h_0$) but with bottom slope (h_0/l_0)? Verify your answer for $h_1 = 5h_0$, $l_1 = 5l_0$, and A (for case a) $= 0.05$.

5.17 Show that the storm surge for a continental shelf modeled as $h = h_0 [1 - (x/l)]^2$ can be approximated by

$$\eta(x) = \sqrt{h^2 + \frac{2n\tau_{s_x}x}{\rho g}} - h$$

(*Note*: There is another possible solution to the linearized problem; however, it gives infinite surge heights at $x = l$.)

5.18 Show that the governing linearized momentum equation for long waves forced by an atmospheric pressure anomaly with linear friction present is

$$h\frac{\partial u}{\partial t} = -gh\frac{\partial \eta}{\partial x} - \frac{h}{\rho}\frac{\partial p_0}{\partial x} - AU$$

and that the solution depends on the wave number (k) of the forcing disturbance and that the solution in terms of the ratio of the modulus of the dynamic to static water surface displacements is

$$\frac{|F_\eta(\sigma)|_{\text{dyn}}}{|F_\eta(\sigma)|_{\text{stat}}} = \frac{1}{\sqrt{(U^2/gh - 1)^2 + (AU/khgh)^2}}$$

[*Note*: There are at least two ways of approaching this problem. One is to represent the traveling pressure and water surface displacements as

$$p = P_R \cos(\sigma t - kx)$$

$$\eta = N_R \cos(\sigma t - kx - \alpha)$$

and to substitute these in the governing equation above and solve for N_R and α. The second (equivalent) method is to represent p and η as Fourier integrals

$$p(x, t) = \frac{1}{\sqrt{2\pi}} \int_{-\infty}^{\infty} F_p(\sigma)e^{i(\sigma t - kx)}\, d\sigma$$

$$\eta(x, t) = \frac{1}{\sqrt{2\pi}} \int_{-\infty}^{\infty} F_\eta(\sigma)e^{i(\sigma t - kx)}\, d\sigma$$

in which $F_p(\sigma)$ and $F_\eta(\sigma)$ are complex amplitude spectra (i.e., they contain phases). The latter approach is the simpler of the two, algebraically.]

6

Wavemaker Theory

Dedication
SIR THOMAS HENRY HAVELOCK

Sir Thomas Henry Havelock (1877–1968) pursued a variety of water wave areas, including ship wave problems and the generation of waves by wavemakers, the subject of this chapter.

Havelock was born in Newscastle upon Tyne. He obtained his education at Armstrong College, University of Durham, and St. Johns College, University of Cambridge. Returning to Durham, he became a lecturer and then professor of mathematics. He received knighthood for his scientific works in 1957 and accepted honorary doctorates from University of Durham and University of Hamburg. He received the first William Froude Gold Medal in 1956 for his work in naval architecture.

Havelock was a Fellow of the Royal Society and a corresponding member of the Academy of Science, Paris.

6.1 INTRODUCTION

To date most laboratory testing of floating or bottom-mounted structures and studies of beach profiles and other related phenomena have utilized wave tanks, which are usually characterized as long, narrow enclosures with a wavemaker of some kind at one end; however, circular beaches have been proposed for littoral drift studies and a spiral wavemaker has been used (Dalrymple and Dean, 1972). For all of these tests, the wavemaker is very important. The wave motion that it induces and its power requirements can be determined reasonably well from linear wave theory.

Wavemakers are, in fact, more ubiquitous than one would expect. Earthquake excitation of the seafloor or human-made structures causes waves which can be estimated by wavemaker theory; in fact, the loading on the structures can be determined (see Chapter 8). Any moving body in a fluid with a free surface will produce waves: ducks, boats, and so on.

6.2 SIMPLIFIED THEORY FOR PLANE WAVEMAKERS IN SHALLOW WATER

In shallow water, a simple theory for the generation of waves by wavemakers was proposed by Galvin (1964), who reasoned that the water displaced by the wavemaker should be equal to the crest volume of the propagating wave form. For example, consider a piston wavemaker with a stroke S which is constant over a depth h. The volume of water displaced over a whole stroke is Sh (see Figure 6.1). The volume of water in a wave crest is $\int_0^{L/2} (H/2) \sin kx \, dx$ $= H/k$. Equating the two volumes,

$$Sh = \frac{H}{k} = \frac{H}{2}\left(\frac{L}{2}\right)\frac{2}{\pi}$$

in which the $2/\pi$ factor represents the ratio of the shaded area to the area of the enclosing rectangle (i.e., an area factor). This equation can also be expressed

$$\left(\frac{H}{S}\right)_{\text{piston}} = kh \tag{6.1}$$

where H/S is the height-to-stroke ratio. This relationship is valid in the shallow water region, $kh < \pi/10$. For a flap wavemaker, hinged at the bottom, the volume of water displaced by the wavemaker would be less by a factor of 2.

$$\left(\frac{H}{S}\right)_{\text{flap}} = \frac{kh}{2} \tag{6.2}$$

These two relationships are shown as the straight dashed lines in Figure 6.2.

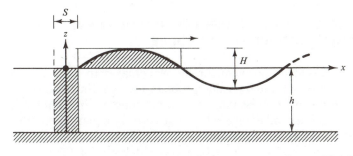

Figure 6.1   Simplified shallow water piston-type wavemaker theory of Galvin.

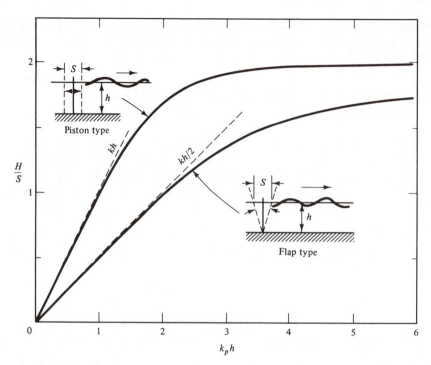

Figure 6.2 Plane wavemaker theory. Wave height to stroke ratios versus relative depths. Piston and flap type wavemaker motions.

Another type of wavemaker is the plunger wavemaker. This could be, as an example, a horizontal cylinder moving vertically about the mean water level. If the cylinder has a radius R and a stroke R, then the cylinder position ranges from fully emerged to half submerged at full stroke. If waves are generated in each direction normal to the cylinder axis, then for shallow water conditions the wave height-to-stroke ratio can be easily shown to be

$$\frac{H}{R} = \frac{\pi(kR)}{4} \tag{6.3}$$

6.3 COMPLETE WAVEMAKER THEORY FOR PLANE WAVES PRODUCED BY A PADDLE

The boundary value problem for the wavemaker in a wave tank follows directly from the boundary value problem for two-dimensional waves propagating in an incompressible, irrotational fluid, as in Chapter 3. For the geometry depicted in Figure 6.1, the governing equation for the velocity potential is the Laplace equation,

$$\frac{\partial^2 \phi}{\partial x^2} + \frac{\partial^2 \phi}{\partial z^2} = 0 \tag{6.4}$$

The linearized forms of the dynamic and kinematic free surface boundary conditions are the same as before.

$$\eta = \frac{1}{g}\frac{\partial \phi}{\partial t}, \qquad z = 0 \tag{6.5}$$

$$-\frac{\partial \phi}{\partial z} = \frac{\partial \eta}{\partial t}, \qquad z = 0 \tag{6.6}$$

The bottom boundary condition is the usual no-flow condition

$$-\frac{\partial \phi}{\partial z} = 0, \qquad z = -h \tag{6.7}$$

The only conditions that change are the lateral boundary conditions. In the positive x direction, as x becomes large, we require that the waves be outwardly propagating, imposing the radiation boundary condition (Sommerfeld, 1964). At $x = 0$, a kinematic condition must be satisfied on the wavemaker. If $S(z)$ is the stroke of the wavemaker, its horizontal displacement is described as

$$x = \frac{S(z)}{2}\sin \sigma t \tag{6.8}$$

where σ is the wavemaker frequency.

The function that describes the surface of the wavemaker is

$$F(x, z, t) = x - \frac{S(z)}{2}\sin \sigma t = 0 \tag{6.9}$$

The general kinematic boundary condition is Eq. (3.6).

$$\mathbf{u} \cdot \mathbf{n} = -\frac{\partial F(x, z, t)/\partial t}{|\nabla F|} \qquad \text{on } F(x, z, t) = 0 \tag{6.10}$$

where $\mathbf{u} = u\mathbf{i} + w\mathbf{k}$ and $\mathbf{n} = \nabla F/|\nabla F|$. Substituting for $F(x, z, t)$ yields

$$u - \frac{w}{2}\frac{dS(z)}{dz}\sin \sigma t = \frac{S(z)}{2}\sigma \cos \sigma t \qquad \text{on } F(x, z, t) = 0 \tag{6.11}$$

For small displacements $S(z)$ and small velocities, we can linearize this equation by neglecting the second term on the left-hand side.

As at the free surface, it is convenient to express the condition at the moving lateral boundary in terms of its mean position, $x = 0$. To do this we expand the condition in a truncated Taylor series.

$$\left[u - \frac{S(z)}{2}\sigma \cos \sigma t \right]_{x=[S(z)/2]\sin \sigma t} = \left[u - \frac{S(z)}{2}\sigma \cos \sigma t \right]_{x=0} \tag{6.12}$$

$$+ \frac{S(z)}{2}\sin \sigma t \frac{\partial}{\partial x}\left[u - \frac{S(z)}{2}\sigma \cos \sigma t \right]_{x=0} + \cdots$$

Clearly, only the first term in the expansion is linear in u and $S(z)$; the others are dropped, as they are assumed to be very small. Therefore, the final lateral boundary condition is

$$u(0, z, t) = \frac{S(z)}{2} \sigma \cos \sigma t \qquad (6.13)$$

Now that the boundary value problem is specified, all the possible solutions to the Laplace equation are examined as possible solutions to determine those that satisfy the boundary conditions. Referring back to Table 3.1, the following general velocity potential, which satisfies the bottom boundary condition, is presented.

$$\phi(x, z, t) = A_p \cosh k_p (h + z) \sin (k_p x - \sigma t) + (Ax + B) \qquad (6.14)$$
$$+ Ce^{-k_s x} \cos k_s(h + z) \cos \sigma t$$

The subscripts on k indicate that that portion of ϕ is associated with a progressive or a standing wave. For the wavemaker problem, A must be zero, as there is no uniform flow possible through the wavemaker and B can be set to zero without affecting the velocity field. The remaining terms must satisfy the two linearized free surface boundary conditions. It is often useful to employ the combined linear free surface boundary condition, made up of both conditions. This condition is

$$\frac{\partial \phi}{\partial z} - \frac{\sigma^2 \phi}{g} = 0, \qquad z = 0 \qquad (6.15)$$

which can be obtained by eliminating the free surface η from Eqs. (6.5) and (6.6). Substituting our assumed solution into this condition yields

$$\sigma^2 = gk_p \tanh k_p h \qquad (6.16)$$

and

$$\sigma^2 = -gk_s \tan k_s h \qquad (6.17)$$

The first equation is the dispersion relationship for progressive waves, as obtained in Chapter 3, while the second relationship, which relates k_s to the frequency of the wavemaker, determines the wave numbers for standing waves with amplitudes that decrease exponentially with distance from the wavemaker. Rewriting the last equation as

$$\frac{\sigma^2 h}{gk_s h} = -\tan k_s h \qquad (6.18)$$

the solutions to this equation can be shown in graphical form (see Figure 6.3).

There are clearly an infinite number of solutions to this equation and all are possible. Each solution will be denoted as $k_s(n)$, where n is an integer. The

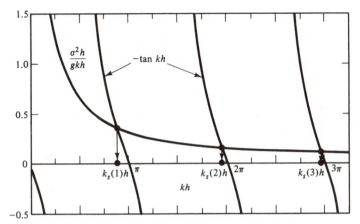

Figure 6.3 Graphical representation of the dispersion relationship for the standing wave modes, showing three of the infinite numbers of roots, $k_s(n)$. Here, $\sigma^2 h/g = 1.0$.

final form for the boundary value problem is proposed as

$$\phi = A_p \cosh k_p (h + z) \sin (k_p x - \sigma t) \tag{6.19}$$

$$+ \sum_{n=1}^{\infty} C_n e^{-k_s(n)x} \cos [k_s(n)(h + z)] \cos \sigma t$$

Again, the first term represents a progressive wave, made by the wavemaker, while the second series of waves are standing waves which decay away from the wavemaker. To determine how rapidly the exponential standing waves decrease in the x direction, let us examine the first term in the series, which decays the least rapidly. The quantity $k_s(1)h$, from Figure 6.3, must be greater than $\pi/2$, but for conservative reasons, say $k_s(1)h = \pi/2$, therefore, the decay of standing wave height is greater than $e^{-(\pi/2)(x/h)}$. For $x = 2h$, $e^{-(\pi/2)(x/h)} = 0.04$, for $x = 3h$, it is equal to 0.009. Therefore, the first term in the series is virtually negligible two to three water depths away from the wavemaker.

For a complete solution, A_p and the C_n's need to be determined. These are evaluated by the lateral boundary condition at the wavemaker.

$$u(0, z, t) = \frac{S(z)}{2} \sigma \cos \sigma t = -\frac{\partial \phi}{\partial x} (0, z, t)$$

$$= -A_p k_p \cosh k_p(h + z) \cos \sigma t$$

$$+ \sum_{n=1}^{\infty} C_n k_s(n) \cos [k_s(n)(h + z)] \cos \sigma t$$

or

$$\frac{S(z)}{2} \sigma = -A_p k_p \cosh k_p(h + z) + \sum_{n=1}^{\infty} C_n k_s(n) \cos [k_s(n)(h + z)] \tag{6.20}$$

Now we have a function of z equal to a series of trigonometric functions of z on the right-hand side, similar to the situation for the Fourier series. In fact, the set of functions, $\{\cosh k_p(h + z),\ \cos [k_s(n)(h + z)],\ n = 1,\ \infty\}$ form a complete harmonic series of orthogonal functions and thus any continuous function can be expanded in terms of them.[1] Therefore, to find A_p, the equation above is multiplied by $\cosh k_p(h + z)$ and integrated from $-h$ to 0. Due to the orthogonality property of these functions there is no contribution from the series terms and therefore

$$A_p = \frac{-\displaystyle\int_{-h}^{0} \frac{S(z)}{2}\, \sigma \cosh k_p(h + z)\, dz}{k_p \displaystyle\int_{-h}^{0} \cosh^2 k_p(h + z)\, dz} \tag{6.21}$$

Multiplying Eq. (6.20) by $\cos \{k_s(m)(h + z)\}$ and integrating over depth yields

$$C_m = \frac{\displaystyle\int_{-h}^{0} \frac{S(z)}{2}\, \sigma \cos [k_s(m)(h + z)]\, dz}{k_s(m) \displaystyle\int_{-h}^{0} \cos^2 [k_s(m)(h + z)]\, dz} \tag{6.22}$$

Depending on the functional form of $S(z)$, the coefficients are readily obtained. For the simple cases of piston and flap wavemakers, the $S(z)$ are specified as

$$S(z) = \begin{cases} S, & \text{piston motion} \\[2mm] S\left(1 + \dfrac{z}{h}\right), & \text{flap motion} \end{cases} \tag{6.23}$$

The wave height for the progressive wave is determined by evaluating η far from the wavemaker.

$$\eta = \frac{1}{g}\frac{\partial \phi}{\partial t}\bigg|_{z=0} = -\frac{A_p}{g}\, \sigma \cosh k_p h \cos (k_p x - \sigma t)$$

$$= \frac{H}{2} \cos (k_p x - \sigma t) \qquad x \gg h \tag{6.24}$$

---

[1]This follows from the Sturm–Liouville theory. Proof of the orthogonality can be obtained by showing that the integrals below are zero, that is,

$$\int_{-h}^{0} \cosh k_p(h + z) \cos [k_s(n)(h + z)]\, dz = 0; \qquad \int_{-h}^{0} \cos [k_s(m)(h + z)] \cos [k_s(n)(h + z)]\, dz = 0$$

for $m \neq n$ using the dispersion relation and Eq. (6.17), Problem 6.8.

Substituting for A_p, we can find the ratio of wave height to stroke as

$$\frac{H}{S} = 4\left(\frac{\sinh k_p h}{k_p h}\right)\frac{k_p h \sinh k_p h - \cosh k_p h + 1}{\sinh 2k_p h + 2k_p h}, \qquad \text{flap motion} \qquad (6.25)$$

$$\frac{H}{S} = \frac{2(\cosh 2k_p h - 1)}{\sinh 2k_p h + 2k_p h}, \qquad \text{piston motion} \qquad (6.26)$$

In Figure 6.2, the wave height-to-stroke ratio is plotted for both flap and piston wavemaker motions for different water depths. This graph enables the rapid prediction of wave height given the stroke of the wavemaker. The reader is referred to Ursell et al. (1960) for further details.

The power required to generate these waves can be easily obtained by determining the energy flux away from the wavemaker.

$$P = ECn \qquad (6.27)$$

where E is proportional to the propagating wave height, as obtained from the preceding equation. The power necessary to generate waves in various water depths is shown in Figure 6.4. By examining Figures 6.2 and 6.4, it can be seen that to generate a wave of the same height, in shallow water, it is easier to generate it with a piston wavemaker motion, as the piston motion more closely resembles the water particle trajectories under the waves, while in deeper water, the flap generator is more efficient.

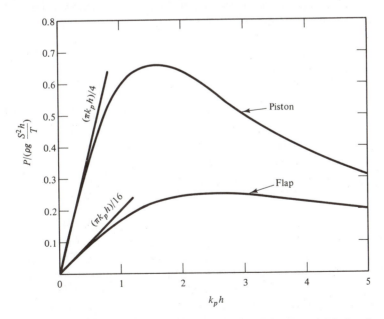

Figure 6.4 Dimensionless mean power as a function of water depth for piston and flap wavemakers.

The wavemaker theory has been developed assuming both small-amplitude motions of the paddle and small wave heights. There are singificant nonlinear effects that occur when the wavemaker moves with large displacements; in fact, the waves that result are of different size and shape at different locations away from the wavemaker (see, e.g., Madsen, 1971, and Flick and Guza, 1980).

6.3.1 Planar Wave Energy Absorbers

Energy may be removed from waves by moving paddles as well as added, as in the preceding section. One means to extract wave energy from waves under various conditions has been discussed by Milgram (1970).

The principle behind the wave absorber is that incident waves onto the paddle are absorbed by the paddle moving in a manner so as to be invisible to the waves. In other words, while in the wavemaking problem, the paddle is pushed forward to make a wave crest, in this case the paddle will move backward as a wave crest impinges on it (thus making waves on the other side of the paddle, if there is water), making it appear that the waves have passed through.

The most efficient absorption of the waves, of course, is dependent on moving the paddle in just the "right" motions, which can be determined theoretically. The mathematical formulation involves examining the waves on the opposite side of the paddle from the previous analysis. The velocity potential remains the same, except for the x dependency of the standing wave terms.

$$\phi_{incident} = \frac{Hg \cosh k_p(h + z)}{2\sigma \cosh k_p h} \sin (k_p x - \sigma t) \tag{6.28}$$

$$+ \sum_{n=1}^{\infty} C_n e^{+k_s(n)x} \cos [k_s(n)(h + z)] \cos \sigma t \qquad \text{for } x \leqslant 0$$

The value of wave absorber stroke S must be found for a given incident wave. To do this we use the boundary condition at $x = 0$.

$$u(z) = \frac{S(z)\sigma \cos \sigma t}{2} = - \frac{\partial \phi}{\partial x} \qquad \text{at } x = 0 \tag{6.29}$$

Following the same procedure as before, the same relationship [Eqs. (6.25) and (6.26)] results for H/S. Therefore, for a given incident wave height, the stroke necessary to absorb the waves can be determined. There still are, of course, the standing waves that are set up to account for the fact that the paddle velocities do not exactly match those of the incident wave. In addition, the velocity of the wavemaker motion must have exactly the same phase as that of the horizontal velocity of the incoming wave.

6.3.2 Three-Dimensional Wavemakers

The "snake" wavemaker.   By using an articulated long wavemaker in a wave basin, it is possible to make waves propagating in different directions depending on the motion of the wavemaker. To study this case, consider a wavemaker located on the y axis, making waves that propagate in the x-y plane. For simplicity the wavemaker will be assumed to be infinitely long. The motion of the wavemaker at $x = 0$ generates velocities in the x direction, $u(y, z; t)$, which in the simplest case may be written

$$u(y, z; t) = U(z) \cos (\lambda y - \sigma t) \qquad \text{on } x = 0 \qquad (6.30)$$

This represents a horizontal velocity at the wavemaker which consists of periodic motion, propagating in the $+y$ direction.

The boundary value problem which must be solved is

$$\frac{\partial^2 \phi}{\partial x^2} + \frac{\partial^2 \phi}{\partial y^2} + \frac{\partial^2 \phi}{\partial z^2} = 0 \qquad \text{in} \begin{cases} 0 \leqslant x < \infty \\ -\infty < y < \infty \\ -h \leqslant z \leqslant 0 \end{cases} \qquad (6.31)$$

At the horizontal bottom of the basin, the bottom boundary condition must be met. At the surface, the linearized kinematic and dynamic conditions apply, as before.

Using separation of variables a solution is assumed which satisfies the bottom boundary condition.

$$\phi = A_p \cosh k_p(h + z) \sin \left(\sqrt{k_p^2 - \lambda^2} \, x + \lambda y - \sigma t \right) \qquad (6.32)$$

$$+ \sum_{n=1}^{\infty} C_n \cos [k_s(n)(h + z)] \exp \left[-\sqrt{k_s^2(n) + \lambda^2} \, x \right] \cos (\lambda y - \sigma t)$$

where $\sigma^2 = g k_p \tanh k_p h$; $\sigma^2 = -g k_s(n) \tan k_s(n) h$.

It can be shown, by examining all other possible solutions, that only this form provides for a propagating wave in the x direction with the usual $\cosh k_p(h + z)$ depth dependency. Further, this imposes a restriction that $k_p \geqslant \lambda$.

Invoking the wavemaker boundary condition at $x = 0$,

$$U(z) \cos (\lambda y - \sigma t) = - \left. \frac{\partial \phi}{\partial x} \right|_{x=0} \qquad (6.33)$$

$$= -A_p \sqrt{k_p^2 - \lambda^2} \cosh k_p (h + z) \cos (\lambda y - \sigma t)$$

$$+ \sum_{n=1}^{\infty} C_n \sqrt{k_s^2(n) + \lambda^2} \cos \{k_s (n)(h + z)\} \cos (\lambda y - \sigma t)$$

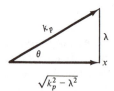

$$\sqrt{k_p^2 - \lambda^2}$$ **Figure 6.5**   Definition for θ.

Examining only the propagating mode (in the x direction), and utilizing the orthogonal properties of $\{\cosh k_p(h + z); \cos [k_s(n)(h + z)], \quad n = 1, 2, \ldots, \infty,$ we have

$$A_p = -\frac{4k_p \int_{-h}^{0} U(z) \cosh k_p(h + z) \, dz}{\sqrt{k_p^2 - \lambda^2} \, (\sinh 2k_ph + 2k_ph)} \tag{6.34}$$

which is nearly the same as before.

If we introduce a directional angle θ made by the wave orthogonal to the x axis as in Figure 6.5, where λ is the wave number in the y direction and $\sqrt{k_p^2 - \lambda^2}$ is the wave number in the x direction, we see that k_p represents the wave number in the propagation direction. Further, $\sqrt{k_p^2 - \lambda^2} = k_p \cos \theta$ and $\lambda = k_p \sin \theta$. This latter expression requires that the wavelength λ of the wavemaker displacement be related to the desired wave angle. Substituting, the velocity potential of the propagating wave can be written

$$\phi_p(x, y, z; t) = A_p \cosh k_p(h + z) \cos ((k_p \cos \theta)x + (k_p \sin\theta)y - \sigma t) \tag{6.35}$$

where A_p is given by Eq. 6.34 and is related to the planar value [Eq. (6.21)] by $(\cos \theta)^{-1}$. To make waves in the opposite $-\theta$ direction, the wave displacement must propagate in the opposite direction

$$u(z, y; t) \propto \cos (\lambda y + \sigma t)$$

In order to generate a realistic sea state in a wave basin, numerous wavemaker motions can be superimposed due to the linearity of the problem.

6.4 CYLINDRICAL WAVEMAKERS

Although not in common use, the wavemaker theory for water waves generated by moving vertical cylinders follows directly from plane wavemaker theory, the only exception being that the problem is worked in polar coordinates (see Chapter 2).

The fluid motion can be described by a velocity potential which is governed by the Laplace equation with the usual linearized free surface and bottom boundary conditions.

$$\nabla^2\phi = \frac{\partial^2\phi}{\partial r^2} + \frac{1}{r}\frac{\partial\phi}{\partial r} + \frac{1}{r^2}\frac{\partial^2\phi}{\partial\theta^2} + \frac{\partial^2\phi}{\partial z^2} = 0 \tag{6.36}$$

where r and θ are the polar coordinates of the horizontal plane.

$$\eta = \frac{1}{g}\frac{\partial \phi}{\partial t} \quad \text{on} \quad z = 0$$

$$-\frac{\partial \phi}{\partial z} = \frac{\partial \eta}{\partial t} \quad \text{on} \quad z = 0 \tag{6.37}$$

$$-\frac{\partial \phi}{\partial z} = 0 \quad \text{on} \quad z = -h$$

Additionally, a radiation boundary condition is imposed at large r to ensure outgoing waves and a kinematic condition must be applied to the moving wall of the cylinder.

There are several possibilities for cylindrical wavemakers, which will be denoted by different types. Type I will be a vertical cylinder (located at $r = 0$, with radius a) moving in piston or flap motion in a fixed vertical plane, taken as $\theta = 0$ or π. Applying a kinematic condition that the fluid at the cylinder wall follows the cylinder's motion, we have in linearized form

$$v_r = -\frac{\partial \phi}{\partial r} = \frac{S(z)\sigma}{2}\cos m\theta \cos \sigma t = \text{Re}\left\{\frac{S(z)\sigma}{2}\cos m\theta\, e^{i\sigma t}\right\} \tag{6.38}$$

where Re{ } denotes real part,[2] m is an integer equal to unity and $S(z)$ is the vertical variation of the displacement of the cylinder. The Type II wavemaker is a pulsating cylinder, which expands and contracts radially with no θ dependency. The corresponding linear kinematic condition is

$$-\frac{\partial \phi}{\partial r} = \frac{S(z)\sigma}{2}\cos \sigma t = \text{Re}\left\{\frac{S(z)\sigma}{2}\cos m\theta\, e^{i\sigma t}\right\} \quad \text{at } r = a \tag{6.39}$$

with $m = 0$. Finally, the Type III wavemaker is a spiral wavemaker discussed by Dalrymple and Dean (1972), who advocate its use in littoral drift studies. In a circular basin the spiral wavemaker generates waves which impinge on a circular beach everywhere at the same angle, thus resulting in an "infinite" beach ideal for sediment transport studies. [In some cases, the spiral wave shoals in a manner differently than plane waves (Mei, 1973).] The cylinder motion can be visualized by placing a pencil point down on a table and rotating the top in a small circular path. The linearized kinematic boundary condition becomes

$$-\frac{\partial \phi}{\partial r} = \frac{S(z)\sigma}{2}\cos (m\theta - \sigma t) = \text{Re}\left\{\frac{S(z)\sigma}{2}e^{i(m\theta - \sigma t)}\right\} \quad \text{on } r = a \tag{6.40}$$

where $m = 1$ for the case of the rotating pencil, but could be greater than unity for a lobe-shaped cylinder.

---

[2]See page 190.

The solution for the velocity potential is obtained by separation of variables, in the same manner as before (see Problem 6.9). The solutions that satisfy all the boundary conditions with the exception of the kinematic condition on the cylinder are

$$\phi(r, \theta, z, t) = A_p H_m^{(1)} (k_p r) \cosh k_p (h + z) \begin{cases} \cos m\theta \, e^{i\sigma t}, \text{ Type I, II} \\ e^{i(m\theta - \sigma t)}, \text{ Type III} \end{cases} \tag{6.41}$$

$$+ \sum_{n=1}^{\infty} C_n K_m(k_s(n)r) \cos [k_s(n)(h + z)] \begin{cases} \cos m\theta \, e^{i\sigma t}, \text{ Type I, II} \\ e^{i(m\theta - \sigma t)}, \text{ Type III} \end{cases}$$

where $H_m^{(1)}(k_p r)$ is the Hankel function of the first kind, defined as $H_m^{(1)}(k_p r) = J_n(k_p r) + i Y_n(k_p r)$, a complex number formed by the Bessel functions, and $K_m(k_s(n)r)$ is the modified Bessel function of the second kind. Associated with these solutions are the dispersion relationships relating the angular frequency to the wave number(s),

$$\sigma^2 = g k_p \tanh k_p h$$

and

$$\sigma^2 = -g k_s(n) \tan k_s(n)h, \qquad n = 1, 2, \ldots, \infty \tag{6.42}$$

The unknown coefficients in the series for the velocity potential are obtained by satisfying the remaining boundary condition at the cylindrical wall using the orthogonality of the depth-dependent functions, with the result that

$$A_p = - \frac{\displaystyle\int_{-h}^{0} \frac{S(z)\sigma}{2} \cosh k_p(h + z) \, dz}{k_p [H_m^{(1)}(k_p a)]' \displaystyle\int_{-h}^{0} \cosh^2 k_p(h + z) \, dz} \tag{6.43}$$

and

$$C_n = - \frac{\displaystyle\int_{-h}^{0} \frac{S(z)\sigma}{2} \cos [k_s(n) (h + z)] \, dz}{k_s(n) [K_m(k_s(n)a)]' \displaystyle\int_{-h}^{0} \cos^2 [k_s(n) (h + z)] \, dz} \tag{6.44}$$

The [·]' denotes the derivative with respect to the argument of the function. The coefficients A_p and C_n are. the same for all three types of cylinder wavemaker and are similar to the coefficients for the planar wavemaker, differing due to the presence of the derivative of a Bessel function in the denominator. These terms in the velocity potential account for the radial decay of the waves away from the wavemaker.

Far from the wavemaker, the water surface displacement η may be determined from the linear dynamic free surface boundary condition and the Hankel function term as the others become negligible several water depths from the wavemaker. Using the asymptotic form for the Hankel function, we have

$$\eta(r, \theta, t) = \text{Re}\left\{ -i\sigma A_p \sqrt{\frac{2}{k_p r}} \cosh k_p h \right.$$ (6.45)

$$\left. \begin{cases} \cos m\theta \exp\left\{ i\left[k_p r - \sigma t - \frac{(2m + 1)\pi}{4} \right] \right\}, \text{ Type I, II} \\[2ex] \exp\left\{ i\left[k_p r + m\theta - \sigma t - \frac{(2m + 1)\pi}{4} \right] \right\}, \text{ Type III} \end{cases} \right\}$$

Using the relationship between strokes $S(z)$ and A_p and the last equation, the wave height-to-stroke ratio can be determined. This is shown in Figure 6.6 for

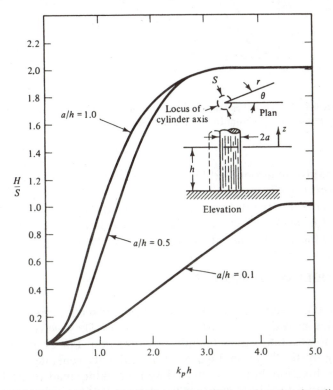

Figure 6.6 Dimensionless progressive wave amplitude evaluated at the cylinder for piston or circular motion of the wavemaker. $m = 1$. (From Dalrymple and Dean, 1972.)

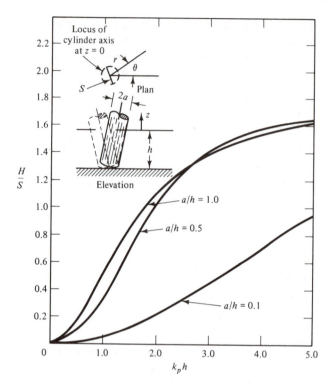

Figure 6.7 Dimensionless progressive wave amplitude evaluated at cylinder, sway motion. $m = 1$. (From Dalrymple and Dean, 1972.)

the case of piston motion and Figure 6.7 for sway motion for Type I and III wavemakers ($m = 1$).

Power requirements to generate these radial waves, energy flux, and the direction (for spiral waves) can be determined fairly simply; the reader is referred to the original paper by Dalrymple and Dean (1972) for details.

6.5 PLUNGER WAVEMAKERS

Plunger wavemakers with a wedge-shaped cross section are often used in laboratories instead of piston or flap-type paddles. These wavemakers can be designed to generate waves in only one direction. For example, a wedge oscillating vertically as in Figure 6.6 would only generate waves in the positive x direction. For an immersed wedge making small vertical motions and for small β, the linear theory is the same as that for piston wavemakers; for larger vertical strokes and for large β, as well as for other shapes the reader is referred to Wang (1974), who solved the plunger problem using a conformal

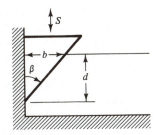

Figure 6.8 Schematic of wedge-shaped plunger wavemakers.

transformation. He presents figures of amplitude/stroke ratios versus dimensionless geometrical parameters for wedge-shaped wavemakers as shown in Figure 6.8.

REFERENCES

DALRYMPLE, R. A., and R. G. DEAN, "The Spiral Wavemaker for Littoral Drift Studies," *Proc. 13th Conf. Coastal Eng., ASCE*, 1972.

FLICK, R. E., and R. T. GUZA, "Paddle Generated Waves in Laboratory Channels," *J. Waterways, Ports, Coastal Ocean Div., ASCE*, Vol. 106, Feb. 1980.

GALVIN, C. J., Jr., "Wave–Height Prediction for Wave Generators in Shallow Water," Tech. Memo 4, U.S. Army, Coastal Engineering Research Center, Mar. 1964.

MADSEN, O. S., "On the Generation of Long Waves," *J. Geophys. Res.*, Vol. 76, No. 36, 1971.

MADSEN, O. S., "A Three-Dimensional Wavemaker, Its Theory and Application," *J. Hydraulics Res.*, Vol. 12, No. 2, 1974.

MEI, C. C., "Shoaling of Spiral Waves in a Circular Basin," *J. Geophys. Res.*, Vol. 78, No. 6, 1973.

MILGRAM, J. H., "Active Water–Wave Absorbers," *J. Fluid Mech.*, Vol. 43, Pt. 4, 1970

SOMMERFELD, A., *Mechanics of Deformable Bodies*, Vol. 2 of *Lectures on Theoretical Physics*, Academic Press, New York, 1964.

URSELL, F., R. G. DEAN, and Y. S. YU, "Forced Small Amplitude Water Waves: A Comparison of Theory and Experiment," *J. Fluid Mech.*, Vol. 7, Pt.. 1, 1960.

WANG, S., "Plunger-Type Wavemakers: Theory and Experiment," *J. Hydraulics Res.*, Vol. 12, No. 3, 1974.

PROBLEMS

6.1 A piston wavemaker operates over only half the water depth and oscillates with frequency σ and a maximum velocity U_0.

(a) Determine the wave height away from the wavemaker in terms of U_0 if the wavemaker operates over the top half of the water column.

(b) An alternative design is to operate the wavemaker over the bottom half of the water column. Plot the ratio of wave heights (away from the wavemaker), H_{top}/H_{bottom} as a function of kh, where H_{top} indicates the wave height in part (a). Which wavemaker is more efficient and why?

(c) Calculate the ratio H/S for shallow water using the simplified approach and compare with the results developed in parts (a) and (b).

6.2 Show, using the simplified shallow water approach, that the ratio of wave height near the cylinder to wave height stroke for a vertical cylinder, oscillating vertically with a stroke d and generating circular waves, is

$$\frac{H}{d} = \frac{kR}{2}$$

where R is the radius of the cylinder.

6.3 What are the stroke and power necessary to generate a 2-s period 20-cm-high wave in 2 m of water for both flap and piston wavemakers.

6.4 A long rectangular barge with draft d in shallow water is heaving (moving vertically) with a velocity $V_0 \cos \sigma t$.

(a) Determine the amplitude of the waves generated by this motion if the barge width is given.

(b) Determine the damping of the barge motion due to wave generation. (*Hint*: It is easiest to use energy arguments here.)

6.5 Determine the equations for instantaneous and mean power required for wavemakers using the wave-induced pressure on the wavemaker. Determine, for a wavemaker with a displacement of $S(z) = S \cosh k(h + z)$, the instantaneous and mean power required. Why might it be advantageous to incorporate a flywheel into the generating mechanism?

6.6 Using conservation of energy flux, show how the waves due to a circular wavemaker (see Problem 6.2), would decay in height with radial distance.

6.7 Examine the energy flux at the wavemaker due to the progressive and standing wave mode components. Discuss your results.

6.8 Show that the set $\{\cosh k_p (h + z), \cos [k_s (n)(h + z)], n = 1, 2, \ldots, \infty)\}$ are orthogonal over the range $-h < z < 0$, given the dispersion relationships for σ, k_p, and $k_s(n)$.

6.9 Develop the theory for waves made by a circular cylinder wavemaker with vertical axis moving in piston motion.

6.10 Develop dimensionless expressions for the maximum total forces on piston and flap-type wavemakers.

6.11 Develop the three-dimensional wavemaker theory for waves in a long wave tank. The side walls are located at $|y| = l$, and the waves are made by a paddle with a mean position of $x = 0$, yet which varies in stroke over the vertical and across the tank width (Madsen, 1974).

7

Wave Statistics and Spectra

Dedication
LORD RAYLEIGH

John William Strutt (1842–1919), the third Baron Rayleigh, for whom the Rayleigh probability distribution is named, received (with Sir William Ramsey) the Nobel Prize in 1905 for the discovery of argon.

He was born in Langford Grove, Essex, England, and entered Trinity College, Cambridge, in 1861, becoming a Fellow in 1866.

Over his career, Rayleigh wrote 446 papers that ranged from his noted *Treatise on the Theory of Sound*, published in 1877, to works in electromagnetism and physical optics. These works have been collected in *Scientific Papers*. His research interests included electricity and psychic phenomena and theoretical/experimental work on the explanation of the sky's color.

In 1879 he gained appointment as the second Cavendish Professor and in 1884 became the director of the Cavendish Laboratory at Cambridge University. In 1894 he retired from these positions to do research in his private laboratory in Terling Place, Witham, Essex, where he was Baron (after the death of his father in 1873).

In 1908 he became the Chancellor of Cambridge University. Rayleigh died in 1919 and was buried in Westminster Abbey.

7.1 INTRODUCTION

Previous chapters have discussed waves that are monochromatic (i.e., they have only one frequency). (The term "monochromatic" derives from the analogy of water waves to light waves and the relation of color to frequency.) However, by simply looking at the actual sea surface, one sees that the surface is composed of a large variety of waves moving in different directions and with different frequencies, phases, and amplitudes. For an adequate description of the sea surface, then, a large number of waves must be superimposed to be realistic (as mentioned in Chapter 1). This chapter discusses the methods by which this is done and the characteristics of the sea surface.

7.2 WAVE HEIGHT DISTRIBUTIONS

Designing in the ocean requires an adequate knowledge of possible wave heights. For example, in the design of a structure, the engineer may be faced with designing for the maximum expected wave height, the "highest possible" waves, or some other equivalent wave height. Historically, several wave heights have become popular as characterizing the sea state. These are the $H_{1/3}$ (the *significant* wave height) and the H_{max} wave heights. To envision what these definitions mean, consider a group of N wave heights measured at a point. Ordering these waves from the largest to the smallest and assigning to them a number from 1 to N, two statistical measures may be obtained. First, $H_{1/3}$ is defined as the average of the first (highest) $N/3$ waves. Correspondingly, H_p would be defined as the average of the first pN waves, with $p \leq 1$. (H_1 would be the average wave height.) Second, the probability that the wave height is greater than or equal to an arbitrary wave height $\hat{H}$ is

$$P(H > \hat{H}) = \frac{n}{N} \tag{7.1}$$

where n is the number of waves higher than $\hat{H}$. We note for later use that $P(H \leq \hat{H}) = 1 - n/N$.

The root-mean-square wave height for our group of waves, H_{rms}, is defined as

$$H_{rms} = \sqrt{\frac{1}{N}\sum_{i=1}^{N} H_i^2} \tag{7.2}$$

which is always larger than H_1 in a real sea.

7.2.1 Single Wave Train

It is clear that for the sea surface described by a single sinusoid wave, $\eta(t) = (H_0/2) \cos \sigma t$, the waves are all of the same height and that $H_p = H_0$ for any p and $H_{rms} = H_0$.

7.2.2 Wave Groups

To make the sea surface somewhat more realistic, another wave train is added, with slightly different frequency, in order to make wave groups, as was done in Chapter 4.

$$\eta = \frac{H_0}{2} \cos\left(\sigma - \frac{\Delta\sigma}{2}\right)t + \frac{H_0}{2} \cos\left(\sigma + \frac{\Delta\sigma}{2}\right)t$$

$$= H_0 \cos \sigma t \, \cos \frac{\Delta\sigma}{2}t = \frac{H(t)}{2} \cos \sigma t \tag{7.3}$$

which represents a propagating wave system evaluated at $x = 0$.

The resulting wave system has a carrier wave at frequency σ and a slowly modulated wave height $2H_0 \cos (\Delta\sigma/2)t$ (see Figure 4.12). Therefore, to examine the wave height distribution for the wave system, we need only to look at the envelope from $t = 0$ to $\pi/\Delta\sigma$ (or from the antinode to the first node).

To determine H_p, we average the wave height envelope from $t = 0$ to $p\pi/\Delta\sigma$, since the wave heights decrease monotonically from the maximum to the minimum.

$$H_p = \frac{1}{p\pi/\Delta\sigma} \int_0^{p\pi/\Delta\sigma} 2H_0 \cos \frac{\Delta\sigma}{2}t \, dt \tag{7.4}$$

$$H_p = 4\frac{H_0}{p\pi} \sin \frac{p\pi}{2} \tag{7.5}$$

The rms wave height can be derived:

$$H_{\text{rms}}^2 = \frac{1}{\pi/\Delta\sigma} \int_0^{\pi/\Delta\sigma} 4H_0^2 \cos^2 \frac{\Delta\sigma}{2}t \, dt \tag{7.6}$$

or

$$H_{\text{rms}} = \sqrt{2}H_0$$

We can therefore express the H_p wave height in terms of H_{rms}, which will be a more definable wave height for real seas.

$$H_p = \frac{2\sqrt{2}\,H_{\text{rms}}}{p\pi} \sin \frac{p\pi}{2} \tag{7.7}$$

and since H_{max} must be equal to $2H_0$, we have,[1] from Eq. (7.6),

$$H_{\text{max}} = \sqrt{2}\,H_{\text{rms}} \tag{7.8}$$

Example 7.1

A wave group consisting of two sinusoids of equal height and slightly different periods is generated in a laboratory wave tank and recorded by a fixed wave gage. What are the values of H_{max}, $H_{1/10}$, $H_{1/3}$, and H_1 in terms of H_{rms}?

---

[1]Alternatively, this can be obtained from Eq. (7.7) as in the limit as $p \to 0$.

Solution.

$$H_{max} = \sqrt{2}\, H_{rms} = 1.414 H_{rms}$$

$$H_{1/10} = \frac{20\sqrt{2}\, H_{rms}}{\pi} \sin \frac{\pi}{20} = 1.408 H_{rms}$$

$$H_{1/3} = \frac{6\sqrt{2}\, H_{rms}}{\pi} \sin \frac{\pi}{6} = 1.350 H_{rms}$$

$$H_1 = \frac{2\sqrt{2}\, H_{rms}}{\pi} = 0.90 H_{rms}$$

7.2.3 Narrow-Banded Spectra: The Rayleigh Distribution

For a more realistic case, we assume that the sea surface is composed of a large number of sinusoids, but with their frequencies near a common value, σ. This is referred to as a narrow-banded sea (in that all the frequencies are in a narrow frequency band about σ). Therefore, for M component frequencies

$$\eta(t) = \sum_{m=1}^{M} \frac{H_m}{2} \cos(\sigma_m t - \epsilon_m) \tag{7.9}$$

or equivalently, in complex notation,[2]

$$\eta(t) = \text{Re} \left\{ \sum_{m=1}^{M} \frac{H_m}{2} e^{i(\sigma_m t - \epsilon_m)} \right\} \tag{7.10}$$

The notation $\text{Re}\{\cdot\}$ refers to taking only the real part, $\text{Re}(e^{i\sigma t}) = \cos \sigma t$. Factoring out the carrier wave of frequency σ yields

$$\eta(t) = \text{Re} \left\{ e^{i\sigma t} \sum_{m=1}^{M} \frac{H_m}{2} e^{i[(\sigma_m - \sigma)t - \epsilon_m]} \right\} \tag{7.11}$$

Again, to define the wave height distribution, we need only to examine the statistics of the slowly varying envelope, $B(t)$:

$$B(t) = \sum_{m=1}^{M} \frac{H_m}{2} e^{i[(\sigma_m - \sigma)t - \epsilon_m]} \tag{7.12}$$

---

[2]From complex variable theory, $e^{i\sigma t} = \cos \sigma t + i \sin \sigma t$, where $i = \sqrt{-1}$. These formulas can be readily derived if we express e^{ix} as a Maclaurin series.

$$e^{ix} = 1 + ix + \frac{(ix)^2}{2!} + \frac{(ix)^3}{3!} + \frac{(ix)^4}{4!}$$

$$= \left(1 - \frac{x^2}{2!} + \frac{x^4}{4!} \cdots \right) + i \left(x - \frac{x^3}{3!} + \cdots \right)$$

The terms in the two sets of large parentheses are the power series expansion for cosine and sine.

From statistical theory, it can be shown (e.g., Longuet-Higgins, 1952) that if the individual components of B are statistically independent and a large number M is used, then the probability of the wave height being greater than or equal to an arbitrary wave height ($\hat{H}$) is given by

$$P(H \geqslant \hat{H}) = e^{-(\hat{H}/H_{rms})^2} \tag{7.13}$$

which is called the Rayleigh distribution.

This theoretical probability can be compared to our rank-ordered group of waves, N, Eq. (7.1):

$$P(H \geqslant \hat{H}) = \frac{n}{N} \tag{7.14}$$

or equating,

$$\frac{n}{N} = e^{-(\hat{H}/H_{rms})^2} \tag{7.15}$$

This expression provides a means to determine the number of waves out of the total number N which have a height greater than or equal to a certain height $\hat{H}$. Alternatively, we can solve this expression to determine the height $\hat{H}$ which is exceeded by n waves in our group of N. By taking the natural logarithms of both sides, we find that

$$\hat{H} = H_{rms} \sqrt{\ell n \frac{N}{n}} \tag{7.16}$$

The height that is exceeded by pN of the waves is therefore

$$\hat{H} = H_{rms} \sqrt{\ell n \frac{1}{p}} \tag{7.17}$$

Example 7.2

At a pier in Atlantic City, New Jersey, 400 consecutive wave heights are measured. The H_{rms} is determined by Eq. (7.2). (a) Assuming that the sea state is narrow-banded, determine how many waves are expected to exceed $H = 2H_{rms}$. (b) What height is exceeded by half the waves? (c) What height is exceeded by only one wave?

Solution.
(a) To answer the first part, Eq. (7.15) is used.

$$n = Ne^{-(2)^2}$$

$$= 7.3 \approx 7 \text{ waves}$$

Approximately seven waves, less than 2% of the total number, exceed $2H_{rms}$.
(b) The height $\hat{H}$ exceeded by half the waves ($n = N/2$, or $p = \frac{1}{2}$) is

$$\hat{H} = \sqrt{\ell n(2)}\, H_{rms} = 0.833 H_{rms}$$

For $H_{1/N}$ we have $p = 1/N$ or $H_{1/N} = \sqrt{\ell n \, N} \, H_{rms} = 2.45H_{rms}$. It is perhaps not too surprising that the more waves present in the group (i.e., large N), the higher the maximum wave will be. This is due to the fact that the Rayleigh probability function decays asymptotically to zero for large H, but never reaches zero. Thus all wave heights are statistically but not necessarily physically possible.

7.2.4 The Rayleigh Probability Density Function

The wave height probability density function f_H follows from the Rayleigh probability distribution $P(H < \hat{H})$:

$$f_H = \frac{d}{dH}(P(H \leqslant \hat{H})) = \frac{d}{d\hat{H}}(1 - e^{-(\hat{H}/H_{rms})^2}) = \frac{2He^{-(H/H_{rms})^2}}{H_{rms}^2} \qquad (7.18)$$

This function is plotted in Figure 7.1. Maximizing with respect to H yields the maximum probability for $\hat{H}/H_{rms} = 1/\sqrt{2}$, or the most frequent wave is $H = 0.707H_{rms}$.

From statistical theory we can obtain important relationships using the distribution function for the wave height.

The mean wave height is defined as

$$\overline{H} = \frac{\displaystyle\int_0^\infty H f_H(H) \, dH}{\displaystyle\int_0^\infty f_H(H) \, dH} = \frac{2}{H_{rms}^{-1}} \int_0^\infty \frac{H^2 e^{-(H/H_{rms})^2}}{H_{rms}^2} \, d\left(\frac{H}{H_{rms}}\right)$$

$$= \frac{\sqrt{\pi}}{2} H_{rms} = 0.886H_{rms} \qquad (7.19)$$

To find the average height of the highest pN waves, we first recall that

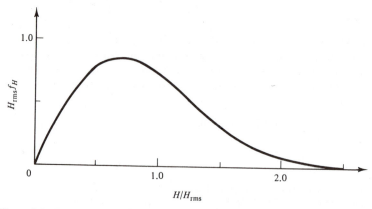

Figure 7.1 The Rayleigh probability distribution function. The area under the curve is unity.

the height $\hat{H}$ exceeded by the pN waves is

$$\frac{H_p}{H_{\text{rms}}} = \sqrt{\ell\text{n}\frac{1}{p}}$$

Next,

$$\bar{H}_p = \frac{\int_{\hat{H}_p}^{\infty} H f_H(H)\, dH}{\int_{\hat{H}_p}^{\infty} f_H(H)\, dH} = H_{\text{rms}} \frac{\int_{\hat{H}_p/H_{\text{rms}}}^{\infty} x^2 e^{-x^2}\, dx}{\int_{\hat{H}_p/H_{\text{rms}}}^{\infty} x e^{-x^2}\, dx} \qquad (7.20)$$

where x is a dummy variable. Integrating by parts, we get

$$\frac{\bar{H}_p}{H_{\text{rms}}} = \frac{\hat{H}_p/H_{\text{rms}} e^{-(\hat{H}_p/H_{\text{rms}})^2} + \int_{\hat{H}_p/H_{\text{rms}}}^{\infty} e^{-x^2} dx}{e^{-(\hat{H}_p/H_{\text{rms}})^2}} \qquad (7.21)$$

$$= \sqrt{\ell\text{n}\frac{1}{p}} + \frac{\sqrt{\pi}}{2p}\, \text{erfc}\left(\sqrt{\ell\text{n}\frac{1}{p}}\right)$$

where erfc (x) is the complementary error function (see Abramowitz and Stegun, 1965).

In Table 7.1 various values of H_p/H_{rms} are presented. It is clear that as p becomes smaller, there is a significant change from the results obtained by the simple wave group model (see Example 7.1).

TABLE 7.1   Relationship of H_p to H_{rms} using the Rayleigh Distribution

$$H_{1/10} = 1.80 H_{\text{rms}}$$
$$H_{1/3} = 1.416 H_{\text{rms}}$$
$$H_1 = 0.886 H_{\text{rms}}$$

Forristall (1978) has shown that for real seas of large magnitude, the Rayleigh distribution tends to overpredict the larger wave heights. This is presumably due to the breaking phenomenon "trimming" these larger heights.

7.3 THE WAVE SPECTRUM

The waves recorded at a wave staff generally are composed of components of many frequencies σ_n and amplitudes a_n with different phases ϵ_n:

$$\eta(t) = \sum_{n=0}^{\infty} a_n \cos(\sigma_n t - \epsilon_n) \qquad (7.22)$$

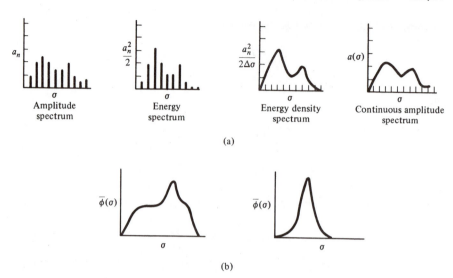

Figure 7.2   (a) Types of spectra; (b) broad versus narrow-banded energy spectrum.

If the amplitudes a_n are plotted versus frequency, an amplitude spectrum results. More commonly used, however, is the energy spectrum, which is a plot of a_n^2. Both of these spectra are line or discrete spectra in that each frequency component is discrete. The energy *density* spectrum, on the other hand, is a plot of $a_n^2/\Delta\sigma$ versus σ, which is more popular, as the area under the curve is a measure of the total energy in the wave field. It is more likely in nature that the spectrum be comprised of a continuous range of frequencies or

$$\eta(t) = \text{Re}\left\{ \int_0^\infty a(\sigma)e^{i[\sigma t - \epsilon(\sigma)]}\, d\sigma \right\} \tag{7.23}$$

where $a(\sigma)\, d\sigma$ is the amplitude of each wave and $a(\sigma)$ might be called the amplitude density function. Examples of these spectra are shown in Figure 7.2a. The shape of the spectrum varies with the types of seas and whether it is broad- or narrow-banded (Figure 7.2b).

7.3.1 Spectral Analysis

The procedure of extracting spectra from wave records is an evolving field and a complete presentation of spectral analysis is beyond the scope of this book. However, some rudimentary aspects of it will be discussed. Of primary importance is the fact that the use of computers in time-series analysis has made it far more convenient to deal with digitized data[3] and spectral analysis is usually done by the fast Fourier transform (FFT) tech-

---

[3] The time series of $\eta(t)$ digitized at an interval of Δt is the sequence of numbers: $\eta(\Delta t)$, $\eta(2\,\Delta t)$, $\eta(3\,\Delta t)$, and so on.

nique, popularized by Cooley and Tukey (1965). It should be noted parenthetically that almost all our knowledge about spectral analysis comes to the ocean engineers via the electronic and communications fields.

7.3.2 Fourier Analysis

The basis for spectral analysis is the Fourier series, named for Joseph Fourier (1768–1830). The premise of Fourier analysis follows from the fact that any (piecewise continuous) function $f(t)$ can be represented over an interval of time (t to $t + T$) as a sum of sines and cosines, where t is arbitrary and $f(t)$ is assumed to be (or is) periodic over the time period, T. The Fourier series is written as

$$f(t) = \sum_{n=0}^{\infty} (a_n \cos n\sigma t + b_n \sin n\sigma t) \tag{7.24}$$

where $\sigma = 2\pi/T$ and $b_0 = 0$ as sin $(0) \equiv 0$, and a_0 is simply the mean of the record. The coefficients a_n and b_n can be obtained by minimizing the mean squared error of the function E, which is defined as

$$E \equiv \frac{1}{T} \int_t^{t+T} \left[f(t) - \sum_{n=0}^{\infty} (a_n \cos n\sigma t + b_n \sin n\sigma t) \right]^2 dt \tag{7.25}$$

Minimizing yields

$$\frac{\partial E}{\partial a_m} = 0; \qquad \frac{\partial E}{\partial b_m} = 0 \tag{7.26a}$$

Expressing these equations fully, we have

$$\int_t^{t+T} \left[f(t) - \sum_{n=0}^{\infty} (a_n \cos n\sigma t + b_n \sin n\sigma t) \right] \cos m\sigma t \, dt = 0$$
$$\int_t^{t+T} \left[f(t) - \sum_{n=0}^{\infty} (a_n \cos n\sigma t + b_n \sin n\sigma t) \right] \sin m\sigma t \, dt = 0 \tag{7.26b}$$

Using the following orthogonality properties of the trigonometric functions:

$$\int_t^{t+T} \sin n\sigma t \, \sin m\sigma t \, dt = \begin{cases} T/2 & m = n \neq 0 \\ 0, & m \neq n \end{cases}$$

$$\int_t^{t+T} \sin n\sigma t \, \cos m\sigma t \, dt = 0$$

$$\int_t^{t+T} \cos n\sigma t \, \cos m\sigma t \, dt = \begin{cases} T, & m = n = 0 \\ T/2 & m = n \neq 0 \\ 0, & m \neq n \end{cases}$$

and carrying out the integration following from Eqs. (7.26b), we obtain

$$a_0 = \frac{1}{T} \int_t^{t+T} f(t)\, dt \tag{7.27a}$$

$$a_n = \frac{2}{T} \int_t^{t+T} f(t) \cos n\sigma t\, dt \qquad \text{for } n = 1, 2, \ldots, \infty \tag{7.27b}$$

$$b_n = \frac{2}{T} \int_t^{t+T} f(t) \sin n\sigma t\, dt \qquad \text{for } n = 1, 2, \ldots, \infty \tag{7.27c}$$

Example 7.3

A square wave centered about $t = 0$, with an amplitude of unity and a period of 4 s, can be described in the interval $|t| < 2$ as

$$f(t) = \begin{cases} 1, & |t| < 1 \\ -1, & |t| > 1 \end{cases} \tag{7.28}$$

(see Figure 7.3).

Since the function is an even function, that is $f(t) = f(-t)$, all the b_n's are identically zero. (Try it if you do not believe it.) Solving, then, solely for the a_n's, using Eq. (7.27b), we get

$$a_n = \frac{4}{T} \int_0^1 (1) \cos n\left(\frac{2\pi}{4}\right) t\, dt + \frac{4}{T} \int_1^2 (-1) \cos n\left(\frac{2\pi}{T}\right) t\, dt$$

or

$$a_n = \frac{4}{n\pi} \sin \frac{n\pi}{2} \tag{7.29}$$

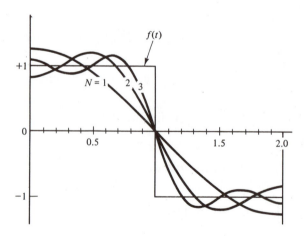

Figure 7.3 Fourier series fit to a square wave. As the results are symmetric about the origin, only the positive axis has been shown. The parameter N denotes the number of terms in the Fourier series.

For n an even number, $a_n = 0$, and for n odd, $a_n = (-1)^{n+1}(4/n\pi)$ for $n = 1, 3, 5, \ldots$; thus

$$f(t) = \frac{4}{\pi} \cos \frac{2\pi t}{T} - \frac{4}{3\pi} \cos \frac{6\pi t}{T} + \frac{4}{5\pi} \cos \frac{10\pi t}{T} - \frac{4}{7\pi} \cos \frac{14\pi t}{T} + \cdots \qquad (7.30)$$

Figure 7.3 shows the fit of the series to the function for one, two, and three terms. For a good representation to a function, it is necessary that a sufficient number of terms N be taken in the summation (practicality dictates that N not be infinite). How large N should be can be determined by finding the mean square value of the function $f(t)$.

$$\frac{1}{T} \int_t^{t+T} f^2(t)\, dt = \frac{1}{T} \int_t^{t+T} \left[a_0 + \sum_{n=1}^{N} (a_n \cos n\sigma t + b_n \sin n\sigma t) \right]^2 dt \qquad (7.31)$$

$$= a_0^2 + \frac{1}{2} \sum_{n=1}^{N} (a_n^2 + b_n^2)$$

This is referred to as Parseval's theorem, and it implies that if one-half the sum of the squares of the coefficients does not approximately equal the average mean square value of $f(t)$, more terms should be taken (N larger). It is often more meaningful in this comparison to subtract out the mean of $f(t)$ prior to using Parseval's theorem, as a_0^2 can dominate the summation.

For the square wave in the example,

$$\frac{1}{T} \int_{-2}^{2} f^2(t)\, dt = 1$$

For various values of n we have [from Eq. (7.29)]

$$\sum_{n=0}^{N} \frac{a_n^2 + b_n^2}{2} = \begin{cases} 0.811, & N = 1 \\ 0.900, & N = 3 \\ 0.933, & N = 5 \\ 0.950, & N = 7 \\ \vdots \\ 1.00, & N = \infty \end{cases}$$

7.3.3 Complex Series Representations

The *exponential* form of the Fourier series is obtained from the Euler identities

$$e^{in\sigma t} = \cos n\sigma t + i \sin n\sigma t$$
$$e^{-in\sigma t} = \cos n\sigma t - i \sin n\sigma t \qquad (7.32)$$

where $i \equiv \sqrt{-1}$. By adding and subtracting these two relationships, we have the identities

$$\cos n\sigma t = \frac{e^{in\sigma t} + e^{-in\sigma t}}{2}$$

$$\sin n\sigma t = \frac{e^{in\sigma t} - e^{-in\sigma t}}{2i} = -i\left(\frac{e^{in\sigma t} - e^{-in\sigma t}}{2}\right)$$

These expressions are then substituted into the Fourier series as represented in Eq. (7.24):

$$f(t) = \sum_{n=0}^{N} \left\{ a_n\left(\frac{e^{in\sigma t} + e^{-in\sigma t}}{2}\right) - b_n i\left(\frac{e^{in\sigma t} - e^{-in\sigma t}}{2}\right) \right\}$$

$$= \sum_{n=0}^{N} \left\{ \left(\frac{a_n - ib_n}{2}\right)e^{in\sigma t} + \left(\frac{a_n + ib_n}{2}\right)e^{-in\sigma t} \right\}$$

(7.33)

If the dummy subscript in the term modifying $e^{-in\sigma t}$ is changed to $-n$, we can write

$$f(t) = \sum_{n=-N}^{N} F(n)e^{in\sigma t}$$

(7.34)

where

$$F(n) = \begin{cases} \dfrac{a_n - ib_n}{2} & \text{for } n \geqslant 0 \\[2mm] \dfrac{a_n + ib_n}{2} & \text{for } n < 0 \end{cases}$$

(7.35)

Since $F(-n) = F^*(n)$, where the asterisk means complex conjugate, the right-hand side of Eq. (7.34) is real. The $F(n)$ may be obtained equivalently from the time series by

$$F(n) = \frac{1}{T}\int_t^{t+T} f(t)e^{-in\sigma t} \, dt$$

(7.36)

using Eqs. (7.35), (7.27b), and (7.27c).

Equations (7.34) and (7.36) constitute a Fourier transform pair. For discrete data, obtained at I points, the Fourier transform pair must be replaced by sums or

$$F(n) = \frac{1}{I}\sum_{m=1}^{N} f(m\,\Delta t)e^{-2\pi imn/I}$$

(7.37)

where $T = I\,\Delta t$ and Δt is the time between samples, and

$$f(m\,\Delta t) = \sum_{n=-I/2}^{I/2} F(n)e^{2\pi imn/I}$$

(7.38)

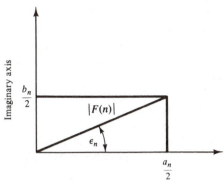

Figure 7.4   Argand diagram for $F(n)$.   Real axis

Any complex number such as $F(n)$ can be expressed in terms of an amplitude and a phase, using an Argand diagram (see Figure 7.4), which shows the real number along the abscissa and the imaginary numbers on the ordinate.

$$F(n) = |F(n)| \, e^{-i\epsilon_n} \tag{7.39}$$

where

$$|F(n)| = \frac{1}{2} \sqrt{a_n^2 + b_n^2} \, .$$

and

$$\epsilon_n = \tan^{-1}\frac{b_n}{a_n}$$

The phase ϵ_n gives the relationship of each particular harmonic term to the origin. For example, if the function $f(t)$ is even, then all the b_n's are zero and the phases are either $0°$ or $180°$. If the function is odd, the ϵ_n values are either $\pi/2$ or $3\pi/2$ for all n. If the $f(t)$ is translated with respect to the origin, the phases change, but $|F(n)|$ remains the same. Thus the $|F(n)|$'s provide a good characterization of a function.

7.3.4 Covariance Function

The covariance function, or the correlation function of two time-varying quantities $f_i(t)$ and $f_j(t)$, can be defined as

$$C_{ij}(\tau) = \frac{1}{T}\int_t^{t+T} f_i(t)f_j(t + \tau) \, dt \tag{7.40}$$

where τ is a time lag. If $i = j = 1$, then $C_{ij}(\tau)$ is the autocorrelation function, while if $i \neq j$, this quantity is the cross-correlation or cross-covariance function.

There are two important uses of the autocovariance function. The first is to identify periodicity within the time series $f_1(t)$. For periodic data, $C_{11}(\tau)$ will be periodic with the same period as $f_1(t)$. The second utility for a covariance function is that it is related directly to the energy spectrum, as will be shown shortly.

It can be shown that $C_{11}(\tau) = C_{11}(-\tau)$, that is, $C_{11}(\tau)$ is symmetric about the origin, and that the covariance is independent of the phase angles of the components of $f_1(t)$.

If we now substitute the Fourier series representation for $f_j(t + \tau)$ into the equation for the covariance, we obtain

$$C_{ij}(\tau) = \frac{1}{T} \int_t^{t+T} f_i(t) \sum_{n=-N/2}^{N/2} F_j(n) e^{in\sigma(t+\tau)} \, dt$$

$$= \frac{1}{T} \sum_{n=-N/2}^{N/2} \int_t^{t+T} f_i(t) e^{in\sigma t} \, dt \, F_j(n) e^{in\sigma\tau} \tag{7.41}$$

$$= \sum_{n=-N/2}^{N/2} F_j(n) F_i^*(n) e^{in\sigma\tau} = \sum_{n=-N/2}^{N/2} |F_j(n)| \, |F_i(n)| e^{i(\epsilon_j - \epsilon_i)} e^{in\sigma\tau}$$

where $F_i^*(n)$ is the complex conjugate of the complex Fourier coefficients of $f_i(t)$.

For the autocovariance,

$$C_{11}(\tau) = \sum_{n=-N/2}^{N/2} |F_1(n)|^2 \, e^{in\sigma\tau}$$

$$= \sum_{n=-N/2}^{N/2} |F_1(n)|^2 \cos n\sigma\tau \tag{7.42}$$

since $C_{11}(\tau)$ is symmetric. For the case where the time lag τ is zero,

$$C_{11}(0) = \sum_{n=-N/2}^{N/2} |F_1(n)|^2 = \frac{1}{T} \int_t^{t+T} f_1^2(t) \, dt \tag{7.43}$$

which recovers Parseval's theorem.

7.3.5 Power Spectrum

The Fourier transform of the covariance function is defined as the power spectrum (for $i = j$) or the cross spectrum (for $i \neq j$). For water waves it is more appropriate to call it the energy spectrum ($i = j$), as in the context the components of the spectrum are the squares of the wave amplitude at each frequency which are related to wave energy. Taking the Fourier transform of $C_{11}(\tau)$, we obtain

$$\Phi_{11}(n) = \frac{1}{T} \int_t^{t+T} C_{11}(\tau) e^{-in\sigma\tau} \, d\tau = |F_1 n)|^2 \tag{7.44}$$

for $-N/2 < n \leqslant N/2$, which is the two-sided energy spectrum. In practice, the one-sided energy spectrum is used, which is physically more intuitive as it does not involve negative frequencies, $-n\sigma$.

$$\Phi_{11}'(n) = 2|F_1(n)|^2, \qquad n > 0 \tag{7.45}$$

$$\Phi_{11}'(0) = |F_1(0)|^2, \qquad n = 0$$

for $0 \leqslant n \leqslant N/2$ only.

In the past, the procedure described above to obtain the wave spectrum was the only practical procedure available. This method, called the mean-lagged products method, involved the computation of the covariance function and then its Fourier transform was calculated to obtain the power spectrum. This laborious method was necessary, instead of the more direct technique of just taking the Fourier transform of the wave record to obtain the $F(n)$ coefficients and then finding $|F(n)|^2$, as it was very time consuming to obtain the Fourier transform. However, in the last two decades, with the implementation of the fast Fourier transform (FFT), which drastically reduced the amount of time necessary for computation, the more direct technique is now favored. In fact, most computer library systems have FFT algorithms available.

The cross spectrum $\Phi_{ij}(n)$ (for $i \neq j$) is obtained in a similar manner as Φ_{ii}.

$$\Phi_{ij}(n) = \frac{1}{T} \int_t^{t+T} C_{ij}(\tau) e^{-in\sigma\tau} \, d\tau = F_i^*(n) F_j(n) \tag{7.46}$$

or, it is the product of the Fourier coefficients of time series j and the complex conjugate of the coefficients for series i. The cross spectrum is in general complex, the real part is denoted the cospectrum, and the imaginary, the quad(rature) spectrum, or $\Phi_{ij}(n) = \text{Co}_{ij}(n) + i\text{Quad}_{ij}(n)$.

There are numerous intricacies of spectral estimation, such as stability and resolution of the spectrum, length of time series necessary, digitizing frequency, and so on. The interested reader is referred to other references for this; see, for example, Jenkins and Watts (1968).

7.3.6 The Continuous Spectrum

The amplitude, phase, and energy spectra that have been discussed have been discrete; that is, there are contributions only at discrete frequencies, for example, for the energy spectrum $\Phi_{11}(n)$, and the spacing on the frequency axis is

$$\Delta\sigma = \frac{2\pi}{T} \tag{7.47}$$

The discrete nature of the spectra is a direct result of considering the time series to be periodic. Natural phenomena such as gustiness in the atmosphere or water waves are usually considered to be aperiodic, and therefore there are a number of analytic continuous wave spectra which are used in design.

The formal derivation of aperiodic spectral relationships will not be presented here. It suffices to note that the procedure is one of considering the interval of periodicity T to approach infinity and recognizing that in the limit the contributions are densely packed on the frequency axis [cf. Eq. (7.47)] and thus approach a continuous distribution.

In practice, to represent the periodic energy spectrum as a continuous spectrum, the following simple transformation ensures that the total energy is conserved:

$$|F(\sigma_n)|^2 \, \Delta\sigma = |F(n)|^2 \qquad (7.48)$$

where $\sigma_n = n \, \Delta\sigma$ and it is seen that for the one-sided spectra $|F'(n)|^2$ and $|F'(\sigma)|^2$,

$$\sum_{n=0}^{+\infty} |F'(n)|^2 = \int_0^\infty |F'_1(\sigma)|^2 \, d\sigma = \overline{f^2(t)} \qquad (7.49)$$

7.4 THE DIRECTIONAL WAVE SPECTRUM

During a storm, such as a hurricane, a great number of waves are present on the sea surface, coming from many different directions. To characterize this, a directional wave spectrum is used. This generalizes the frequency spectrum, (7.23), by adding the variable θ, the wave direction, in addition to the wave frequency. Thus for each frequency there may be a number of wave trains from different directions. This directional wave system is expressed as[4]

$$\eta(x, y, t) = \sum_{n=-N/2}^{N/2} \int_0^{2\pi} F(n, \theta)e^{i(n\sigma t - (k_n\cos\theta)x - (k_n\sin\theta)y)} \, d\theta \qquad (7.50)$$

where θ is the angle made by the wave orthogonal and the x axis.

For waves measured at a point, say the origin, as a function of time, this reduces to

$$\eta(t) = \sum_{n=-N/2}^{N/2} \int_0^{2\pi} F(n, \theta)e^{in\sigma t} \, d\theta \qquad (7.51)$$

Measurement of the directional spectrum and its use in design has recently become widespread in the ocean industry. In fact, in relatively deep water, the directional nature of the sea surface during storms is at least as important as the nonlinearities present due to large waves. (For shallow

---

[4]The artifice of negative frequencies is required here to ensure that $\eta(t)$ is real. Note that this requires that $k_n = -k_n$, but that in the depth-dependent terms, $k \rightarrow |k_n|$, to ensure the decay with depth.

water conditions, the nonlinearities are generally much more significant than in deep water.)

As an example of the formulations necessary to develop the directional spectra, we will consider measurements made by a surface-piercing wave gage [Eq. (7.50)] and a two-component current meter, oriented such that it measures the horizontal components (u, v).

The velocities u and v can be represented as

$$u(t) = \sum_{n=-N/2}^{N/2} \int_0^{2\pi} \frac{gk_n \cos\theta}{n\sigma} K_p(z)F(n, \theta)e^{in\sigma t}\, d\theta \tag{7.52}$$

$$v(t) = \sum_{n=-N/2}^{N/2} \int_0^{2\pi} \frac{gk_n \sin\theta}{n\sigma} K_p(z)F(n, \theta)e^{in\sigma t}\, d\theta \tag{7.53}$$

where, as developed in Chapter 4,

$$K_p = \frac{\cosh |k_n|\,(h + z)}{\cosh |k_n|h}$$

and the associated velocity potential is

$$\Phi(x, y, t) = -i \sum_{n=-N/2}^{N/2} \int_0^{2\pi} F(n, \theta)\, \frac{g}{n\sigma}\, \frac{\cosh |k_n|\,(h + z)}{\cosh |k_n|h} e^{+i(n\sigma t - (k_n\cos\theta)x - (k_n\sin\theta)y)}\, d\theta$$

$$\tag{7.54}$$

The energy density spectrum $\Phi_{\eta\eta}(n)$ is obtained analytically by first determining the covariance function $C_{\eta\eta}(\tau)$.

$$C_{\eta\eta}(\tau) = \frac{1}{T} \int_t^{t+T} \eta(t) \sum_{n=-N/2}^{N/2} \int_0^{2\pi} F(n, \theta)e^{in\sigma(t+\tau)}\, d\theta\, dt \tag{7.55}$$

$$C_{\eta\eta}(\tau) = \sum_{n=-N/2}^{N/2} \int_0^{2\pi} F^*(n, \theta')\, d\theta' \int_0^{2\pi} F(n, \theta)\, d\theta e^{in\sigma\tau}$$

The integrands are periodic functions, and it can be shown (by expanding $F(n, \theta)$ in a complex Fourier series) that $C_{\eta\eta}(\tau)$ can be written as

$$C_{\eta\eta}(\tau) = \sum_{n=-N/2}^{N/2} \int_0^{2\pi} |F(n, \theta)|^2 e^{in\sigma\tau}\, d\theta \tag{7.56}$$

The energy density spectrum of the surface displacement $\Phi_{\eta\eta}(n)$ is the Fourier transform of $C_{\eta\eta}(\tau)$, or

$$\Phi_{\eta\eta}(n) = \frac{1}{T} \int_t^{t+\tau} C_{\eta\eta}(\tau)e^{-in\sigma\tau}\, d\tau$$

$$\tag{7.57}$$

$$= \int_0^{2\pi} |F(n, \theta)|^2\, d\theta \qquad \text{for } -N/2 \leqslant n \leqslant N/2$$

This quantity $\Phi_{\eta\eta}(n)$ is the energy at each frequency σ_n, and it is seen to be the integral over the directions θ. The directional energy density spectrum is $|F(n, \theta)|^2$, which gives the distribution of energy with direction as well as frequency. Alternatively, if we examine the energy density spectra of the horizontal velocities, we obtain

$$\Phi_{uu}(n) = K^2 \int_0^{2\pi} \cos^2 \theta |F(n, \theta)|^2 \, d\theta \tag{7.58}$$

$$\Phi_{vv}(n) = K^2 \int_0^{2\pi} \sin^2 \theta |F(n, \theta)|^2 \, d\theta \qquad \text{for } -N \leqslant n \leqslant N \tag{7.59}$$

where $K = gkK_p(z)/n\sigma$.

Finally, the cross-spectra

$$\Phi_{u\eta}(n) = K \int_0^{2\pi} \cos \theta |F(n, \theta)|^2 \, d\theta \tag{7.60a}$$

$$\Phi_{v\eta}(n) = K \int_0^{2\pi} \sin \theta |F(n, \theta)|^2 \, d\theta \tag{7.60b}$$

$$\Phi_{uv}(n) = K^2 \int_0^{2\pi} \sin \theta \cos \theta \, |F(n, \theta)|^2 \, d\theta \quad \text{for } -N/2 \leqslant n \leqslant N/2 \tag{7.60c}$$

To obtain the directional wave spectrum, a method developed by Longuet-Higgins et al. (1963) may be used. The directional spectrum is expressed as a Fourier series,

$$|F(n, \theta)|^2 = \sum_{m=0}^{\infty} A_m(n) \cos m\theta + \sum_{m=1}^{\infty} B_m(n) \sin m\theta \tag{7.61}$$

Now, A_m and B_m can be evaluated in the foregoing expressions for the energy spectra. Thus

$$\Phi_{\eta\eta}(n) = \pi A_0(n) \tag{7.62a}$$

$$\Phi_{uu}(n) = K^2 \left[A_0(n)\pi + \frac{A_2(n)\pi}{2} \right] \tag{7.62b}$$

$$\Phi_{u\eta}(n) = K[A_1(n)]\pi \tag{7.62c}$$

$$\Phi_{v\eta}(n) = K[B_1(n)]\pi \tag{7.62d}$$

$$\Phi_{uv}(n) = K^2 \frac{B_2(n)\pi}{2} \qquad \text{for } -N \leqslant n \leqslant N \tag{7.62e}$$

From the equations above, the first five harmonics of the directional spectra can be determined in terms of the cross-spectra. The reader should verify that the spectrum $\Phi_{vv}(n)$ would yield an additional but not independent equation in $A_0(n)$ and $A_2(n)$.

Different methods for obtaining the directional spectrum using wave staffs or pressure transducers have been discussed or utilized. Panicker and Borgman (1970) discuss various gage arrays and Borgman (1979) presents a unified approach to arrays using different types of sensors. Seymour and Higgins (1978) have developed the slope array, which uses pressure transducers to provide estimates of the directional spectrum.

Example 7.4: Directional Wave Spectrum from a Linear Array

Pawka (1974) uses a linear array of pressure transducers parallel to shore. Using, instead, wave staffs, a method of determining the directional spectra will be illustrated, differing only in the fact that the pressure response factor is not included for ease of presentation.

Consider three wave gages distributed at $x = 0$, l_1, and l_2 along the x axis with the y axis pointing offshore. For each gage the wave records with time are

$$\eta_0(t) = \sum_{n=-N/2}^{N/2} \int_0^{2\pi} F(n, \theta) \, d\theta e^{-in\sigma t} \tag{7.63a}$$

$$\eta_1(t) = \sum_{n=-N/2}^{N/2} \int_0^{2\pi} F(n, \theta) e^{ik_n \cos \theta l_1} \, d\theta \, e^{-in\sigma t} \tag{7.63b}$$

$$\eta_2(t) = \sum_{n=-N/2}^{N/2} \int_0^{2\pi} F(n, \theta) e^{ik_n \cos \theta l_2} \, d\theta \, e^{-in\sigma t} \tag{7.63c}$$

where k_n is related to $n\sigma$ by the dispersion relationship

$$(n\sigma)^2 = gk_n \tanh k_n h \tag{7.64}$$

and $k_{-n} = -k_n$.

If the cross spectrum between η_0 and η_1 is examined, we find that

$$\Phi_{01}(n) = \int_0^{2\pi} |F(n, \theta)|^2 e^{ik_n \cos \theta l_1} \, d\theta \tag{7.65}$$

for $-N/2 \leq n \leq N/2$. Again expressing the directional spectrum $|F(n, \theta)|^2$ in terms of a Fourier series, as in Eq. (7.61) and substituting into Eq. (7.65), integrals of the following form result:

$$\int_0^{2\pi} \cos m\theta \, e^{ik_n \cos \theta l_1} \, d\theta = \pi i^m J_m(k_n l_1) \tag{7.66}$$

and

$$\int_0^{2\pi} \sin m\theta \, e^{ik_n \cos \theta l_1} \, d\theta = 0 \tag{7.67}$$

where $J_m(k_n l_1)$ is the mth-order Bessel function of the first kind. Therefore,

$$\Phi_{01}(n) = \pi \sum_{m=0}^{M} i^m A_m(n) J_m(k_n l_1) \tag{7.68}$$

The other possible cross-spectra are

$$\Phi_{02}(n) = \pi \sum_{m=0}^{M} i^m A_m(n) J_m(k_n l_2) , \qquad 0 \leqslant n \leqslant \infty \qquad (7.69)$$

$$\Phi_{12}(n) = \pi \sum_{m=0}^{M} i^m A_m(n) J_m(k_n(l_2 - l_1)) \qquad (7.70)$$

The energy spectrum for each gage is

$$\Phi_{00}(n) = \Phi_{11}(n) = \Phi_{22}(n) = A_0(n) 2\pi \qquad (7.71)$$

With three gages we have three cross-spectra and one autospectrum (since the three autospectra are the same) or seven real linear equations for seven real unknown A_m's.

$$\begin{pmatrix} 1 & 0 & 0 & 0 \\ J_0(k_n l_1) & -J_2(k_n l_1) & J_4(k_n l_1) & -J_6(k_n l_1) \\ J_0(k_n l_2) & -J_2(k_n l_2) & J_4(k_n l_2) & -J_6(k_n l_2) \\ J_0(k_n(l_2-l_1)) & -J_2(k_n(l_2-l_1)) & J_4(k_n(l_2-l_1)) & -J_6(k_n(l_2-l_1)) \end{pmatrix} \begin{pmatrix} A_0(n) \\ A_2(n) \\ A_4(n) \\ A_6(n) \end{pmatrix} = \begin{pmatrix} \dfrac{\Phi_{00}}{2\pi} \\ \dfrac{(Co)_{01}}{\pi} \\ \dfrac{(Co)_{02}}{\pi} \\ \dfrac{(Co)_{12}}{\pi} \end{pmatrix}$$

$$(7.72)$$

and

$$\begin{pmatrix} J_1(k_n l_1) & -J_3(k_n l_1) & J_5(k_n l_1) \\ J_1(k_n l_2) & -J_3(k_n l_2) & J_5(k_n l_2) \\ J_1(k_n(l_2-l_1)) & -J_3(k_n(l_2-l_1)) & J_5(k_n(l_2-l_1)) \end{pmatrix} \begin{pmatrix} A_1(n) \\ A_3(n) \\ A_5(n) \end{pmatrix} = \begin{pmatrix} \dfrac{(Quad)_{01}}{\pi} \\ \dfrac{(Quad)_{02}}{\pi} \\ \dfrac{(Quad)_{12}}{\pi} \end{pmatrix}$$

$$(7.73)$$

where $(Co)_{ij}$ and $(Quad)_{ij}$ refer to the real and imaginary parts of the cross-spectrum, Φ_{ij}.

The resulting values of A_0 to A_6 thus define the directional energy spectrum. The fact that the B_m's are not obtained means that the resulting directional spectrum is symmetric about $\theta = 0$. That is, there is an ambiguity in the results in the sense that the sensor array cannot tell if waves are coming from the $+y$ direction or the $-y$ direction and hence the physical reason for the array being parallel to shore, as the assumption can be made that waves do not come from shore (of course, wave reflection or a significant wind generation area behind the array could affect these results).

It is important to notice that if the gages are spaced evenly, that is, $l_2 = 2l_1$, then two of the equations in the matrices are redundant, and only five Fourier coefficients can be obtained instead of seven.

7.5 TIME-SERIES SIMULATION

Simulation refers to the calculation of phenomena of interest to investigate their characteristics or to evaluate the effectiveness of various designs to measure or withstand the phenomena. An example is the simulation of directional waves to investigate the forces caused on a particular structural design. Numerical simulation is feasible through the extremely efficient FFT procedures noted earlier. In principle, simulations for one-dimensional and directional spectra are essentially the same; the procedure will be discussed here for a directional spectrum.

Consider a continuous directional spectrum $|F(\sigma, \theta)|^2$, representing the continuous directional spread of energy over direction θ and frequency σ. For numerical simulation, the water surface displacement η is expressed as

$$\eta(x, y, t) = 2 \sum_{n=0}^{N/2} \sum_{m=1}^{M} \sqrt{|F(\sigma_n, \theta_m|^2 \, \Delta\theta_m \, \Delta\sigma_n} \tag{7.74}$$

$$\cos(n\sigma t - k_{mn_x}x - k_{mn_y}y - \epsilon_{mn})$$

in which the above represents a total of $M \times N/2$ wavelets, with M directions at each of $N/2$ frequencies. The phase angles ϵ_{mn} are considered to be random, in accordance with the concept of the generation of a wavelet over a fetch which is long compared to the wavelength. Since the set of ϵ_{mn} is random, any number of simulations can be carried out based on a single spectrum; each simulation is termed a "realization" of the spectrum and is interpreted as one of an infinite number of possible wave systems that could result from a storm that caused the spectrum of interest. Thus statistics can be developed describing the probability of the maximum wave height or force or probability of exceeding design limits, and so on.

In carrying out the simulation, the FFT is generally used due to its speed. Thus it should be recognized that Eq. (7.74) represents a periodic time series and any attempt to apply a simulation for a greater period than the interval of periodicity ($= 2\pi/\Delta\sigma$) would not yield any additional information and probably would be misleading. To apply the FFT to simulation, it is more useful to express Eq. (7.74) as

$$\eta(x, y, t_j) = \sum_{n=-N/2}^{N/2} (a_n - ib_n)e^{2\pi inj/N} \tag{7.75}$$

in which a_n and b_n depend on x and y and include the contributions from all directions at the nth frequency,

$$a_n = \sum_{m=1}^{M} \sqrt{|F(\sigma_n, \theta_m|^2 \, \Delta\theta_m \, \Delta\sigma}} \cos(k_{nm_x}x + k_{nm_y}y + \epsilon_{mn})$$

$$b_n = \sum_{m=1}^{M} \sqrt{|F(\sigma_n, \theta_m|^2 \, \Delta\theta_m \, \Delta\sigma}} \sin(k_{nm_x}x + k_{nm_y}y + \epsilon_{mn})$$

$$\tag{7.76}$$

As an illustration of a simulation, suppose that a wave gage array has been designed to determine the directional spectrum. For selected input directional spectra, simulations could be carried out and from these the directional spectra calculated. The use of various record sampling lengths, various levels of random noise added to the input, and so on, would assist in evaluating both the methodology developed for extracting the directional spectrum and the effectiveness of the array for different directional spectra. A specific example would be one in which the longshore component of energy flux at a particular point is of primary interest. Simulations would assist in the evaluation of the ranking of different array designs for extracting the parameter of interest for a range of directional spectra considered likely to occur.

7.6 EXAMPLE OF USE OF SPECTRAL METHODS TO DETERMINE MOMENTUM FLUX

In Chapter 10 it will be shown that the onshore flux of the longshore component of momentum S_{xy} is given by

$$S_{xy} = \frac{E}{2} \frac{C_G}{C} \sin 2\theta \tag{7.77}$$

in which θ is measured counterclockwise with respect to the x axis and the x axis is directed shoreward, and E is the usual total energy per unit surface area. Equation (7.77) represents the contribution for a particular frequency and wave direction. If measurements of waves are made such that the directional spectrum is obtained, the contribution is given, in terms of the directional spectrum, as

$$S_{xy}(n, \theta_m) = \frac{\gamma |F(n, \theta_m)|^2}{4} \left(1 + \frac{2k_n h}{\sinh 2k_n h} \right) \sin 2\theta_m \, \Delta\theta_m \tag{7.78}$$

where $\gamma = \rho g$, the specific weight of water. The contribution to the momentum flux component on a frequency-by-frequency basis yields

$$S_{xy}(\sigma_n) = \frac{\gamma}{4} \left(1 + \frac{2k_n h}{\sinh 2k_n h} \right) \sum_{m=1}^{M} |F(n, \theta_m)|^2 \sin 2\theta_m \, \Delta\theta_m \tag{7.79}$$

and the total longshore component of the onshore component of momentum flux is

$$S_{xy} = \sum_{n=-N/2}^{N/2} S_{xy}(\sigma_n) \tag{7.80}$$

7.6.1 Measurement of S_{xy} in Shallow Water

If shallow water wave conditions prevail, an interesting and simple application of spectral theory affords a direct determination of the momentum flux component S_{xy}.

The integral counterparts to Eq. (7.74) expressed for the u and v components of water particle velocity are

$$u(z, t) = \sum_{n=-N/2}^{n=N/2} \int_0^{2\pi} F(n, \theta)\sigma \cos \theta \frac{\cosh k(h + z)}{\sinh kh} e^{in\sigma t}\, d\theta \qquad (7.81)$$

$$v(z, t) = \sum_{n=-N/2}^{n=N/2} \int_0^{2\pi} F(n, \theta)\sigma \sin \theta \frac{\cosh k(h + z)}{\sinh kh} e^{in\sigma t}\, d\theta \qquad (7.82)$$

Consider the time average of the product of u and v:

$$\overline{uv} = \sum_{n=-N/2}^{n=N/2} \int_0^{2\pi} |F(n, \theta)|^2 \frac{\sigma^2}{2} \sin 2\theta \frac{\cosh^2 k(h + z)}{\sinh^2 kh}\, d\theta \qquad (7.83)$$

which upon using the dispersion equation (3.44) and shallow water approximations becomes

$$\overline{uv} = \sum_{n=-N/2}^{n=N/2} \int_0^{2\pi} \frac{|F(n, \theta)|^2}{2h} \sin 2\theta\, d\theta \qquad (7.84)$$

which can be shown to be proportional to S_{xy}, that is,

$$S_{xy} = \rho h \overline{uv} \qquad (7.85)$$

Thus the time-averaged product of the output from a biaxial current meter could be used to determine an estimate of the total value of S_{xy}. A running average of this product would provide a useful measure of the longshore forces exerted on the surf zone by the incident waves. The result displayed in Eq. (7.85) should not be surprising since the definition of S_{xy} is

$$S_{xy} = \int_{-h}^0 \rho u v\, dz \qquad (7.86)$$

and for shallow water conditions, u and v are uniform over depth.

REFERENCES

ABRAMOWITZ, M., and I. A. STEGUN. *Handbook of Mathematical Functions*, Dover, New York, 1965.

BORGMAN, L. E., "Directional Spectral Models for Design Use for Surface Waves," *Preprints, Offshore Technol. Conf.*, 1979.

COOLEY, R. J. W., and J. W. TUKEY, "An Algorithm for the Machine Calculation of Complex Fourier Series," *Math. Comput.*, Vol. 19, 1965.

FORRISTALL, G. Z., "On the Statistical Distribution of Wave Heights in a Storm," *J. Geophys. Res.*, Vol. 83, No. C5, 1978.

JENKINS, G. M., and D. G. WATTS, *Spectral Analysis and Its Applications*, Holden-Day, San Francisco, 1968.

LONGUET-HIGGINS, M. S., "On the Statistical Distribution of the Heights of Sea Waves," *J. Mar. Res.*, Vol. 11, pp. 245–266, 1952.

LONGUET-HIGGINS, M. S., D. E. CARTWRIGHT, and N. D. SMITH, "Observations of the Directional Spectrum of Sea Waves Using the Motions of a Floating Buoy," in *Ocean Wave Spectra*, Proceedings of a Conference Held at Easton, Prentice-Hall, Englewood Cliffs, N. J., 1963, pp. 111–131.

PANICKER, N. N., and L. E. BORGMAN, "Directional Spectra from Wave Gage Arrays," *Proc. 12th Conf. Coastal Eng., ASCE*, 1970, pp. 117–136.

PAWKA, S., "Study of Wave Climate in Nearshore Waters," *Proc. Int. Symp. Ocean Wave Meas. Anal., ASCE*, 1974, Vol. 1, pp. 745–760.

PIERSON, W. J., "The Representation of Ocean Surface Waves by a Three-Dimensional Stationary Gaussian Process," New York University, New York, 1954.

SEYMOUR, R. J., and A. L. HIGGINS, "Continuous Estimation of Longshore Sand Transport," *Proc. Coastal Zone '78, ASCE*, Vol. 3, 1978.

PROBLEMS

7.1 In a wave train consisting of 600 waves with a rms wave height H_{rms} of 4 m, what is the probability that the height of *a particular wave* will exceed 6 m? What is the probability that the height of at least one of the 600 waves will exceed 6 m?

7.2 Recognizing that the total area under a spectrum is $\overline{\eta^2}$, that for a single sinusoid $\overline{\eta^2} = H^2/8$, and that for a Rayleigh distribution $H_{1/3} = 1.416H_{rms}$, develop a realtionship between $H_{1/3}$ and the square root of the area under the spectrum η_{rms}.

7.3 For the time functions below: (a) determine the Fourier coefficients a_n and b_n; (b) the phase angles ϵ_n; (c) the complex Fourier coefficients; (d) the two-sided energy spectra; (e) the cross-spectrum.

$$f_1(t) = 1 + 2 \cos \sigma t + 2 \sin \sigma t - 3 \cos 3\sigma t$$

$$f_2(t) = 2 + 3 \sin \left(\sigma t - \frac{\pi}{4} \right) + 4 \cos 4\sigma t$$

7.4 The cross-correlation function $C_{12}(\tau)$ associated with a pair of time functions $f_1(t)$ and $f_2(t)$ is given by

$$C_{12}(\tau) = 3 \cos^2 \sigma t \sin \sigma t$$

If $f_1(t)$ is given as

$$f_1(t) = \tfrac{1}{2} + \tfrac{1}{4} \cos \sigma t + \tfrac{1}{2} \sin 2\sigma t - \tfrac{3}{2} \sin 3\sigma t + 4 \cos 4\sigma t$$

find $f_2(t)$.

7.5 Demonstrate that an arbitrary shift of the time origin by an amount t' changes the individual values of a_n and b_n but does not change $\sqrt{a_n^2 + b_n^2}$.

7.6 Using two wave gages located a distance l apart, show that the wave direction for a sea that has a unique direction for each frequency is

$$\theta(\sigma) = \tan^{-1} \frac{\text{Quad}(\sigma)}{\text{Co}(\sigma)}$$

7.7 For a directional wave system as expressed by Eq. (7.51), derive the following cross-spectra:

$$\Phi_{\eta\eta}(n), \quad \Phi_{\eta(\partial\eta/\partial x)}(n), \quad \Phi_{\eta(\partial\eta/\partial y)}(n), \quad \Phi_{(\partial\eta/\partial x)(\partial\eta/\partial x)}(n), \quad \Phi_{(\partial\eta/\partial y)(\partial\eta/\partial y)}(n)$$

Develop the counterparts to Eq. (7.62) for the coefficients of the directional spectrum.

7.8 Develop the first five harmonics of a directional spectrum based on records of the water surface and the surface slopes, that is,

$$\eta, \quad \frac{\partial\eta}{\partial x}, \quad \frac{\partial\eta}{\partial y}$$

7.9 Compare the values of $H_{1/10}$, $H_{1/3}$, and H_1 obtained by the Rayleigh distribution and by the two-component model. Discuss and develop a reasonable qualitative explanation for the differences. Also compare $H_{\max}$ obtained from the two approaches.

8

Wave Forces

Dedication
WILLIAM FROUDE

William Froude (1810–1879) is well known for the dimensionless parameter that bears his name. This parameter, utilized in model testing involving a liquid free surface, such as would occur in testing of ships, harbor response or wave forces on structures, is a ratio of the inertial forces extant to the gravitational forces.

Froude was born in Dartington, England, and received his bachelor's degree in mathematics from Oriel College, Oxford, in 1832 and his master's degree in 1837. After graduation he worked for Isambard K. Brunel, the well-known civil engineer and naval architect. Brunel asked him in 1856 to study the waves generated by ships. In 1859 he moved to Torquay, an Admiralty establishment, to continue his work in naval architecture. During this time he studied trochoidal waves and developed techniques to reduce ship roll. In 1870 he began a series of experiments to study the resistance of ships using a covered towing tank 76 m long, 10 m wide, and 3 m deep. He used dynamometers to measure the forces of various models of ship hulls and scaled these up to prototype scale.

8.1 INTRODUCTION

An important application of water wave mechanics is the determination of the forces induced by waves on fixed and compliant structures and the motions of floating objects. All objects, whether floating on the sea or attached to the bottom, are subjected to wave forces, and therefore these forces are of central interest to the designer of these structures.

The investigation of wave forces has been under way for a considerable time and numerous studies have been carried out for the case of wave forces on a vertical pile, yet no wave force calculation procedure has been developed to date for this most simple case for which there is uniform agreement. Although for long-crested waves, with a single fundamental period, theories are available which accurately represent the water particle motions in the absence of a pile for a wide range of wave characteristics, at present there is no reliable procedure for calculating the wave interaction with a structure for all conditions of interest. Watching a wave impinge on a vertical pile, the complexity of the problem becomes immediately obvious. As the wave crest approaches the pile, a bow wave forms and run-up occurs on the front of the pile, while a wake develops at the rear. We know from fluid mechanics that the wake signifies separated flow, which is impossible to treat analytically. Moreover at Reynolds numbers of interest, the flow is generally turbulent. As the wave crest passes and the trough reaches the pile, the flow field reverses and the previously formed wake may wash back past the pile as a new wake is formed. All of these phenomena clearly violate our previous assumptions of irrotational flow with small-amplitude waves and small velocities.

Later discussions will describe the wave forces as comprised of an inertia and drag force component. In the case of structures that are large relative to the wave length, the wake effects are not important; the inertia force dominates, and accurate calculation methods exist. For objects that are small, the wake plays a dominant role on both the drag and inertia force components, and the roughness characteristics of the object are also of significance. In the latter case, no reliable analytical approaches are available and experimental results provide the major design basis.

8.2 POTENTIAL FLOW APPROACH

The treatment of ideal flow about a circular cylinder will provide a framework for wave force discussions to follow. If, for convenience, we consider a section of vertical piling far from the free surface, then to obtain a first approximation for the wave force we integrate the pressure distribution around the piling using potential flow. For a circular piling, it is convenient to use polar coordinates (r, θ, z) in the horizontal plane. In this system, the Laplace equation in three dimensions is

$$\nabla^2 \phi = \frac{\partial^2 \phi}{\partial r^2} + \frac{1}{r} \cdot \frac{\partial \phi}{\partial r} + \frac{\partial^2 \phi}{r^2 \, \partial \theta^2} + \frac{\partial^2 \phi}{\partial z^2} = 0 \tag{8.1}$$

and the velocity components are

$$u_r = -\frac{\partial \phi}{\partial r}, \qquad u_\theta = -\frac{1}{r} \frac{\partial \phi}{\partial \theta}, \qquad u_z = -\frac{\partial \phi}{\partial z} \tag{8.2}$$

A solution to this equation, which is uniform in the vertical direction,

$$\phi(r, \theta) = U(t)r\left(1 + \frac{a^2}{r^2}\right)\cos\theta \tag{8.3}$$

At $r = a$, the radius of the pile, there is a no-flow condition in the r direction as expected.

$$u_r(a, \theta) = -\frac{\partial\phi}{\partial r}\bigg|_{r=a} = 0 \tag{8.4}$$

$U(t)$, the far-field velocity, is considered to vary sinusoidally with the wave period T. In plan view the flow around the cylinder is as shown in Figure 8.1. (Note the absence of a wake in potential flow.)

To calculate the pressure distribution around the cylinder, the unsteady form of the Bernoulli equation is applied at the cylinder wall and far upstream at a point where $r = l$, $\theta = 0$, and $l \gg a$:

$$\left[\frac{p(r, \theta)}{\rho} + gz + \frac{u_r^2 + u_\theta^2}{2} - \frac{\partial\phi}{\partial t}\right]_{r=a} = \left[\frac{p(r, \theta)}{\rho} + gz + \frac{u_r^2 + u_\theta^2}{2} - \frac{\partial\phi}{\partial t}\right]_{\substack{r=l \\ \theta=0}} \tag{8.5}$$

The elevation terms cancel, leaving the pressure difference between the free stream pressure in the fluid and that at that cylinder as

$$p(a, \theta) - p(l, 0) = \rho\left[\left(\frac{u_r^2 + u_\theta^2}{2}\right)_{\substack{r=l \\ \theta=0}} - \left(\frac{u_r^2 + u_\theta^2}{2}\right)_{r=a} + \left(\frac{\partial\phi}{\partial t}\right)_{r=a} - \left(\frac{\partial\phi}{\partial t}\right)_{\substack{r=l \\ \theta=0}}\right] \tag{8.6}$$

Substituting from the velocity potential yields

$$p(a, \theta) - p(l, 0) = \rho\left[\frac{U^2(t)}{2}(1 - 4\sin^2\theta) + 2a\frac{dU}{dt}\cos\theta - l\frac{dU}{dt}\right] \tag{8.7}$$

where terms of $O(a^2/l^2)$ have been dropped as extremely small. The pressure term is thus due to two different contributions, the steady flow term, proportional to $U^2(t)$, and an acceleration or inertial term, due to $dU(t)/dt$. Let us examine them term by term.

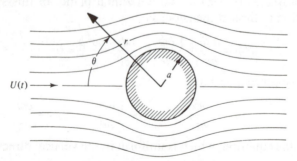

Figure 8-1   Potential flow around a circular cylinder.

8.2.1 Steady Flow Term

The steady pressure contribution as a function of angular position around the pile is

$$p(a, \theta) - p(l, 0) = \frac{\rho U^2(t)}{2}(1 - 4 \sin^2 \theta) \qquad (8.8)$$

This pressure distribution is shown in Figure 8.2. The pressure is symmetrical about the pile and in the absence of a wake, the pressure at the rear of the pile is the same as that at the front. Intuitively, the net pressure force in the downstream direction should be zero. Integrating the pressure around the pile, noting that we use the component of the force in the downstream direction as illustrated in Figure 8.3, yields the steady (drag) force per unit elevation dF_D, where

$$dF_D = \int_0^{2\pi} p(a, \theta)a \cos \theta \, d\theta \qquad (8.9)$$

$$= \int_0^{2\pi} \left[\frac{\rho U^2(t)}{2}(1 - 4 \sin^2 \theta) + p(l, 0)\right]a \cos \theta \, d\theta$$

or

$$dF_D = 0 \qquad (8.10)$$

Therefore, as expected from the pressure symmetry, there is no force on the pile in ideal steady flow. However, this is contrary to the actual results determined from real flows; an experience familiar to all is the force occurring on one's arm when extended out the window of a moving car. This discrepancy has been called D'Alembert's paradox and it puzzled the early hydrodynamicists. The reason for the paradox, as alluded to before, is the unrealistic assumption of potential flow, which precludes the formation of boundary layers and a wake.

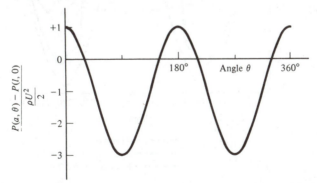

Figure 8.2 Pressure distribution around cylinder for case of ideal flow. Note the low pressure at the sides, $\theta = 90°$, and the symmetry with respect to $\theta = 0°$ and $\theta = 90°$.

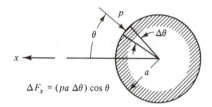

$$\Delta F_x = (pa\,\Delta\theta)\cos\theta$$

Figure 8.3 Calculation of elemental force in x direction. ΔF_x is positive in the downstream $(-x)$ direction.

The real pressure distribution around a cylinder in steady flow is a function of the Reynolds number $\mathbb{R}$, defined as $\mathbb{R} = UD/\nu$, where U is the velocity normal to the cylinder axis, D is the pile diameter, and $\nu = \mu/\rho$, the kinematic viscosity of the fluid, which is the ratio of the dynamic viscosity μ to the fluid density ρ. In Figure 8.4, Goldstein (1938) shows the measured pressure distribution around cylinders for two Reynolds numbers compared to the theoretical ideal flow result. For the upstream portion of the cylinder, with $\theta \leqslant \theta_s$, the separation angle, the pressure may be described approximately by potential flow; however, for $\theta > \theta_s$, which is a function of the Reynolds number, the pressure appears nearly constant. We can therefore approximate the force on a cylinder by using the potential flow solution for $0 \leqslant \theta \leqslant \theta_s$ and using a constant pressure in the wake, as follows:

$$dF_D = 2\int_0^{\theta_s} \frac{\rho U^2(t)}{2}(1 - 4\sin^2\theta)a\cos\theta\,d\theta + 2\int_{\theta_s}^{\pi} p_{\text{wake}}a\cos\theta\,d\theta \qquad (8.11)$$

$$= \rho U^2(t)a\left[\int_0^{\theta_s}(1 - 4\sin^2\theta)\cos\theta\,d\theta + \int_{\theta_s}^{\pi}\frac{p_{\text{wake}}}{\rho U^2(t)/2}\cos\theta\,d\theta\right]$$

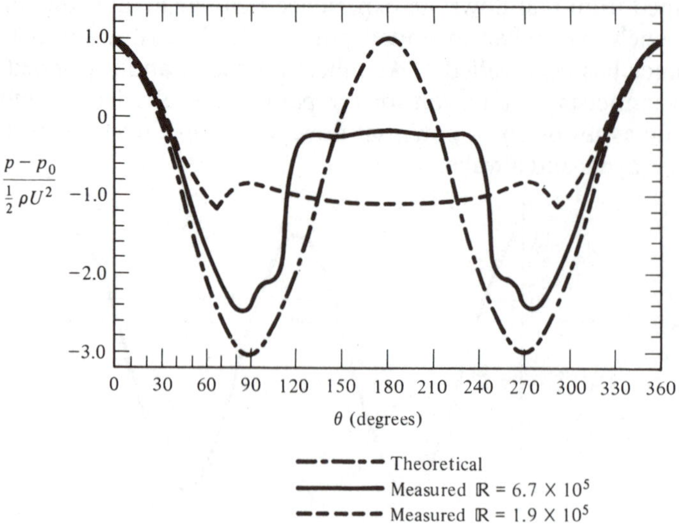

$\dfrac{p - p_0}{\frac{1}{2}\rho U^2}$

θ (degrees)

-·--·-- Theoretical
———— Measured $\mathbb{R} = 6.7 \times 10^5$
----- Measured $\mathbb{R} = 1.9 \times 10^5$

Figure 8.4 Measured pressure distributions around cylinders. (From Goldstein, 1938.)

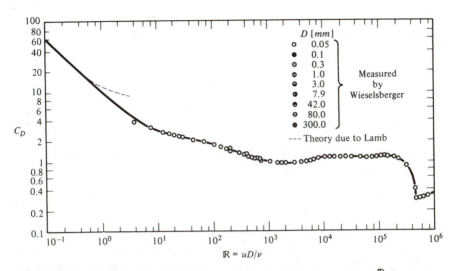

Figure 8.5 Variation of drag coefficient, C_D with Reynolds number $\mathbb{R}$ for a smooth circular cylinder. (From H. Schlichting, *Boundary Layer Theory*. Copyright © 1968 by McGraw-Hill Book Company. Used with the permission of McGraw-Hill Book Company.)

The term within the brackets is a function of Reynolds number $\mathbb{R}$, as both θ_s and p_{wake} vary with Reynolds number. Therefore, the force per unit length, dF, can be related to a function, C_D, which varies with $\mathbb{R}$, allowing us to write the force on the pile per foot of elevation as

$$dF_D = C_D(\mathbb{R})\rho D \, \frac{U^2(t)}{2} = C_D\rho \, \frac{AU^2(t)}{2} \qquad (8.12)$$

where D = piling diameter = $2a$ and for the case of a circular cylinder is equal to A = projected area/unit elevation of the cylinder (i.e., $A = 2a$). The last form of Eq. (8.12) applies to two- and three-dimensional objects, with the stated definition of A. The function C_D is called the "drag coefficient" and its variation with Reynolds number is empirically known for steady flows as shown in Figure 8.5 for a smooth cylinder of circular cross section. In practice, C_D is generally on the order of unity and depends on piling roughness in addition to Reynolds number.

8.2.2 Unsteady Flow

Examining the remaining term in the potential flow expression for the pressure [Eq. (8.7)], we have, integrating the component of force in the downstream direction,

$$dF_I = \int_0^{2\pi} \rho \, \frac{dU(t)}{dt} \, 2a^2 \cos^2 \theta \, d\theta - \int_0^{2\pi} \rho \, \frac{dU(t)}{dt} \, la \cos \theta \, d\theta \qquad (8.13)$$

The second term on the right-hand side integrates to zero, thereby contributing no net force. The first term, however, yields

$$dF_I = \rho a^2 \frac{dU}{dt} 2\pi$$

$$= 2\rho\pi a^2 \frac{dU}{dt}$$

(8.14)

The term πa^2 is the volume V of the pile per unit length, so that the final expression can be written as

$$dF_I = C_M \rho V \frac{dU}{dt}$$

(8.15)

where C_M is defined as the inertia coefficient, which in this case (of potential flow about a circular cylinder) is equal to 2.0. Thus there is a force called the inertial force caused by the fluid accelerating past the cylinder, even in the absence of friction. The general form [Eq. (8.15)] for the inertia force component is valid for two- and three-dimensional objects of arbitrary shapes, except that the inertia coefficient can vary with the flow direction.

The inertia coefficient, in practice, can be discussed meaningfully as the sum of two terms,

$$C_M = 1 + k_m$$

(8.16)

where the second term, k_m, is called the added mass which depends on the shape of the object. The interpretation of the inertia coefficient is that the pressure gradient required to accelerate the fluid exerts a so-called "buoyancy" force on the object, corresponding to the unity term in Eq. (8.16). An additional local pressure gradient occurs to accelerate the neighboring fluid around the cylinder. The force necessary for the acceleration of the fluid around the cylinder yields the added mass term, k_m.

Let us first consider the force on an object due to the unaffected pressure gradient in an accelerating fluid. If the pressure gradient is uniform across the dimension of the object, the knowledge available for vertical buoyancy forces in a hydrostatic fluid can be applied. In the latter case, the hydrostatic buoyancy force F_B on an object of volume V in a fluid of specific weight γ is

$$F_B = \gamma V$$

(8.17)

and for a hydrostatic fluid, the pressure gradient $\partial p/\partial z$ and specific weight γ are related by

$$\gamma = -\frac{\partial p}{\partial z}$$

(8.18)

Therefore,

$$F_B = -\frac{\partial p}{\partial z}V \tag{8.19}$$

Returning to the effect of a horizontal pressure gradient associated with an accelerating fluid, the "buoyancy-like" force component is

$$F_B = -\frac{\partial p}{\partial x}V \tag{8.20}$$

and from the Euler equations, $-\partial p/\partial x$ may be replaced by $\rho\,(du/dt)$, yielding

$$F_B = \rho V \frac{du}{dt} \tag{8.21}$$

and by comparing Eqs. (8.20), (8.15), and (8.16), the origin of the unity term in C_M is clear (i.e., it is due directly to the pressure gradient). The added mass, which is shape dependent, is caused by the disturbance of the flow field. It appears that in all cases, C_M should be greater than unity.

For two-dimensional flow about a cylinder of elliptical cross section, the added mass coefficient k_m can be shown (Lamb, 1945) to be

$$k_m = \frac{b}{a} \tag{8.22}$$

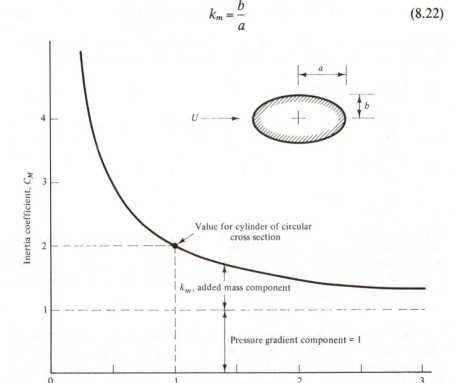

Figure 8.6 Inertia coefficient for a cylinder of ellipsoid cross section.

where a and b are the semiaxes aligned with and transverse to the line of acceleration, respectively. Equation (8.22) is plotted in Figure 8.6, which demonstrates the occurrence of a small k_m for a streamlined body.

Example 8.1

It is instructive to consider the case where a circular cylinder is accelerating through a quiescent ideal fluid. Is the force exerted on the cylinder by the fluid the same as when the fluid accelerates past the cylinder? We expect that since there is no pressure gradient in the fluid, the force would only be due to the added mass coefficient. Therefore, the force should not be the same. To determine this, we write the two-dimensional velocity potential for a moving cylinder as

$$\phi(r, \theta, t) = U(t)\frac{a^2}{r} \cos \theta \tag{8.23}$$

where now, $U(t)$ represents the velocity of the cylinder. It is clear that this equation satisfies the following kinematic boundary condition on the cylinder

$$u_r\big|_{r=a} = U(t) \cos \theta \tag{8.24}$$

The pressure at the wall of the cylinder due to the fluid acceleration is given as

$$p(a, \theta) = p(l, 0) + \rho\left[\left(\frac{\partial \phi}{\partial t}\right)\bigg|_{r=a} - \left(\frac{\partial \phi}{\partial t}\right)\bigg|_{\substack{r=l \\ \theta=0^\circ}}\right] \tag{8.25}$$

where l is as defined before for the case of a stationary cylinder. Integrating the downstream component of the pressure force around the cylinder, we have

$$
\begin{aligned}
dF_I &= \int_0^{2\pi} [p(a, \theta)]a \cos \theta \, d\theta \\
&= \int_0^{2\pi} \rho \frac{dU}{dt} a^2 \cos^2 \theta \, d\theta + \int_0^{2\pi} \rho \frac{dU}{dt} \frac{a^3}{l} \cos \theta \, d\theta + \int_0^{2\pi} p(l, 0)a \cos \theta \, d\theta \\
&= (1)\rho \frac{dU}{dt} a^2 \pi \\
&= k_m \rho \frac{dU}{dt} V
\end{aligned}
\tag{8.26}
$$

In addition to this force by the fluid on the accelerating cylinder, a force is necessary, of course, to accelerate the mass of the cylinder itself. Therefore, the total force required to accelerate a cylinder through water could be greater or less than if the water accelerated past the cylinder depending on whether the mass of the cylinder is greater or less than that of the displaced water.

In interpreting the physics and terminology associated with the added mass concept it is helpful to consider the energetics of the case of a circular cylinder accelerating through a fluid. As noted previously, the force per unit length exerted by an accelerating circular cylinder on the surrounding fluid is, from Eq. (8.26),

$$F_I = \rho k_m \pi a^2 \frac{dU}{\partial t} \tag{8.27}$$

where $k_m = 1$ for a circular cylinder.

Let us now calculate the kinetic energy of the accelerated fluid as a function of time. The radial and angular components of velocity are, from Eq. (8.23),

$$u_r = -\frac{\partial \phi}{\partial r} = U(t)\frac{a^2}{r^2}\cos\theta \tag{8.28}$$

$$u_\theta = -\frac{1}{r}\frac{\partial \phi}{\partial \theta} = U(t)\frac{a^2}{r^2}\sin\theta \tag{8.29}$$

The total fluid kinetic energy KE at any time is

$$\begin{aligned}
\text{KE} &= \int_0^{2\pi}\int_a^\infty \frac{\rho}{2}(u_r^2 + u_\theta^2)r\,dr\,d\theta \\
&= \int_0^{2\pi}\int_a^\infty \frac{\rho}{2}U^2\frac{a^4}{r^3}\,dr\,d\theta = \frac{\rho}{2}U^2\pi a^2
\end{aligned} \tag{8.30}$$

The time rate of change of kinetic energy should equal the product of the force and the velocity, $(F_I \cdot U)$, that is, the rate at which work is being done by the cylinder, which is verified as follows:

$$\frac{\partial(\text{KE})}{\partial t} = \rho\pi a^2 \frac{\partial U}{\partial t}U \tag{8.31}$$

and by comparison with Eq. (8.26), we see that this is exactly equal to $F \cdot U$. Thus the added mass coefficient represents the ratio of the additional mass of fluid that is accelerated with the cylinder to the mass of the fluid displaced by the cylinder.

8.3 FORCES DUE TO REAL FLUIDS

8.3.1 The Morison Equation

Previously, we have treated the inertia and steady-state drag force components independently. However, in a wave field both forces occur and vary continuously with time. Morison et al. (1950) proposed the following formula for the total wave force, which is just the sum of the two forces, drag and inertia.

$$\begin{aligned}
dF &= dF_D + dF_I \\
&= \tfrac{1}{2}C_D\rho Au|u| + C_M\rho V\frac{Du}{Dt}
\end{aligned} \tag{8.32}$$

Equation (8.32) is frequently referred to as the "Morison equation."

It is noted that in Eq. (8.32), an absolute value sign on one of the velocity terms ensures that the drag force is in the direction of the velocity, which changes direction as the wave passes.

8.3.2 Total Force Calculation

To determine the total force on a vertical pile, the force per unit elevation must be integrated over the immersed length of the pile.

$$F = \int_{-h}^{\eta} dF$$

$$= \int_{-h}^{\eta} \tfrac{1}{2} C_D \rho \, D u |u| \, dz + \int_{-h}^{\eta} \rho C_M \frac{\pi D^2}{4} \frac{Du}{Dt} dz \qquad (8.33)$$

In general, C_D and possibly C_M vary over the length of the pile as the Reynolds number surely does. Therefore, we cannot integrate these equations directly. If, however, we take constant values of C_D and C_M and use linear wave theory[1] and consider only the local acceleration term, the integration can be carried out up to the *mean free surface* to give an approximation to the total force.

$$F = \frac{\rho C_D D}{2} \int_{-h}^{0} \left(\frac{H}{2}\right)^2 \sigma^2 \frac{\cosh^2 k(h+z)}{\sinh^2 kh} \cos(kx_1 - \sigma t) \, |\cos(kx_1 - \sigma t)| \, dz$$

$$\qquad (8.34)$$

$$+ \frac{\rho C_M \pi D^2}{4} \int_{-h}^{0} \frac{H}{2} \sigma^2 \frac{\cosh k(h+z)}{\sinh kh} \sin(kx_1 - \sigma t) \, dz$$

$$F = \frac{\rho C_D D H^2 g}{4 \sinh 2kh} \left(\frac{2kh + \sinh 2kh}{4}\right) \cos(kx_1 - \sigma t) \, |\cos(kx_1 - \sigma t)| \qquad (8.35)$$

$$+ C_M \frac{\rho \pi D^2}{4k} \frac{H}{2} \sigma^2 \sin(kx_1 - \sigma t)$$

or

$$F = C_D \, DnE \cos(kx_1 - \sigma t) \, |\cos(kx_1 - \sigma t)| \qquad (8.36)$$

$$+ C_M \pi DE \frac{D}{H} \tanh kh \sin(kx_1 - \sigma t)$$

where x_1 is the location of the pile (conveniently, this can usually be taken as $x_1 = 0$), $E \, (= \tfrac{1}{8} \rho g H^2)$ is the wave energy per unit surface area, and n is the ratio of group velocity C_G to wave celerity C, as given by Eq. (4.82b). The ratio

---

[1] In actual design, a nonlinear theory (see Chapter 11) should be used for horizontal velocity and acceleration.

D/H can be interpreted in terms of the relative importance of the inertia to drag force components. The total moment about the seafloor can be obtained similarly by integrating

$$M = \int_{-h}^{\eta} dM = \int_{-h}^{\eta}(h+z)\,dF$$

$$= \int_{-h}^{\eta}(h+z)\tfrac{1}{2}C_D\rho Du|u|\,dz + \int_{-h}^{\eta}(h+z)\rho C_M \frac{\pi D^2}{4}\frac{Du}{Dt}\,dz \tag{8.37}$$

which yields[2]

$$M = C_D\,DnE\,\cos(kx_1 - \sigma t)\,|\cos(kx_1 - \sigma t)|\left\{h\left[1 - \frac{1}{2n}\left(\frac{\cosh 2kh - 1 + 2(kh)^2}{2kh\,\sinh 2kh}\right)\right]\right\}$$

$$+ C_M\pi DE\frac{D}{H}\tanh kh\,\sin(kx_1 - \sigma t)\left\{h\left[1 - \frac{\cosh kh - 1}{kh\,\sinh kh}\right]\right\} \tag{8.38}$$

in which each of the terms above is recognized as the total force component times the respective lever arm (the lever arms are in the braces, $\{\cdot\}$). The reader should demonstrate that, as expected from physical reasoning, the asymptotes for these lever arms are $h/2$ and h for shallow and deep water conditions, respectively.

8.3.3 Methodology for Determining Drag and Inertia Coefficients

In practice, the reliable determination of drag and inertia coefficients presents a very challenging problem, particularly from field data. The required measurements include the time-varying force F_m at a particular elevation on a pile, and the corresponding instantaneous water particle velocities and accelerations. Given this information, C_D and C_M may be determined by a variety of approaches. Only until the recent development of reliable current meters have the water particle kinematics been available to researchers. Previous investigators have had to rely on calculated kinematics based on measurements of the water surface profile. Even if the kinematics are accurately predicted, which is open to some question, particularly if small-amplitude wave theory is used for large waves, then Morison's equation is only one equation with two unknowns, C_D and C_M. Two methods have been used to surmount this problem. The first is to correlate forces with water particle kinematics only at times when the velocities or accelerations are

[2]The integration again is only carried out to $z = 0$ as opposed to $z = \eta$, for the sake of simplicity of the final result.

zero. For a small-amplitude wave of a single period, this corresponds to times of zero or extreme water surface displacements, respectively. At such times either the drag or inertia term is zero and therefore, there is only one unknown in the equation. For example, at the wave crest, the acceleration (inertia force) is zero and C_D would be found as follows:

$$C_D = \frac{F_m}{\frac{1}{2}\rho A u^2} \tag{8.39}$$

and a similar equation would apply for the inertia coefficient at times when the velocity (drag force component) is zero.

Disadvantages of this approach are that considerable data are not utilized: for instance, the data between the crest and the still water crossing. With real storm-driven waves, the times are not obvious at which zeros of velocities or accelerations occur. This can be seen from Figure 8.7, which represents the largest wave measured during Hurricane Carla in almost 100 ft of water in the Gulf of Mexico (Dean, 1965).

A second method, used by Dean and Aagaard (1970), is to minimize the mean squared error ϵ^2 between measured and predicted forces. This procedure, in order to account for Reynolds number dependency, involves classifying the digitized data into groups with approximately the same Reynolds number. For each group, then, ϵ^2 is minimized with respect to the unknowns, C_D and C_M.

$$\epsilon^2 = \frac{1}{I} \sum_{i=1}^{I} (F_{m_i} - F_{p_i})^2 \tag{8.40}$$

where the lowercase subscripts m and p refer to measured and predicted forces and I is the total number of data points for the data group. The

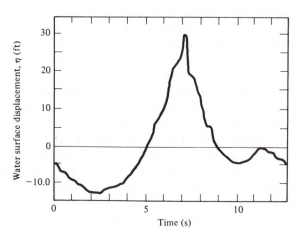

Figure 8.7   Largest measured wave from Hurricane Carla, September 1961. (From Dean, 1965.)

minimization procedure results in two equations in the two unknowns, that is,

$$\frac{\partial \epsilon^2}{\partial C_D} = \frac{2}{I} \sum_{i=1}^{I} (F_{mi} - F_{pi}) \frac{\partial F_{pi}}{\partial C_D} = 0$$

$$\frac{\partial \epsilon^2}{\partial C_M} = \frac{2}{I} \sum_{i=1}^{I} (F_{mi} - F_{pi}) \frac{\partial F_{pi}}{\partial C_M} = 0$$

(8.41)

Multiplying through and simplifying, the equations are

$$\frac{\rho A}{2} C_D \sum_{i=1}^{I} [u|u|]_i^2 + C_M \sum_{i=1}^{I} \left[\rho V \left(\frac{Du}{Dt} \right)_i (u|u|)_i \right] = \sum_{i=1}^{I} F_{mi} (u|u|)_i$$

$$C_D \sum_{i=1}^{I} \left[\frac{\rho A}{2} \left(\frac{Du}{Dt} \right)_i (u|u|)_i \right] + C_M \sum_{i=1}^{I} \rho V \left(\frac{Du}{Dt} \right)^2 = \sum_{i=1}^{I} F_{mi} \left(\frac{Du}{Dt} \right)_i$$

(8.42)

which can be abbreviated as

$$AC_D + BC_M = D$$

$$BC_D + FC_M = G$$

(8.43)

where A, B, D, F, and G are known constants for a given set of data. Eliminating unknowns yields

$$C_D = \frac{GB - DF}{B^2 - AF}$$

and   (8.44)

$$C_M = \frac{DB - GA}{B^2 - AF}$$

Once the coefficients have been obtained, the mean squared error can be found by expanding Eq. (8.40),

$$\epsilon^2 = \sum_{i=1}^{I} F_{m_i}^2 - 2DC_D - 2GC_M + AC_D^2 + 2BC_DC_M + FC_M^2 \qquad (8.45)$$

or

$$\epsilon^2 - \sum_{i=1}^{I} F_{m_i}^2 = AC_D^2 - 2DC_D + 2BC_DC_M - 2GC_M + FC_M^2 \qquad (8.46)$$

It is interesting to note that the last equation is an equation for an ellipse when plotted with C_D and C_M as axes. This is most readily seen for the case of symmetric wave data, in which the constant B would be equal to zero[3] due to the symmetry of the velocity and antisymmetry of the acceleration about the

---

[3]It is interesting to note that most actual data sets approximate this condition of $\overline{\frac{Du}{Dt} u|u|} = 0$.

crest and trough. Rewriting the equation above and completing the square,

$$A\left[C_D^2 - \frac{2C_D D}{A} + \left(\frac{D}{A}\right)^2\right] + F\left[C_M^2 - \frac{2G}{F}C_M + \left(\frac{G}{F}\right)^2\right] \tag{8.47}$$

$$= \epsilon^2 - \sum_{i=1}^{I} F_{m_i} + \frac{D^2}{A} + \frac{G^2}{F}$$

Setting the right-hand side equal to a new constant, J, the equation can be written in a standard ellipse form:

$$\frac{(C_D - D/A)^2}{J/A} + \frac{(C_M - G/F)^2}{J/F} = 1 \tag{8.48}$$

The center of the ellipse is located at $C_D = D/A$ and $C_M = G/F$, which are the values that give the minimum mean squared error for symmetric data [cf. Eq. (8.44) for $B = 0$]. The ratio of the two axes is $\sqrt{F/A}$. The eccentricity of the ellipse is $e = \sqrt{1 - A/F}$ if $A < F$ or $e = \sqrt{1 - F/A}$ if $F < A$. For a perfect circle, $e = 0$; for an extremely flattened ellipse, $e \to 1.0$. The eccentricity of the ellipse is a measure of the conditioning of the data. If the ellipses are as shown in Figure 8.8, the data are well conditioned for the drag coefficient, but poorly conditioned for determination of C_M, as C_M could take on a range of values without changing the error appreciably. Obviously, the best conditioned data for both coefficients occurs when the ellipses become circles, $A = F$. In practice, when the data are grouped by Reynolds numbers, typically the low $\mathbb{R}$ data are poorly conditioned for the drag coefficient, but they yield good results for C_M, while the opposite is true for high Reynolds number data. This is due largely, for example, for the first case, because the drag forces would

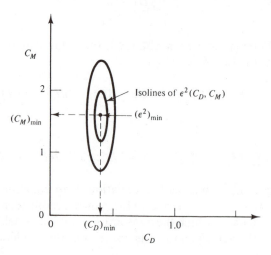

Figure 8.8 Illustration of error surface for data that are well conditioned for determining C_D.

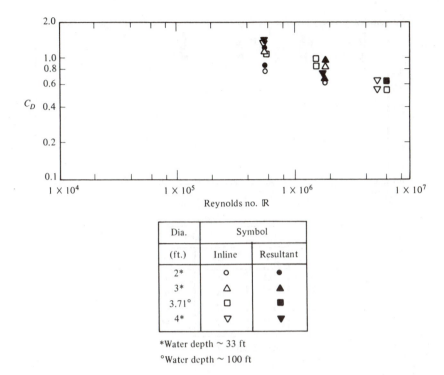

Figure 8.9 Drag coefficient variation with Reynolds number as determined by Dean and Aagaard (1970). Copyright 1970 SPE-AIME.

only be a small portion of the total force. Figures 8.9 and 8.10 show drag and inertia coefficient results as a function of Reynolds number as obtained by Dean and Aagaard (1970). There is a dependency on Reynolds number apparent for the drag coefficient; however, the inertia coefficient appears to be a constant value, 1.33. Note the reduction of k_m to 0.33 from 1.0 for potential flow. Many other data exist for C_D and C_M based on different values of (D/H) and using other parameters. Because of the complexity of the problem no one functional relationship for C_D or C_M presently is known. The values above are recommended for the present for small-diameter vertical piling (say less than 5 ft) when the force is drag dominant, as in most field data for pile-supported platforms.

8.3.4 Wave Forces on Pipelines Resting on the Seafloor

Pipelines are frequently used to convey gas, oil, and other products across the seafloor. A knowledge of the wave forces acting on pipelines resting on the seafloor is essential to a design that will ensure the stability of the

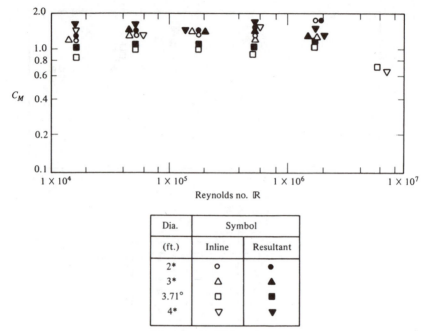

Figure 8.10 Inertia coefficient variation with Reynolds number as determined by Dean and Aagaard (1970). Copyright 1970 SPE-AIME

pipeline. For our purposes here, we will focus on the case of a long-crested wave propagating with its crest parallel to the pipeline (see Figure 8.11).

In earlier sections, we have seen the streamline pattern about a cylinder in an infinite fluid medium. The presence of the plane boundary for the problem being considered here causes interesting streamline patterns and associated forces. Figure 8.12 shows the ideal flow case and it is seen that the streamlines above the cylinder are concentrated, thereby resulting in a maximum lift force coinciding with the time of maximum velocities. If, however, there is a small gap between the cylinder and the seafloor, the concentration of streamlines beneath the cylinder causes a *negative* lift force (i.e., directed downward). This phenomenon has been recognized for many years and was of considerable concern to dirigible pilots landing in a crosswind. As the dirigible would approach the ground a strong downward force would occur, only to change to a positive lift force as the craft "touched down." The problem was solved by winching the dirigible down under conditions of considerable positive buoyancy. If pipelines are not adequately ballasted or anchored, they may experience sufficient lift to be raised off the seafloor, then experience a negative lift due to high velocities between the pipe and bottom, resulting in a possibly damaging oscillation.

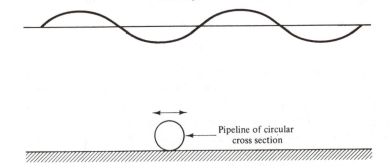

Figure 8.11 Pipeline resting on seafloor subject to oscillating water particle kinematics.

For the case of *ideal* flow of a fluid about a cylinder resting on the bottom, it can be shown that there are inertia forces in the horizontal and vertical directions. In addition, a lift force occurs; but there is no drag force due to the symmetry of the streamline pattern. The inertia forces (per unit length) in the x and z directions and the lift force F_L for the pipeline seated on the seafloor are given by

$$F_x = C_{M_x} \frac{\rho \pi D^2}{4} \frac{Du}{Dt} \tag{8.49}$$

$$F_z = C_{M_z} \frac{\rho \pi D^2}{4} \frac{Dw}{Dt} \tag{8.50}$$

$$F_L = C_L \frac{\rho D}{2} u^2 \tag{8.51}$$

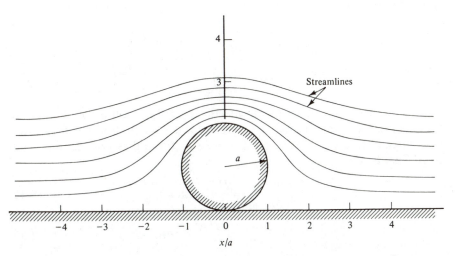

Figure 8.12 Idealized flow field over a cylindrical pipe resting on the seafloor.

in which according to potential flow (Wilson and Reid, 1963)

$$C_{M_x} = C_{M_z} = 3.29 \tag{8.52}$$

$$C_L = 4.493 \tag{8.53}$$

It is noted that the vertical acceleration is very small near the seafloor and under the crest acts in a direction to stabilize the pipeline. Under the trough, both the vertical inertia force and lift forces are directed upward; however, for design waves, the velocities under the trough are generally substantially less than under the crest. Thus the uplift forces under the crest will usually be greater than under the trough.

For the case of real flow fields about a pipeline, both the Reynolds number and relative water particle displacement are of importance. For most design conditions, the flow will be fully separated and if the particle displacement is greater than twice the pipe diameter, drag and inertia coefficients on the order of 1.0 appear reasonable. If the relative displacement (water particle displacement/cylinder diameter) is less than 1.0, experimental data by Wright and Yamamoto (1979) have shown that the potential flow results are applicable. Valuable experimental results are also presented by Sarpkaya (1976).

8.3.5 Relative Importance of Drag and Inertia Force Components

In some situations the drag or inertia force will dominate over the other, thus simplifying the Morison equation. To determine the condition for which this happens, consider the value of the ratio $dF_{I_{max}}/dF_{D_{max}}$. For wave forces on a pile, we know that the maximum velocities occur in the upper portions of the water column. As a reasonable estimate, let us examine the ratio at $z = 0$.

$$\frac{(dF_I)_{max}}{(dF_D)_{max}} = \frac{\left(\rho C_M V \dfrac{\partial u}{\partial t}\right)_{max}}{\left(\dfrac{C_D}{2}\rho A u|u|\right)_{max}} = \frac{\left(C_M \pi D \cdot \dfrac{\partial u}{\partial t}\right)_{max}}{(C_D 2u^2)_{max}} \tag{8.54}$$

The maximum value of the inertia force for small-amplitude waves occurs at the still water crossing, where $\partial u/\partial t$ is a maximum. The maximum drag force occurs at the wave crest. If we substitute these values (for $z = 0$) from Chapter 4:

$$\left(\frac{\partial u}{\partial t}\right)_{max} = \frac{H}{2}\sigma^2 \coth kh \tag{8.55}$$

$$(u^2)_{max} = \left(\frac{H}{2}\right)^2 \sigma^2 \coth^2 kh \tag{8.56}$$

we obtain

$$\frac{(dF_I)_{max}}{(dF_D)_{max}} = \frac{C_M\pi D}{C_D H}\tanh kh \qquad (8.57)$$

In deep water, the ratio equals

$$\frac{(dF_I)_{max}}{(dF_D)_{max}} = \frac{C_M\pi D}{C_D H} \qquad (8.58)$$

In shallow water,

$$\frac{(dF_I)_{max}}{(dF_D)_{max}} = \frac{C_M\pi D(kh)}{C_D H} \qquad (8.59)$$

For the maximum force per unit elevation of the piling then, it is clear that since C_M and C_D are O(1), the ratio D/H is relevant in determining the importance of the inertia force. For large structures, with diameters much greater than the incident wave height, the inertia force will predominate in deep water; in shallower water, where kh becomes small, the importance of the inertia force decreases. To determine which force predominates, we will determine the curve for which the two forces are equal. Equating Eq. (8.57) to 1, we have

$$\frac{H}{D} = \frac{C_M\pi}{C_D}\tanh kh \quad\cdot \qquad (8.60)$$

This curve is shown in Figure 8.13 for $C_D/C_M = 0.5$. For ratios of H/D above the curve, the drag force predominates. Note that in shallow water, the drag force tends to predominate over the inertia force.

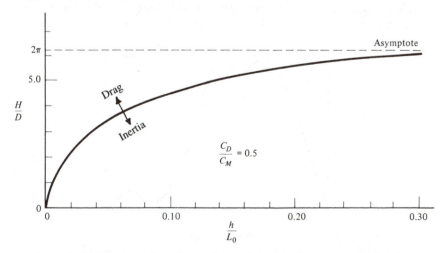

Figure 8.13 H/D versus h/L_0 for condition of equal maximum drag and inertia force components.

It is also instructive to consider the total force expressed in terms of a simple harmonic velocity given by

$$u = u_m \cos \sigma t \qquad (8.61)$$

instead of a form related to the wave height.

The total force on a unit length of cylinder is

$$dF_T = dF_D + dF_I = C_D \frac{\rho}{2} D\, u_m \cos \sigma t \,|u_m \cos \sigma t\,| - C_M \frac{\rho\pi D^2}{4}\, u_m \sigma \sin \sigma t \quad (8.62)$$

The ratio of maximum inertia to drag force component is

$$\frac{(dF_I)_{max}}{(dF_D)_{max}} = \tfrac{1}{2} \frac{C_M \pi D\sigma}{C_D u_m} = \pi^2 \frac{C_M}{C_D} \frac{1}{u_m T/D} \qquad (8.63)$$

and from Eq. (8.61), it can be shown that u_m/σ represents the maximum displacement S of a water particle from its neutral position. Therefore,

$$\frac{(dF_I)_{max}}{(dF_D)_{max}} = \frac{\pi}{2} \frac{C_M}{C_D} \frac{1}{(S/D)} \qquad (8.64)$$

The forms above are interesting because of the background and significance of the parameters $u_m T/D$ and S/D. The parameter $u_m T/D$ was first proposed by Keulegan and Carpenter (1958) and is sometimes referred to as the "Keulegan–Carpenter" parameter or the "period" parameter, while S/D is referred to as the "displacement" parameter. It is noted that, for small and large values of these parameters, the inertia and drag force components dominate, respectively. It is very important, but not surprising, that reliable values of C_D are most readily determined for large values of these parameters and reliable values of C_M are best determined from data for which these parameters are small. Moreover, it is found that if these parameters are

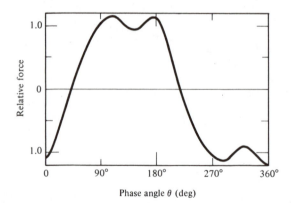

Figure 8.14 Measured force variation for $S/D = 2.5$. (Based on Keulegan and Carpenter, 1958.)

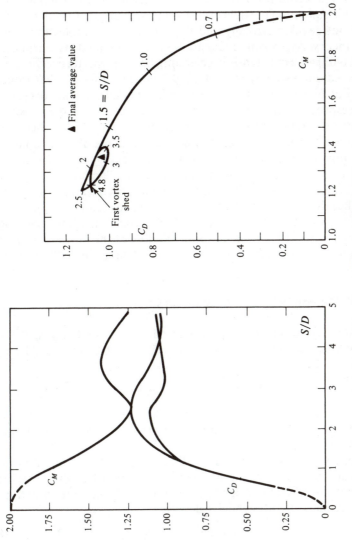

Figure 8.15 Variation of drag and inertia coefficients with the displacement parameter *S/D* for constant acceleration. (From Sarpkaya and Garrison, 1963.)

small, the *form* of the wave force time history is well represented by the theory; however, if these parameters are large, the form may deviate significantly from that predicted by theory (see Figure 8.14 for the form of a measured force record for $S/D \approx 2.5$). The reason for the behavior noted is that if S/D is small, the particle excursion is so small that friction and wake effects do not develop strongly and the flow field resembles that given by potential theory. As shown in Figure 8.15, as presented by Sarpkaya and Garrison (1963) for a constantly accelerating flow, the inertia coefficient C_M is approximated quite well by the potential flow value of 2 for S/D values less than 0.5. For higher values the inertia and drag coefficients decrease and increase, respectively. For S/D values larger than 2.0, the drag and inertia coefficients oscillate with time (S/D), presumably due to eddy shedding.

Figures 8.16 and 8.17 present drag and inertia coefficients obtained by Keulegan and Carpenter (1958) versus the period parameter for forces measured at the node of a standing wave system. (The interpretation of the inertia coefficient being less than unity is that this occurs for a drag-dominant case and that the phasing of the forces are more related to the phases of the near cylinder wake kinematics than to those at far field. Since the drag and inertia coefficients are correlated to the phasing of the far-field kinematics, the inertia force as correlated to the far field is "contaminated" by drag force effects.)

8.3.6 Maximum Total Force on an Object

For an object subjected to simple harmonic oscillations, the time-varying total force can be expressed by Eq. (8.36), which can be abbreviated as

$$F_T = F_D \cos \sigma t \,|\cos \sigma t\,| - F_I \sin \sigma t \qquad (8.65)$$

in which F_D and F_I represent the maxima of the drag and inertia force components, respectively, and can be determined readily by comparing Eqs. (8.36) and (8.65).

It is often of interest to determine the maximum *total* force. Noting that the maximum total force will occur for $\cos \sigma t > 0$, Eq. (8.65) can be written in the following form, from which the maximum can be determined by the normal procedures of differential calculus.

$$F_T = F_D \cos^2 \sigma t - F_I \sin \sigma t \qquad (8.66)$$

$$\frac{dF_T}{dt} = 0 = -2F_D \sigma \cos (\sigma t)_m \sin (\sigma t)_m - F_I \sigma \cos (\sigma t)_m \qquad (8.67)$$

Although not immediately obvious, there are two roots to Eq. (8.67). The first is found by dividing through by $\sigma \cos (\sigma t)_m$, yielding

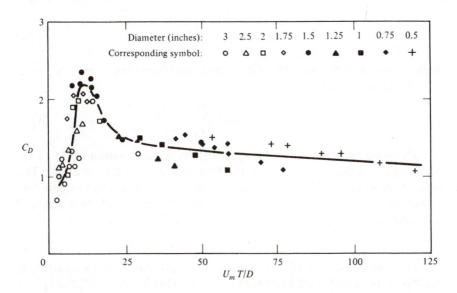

Figure 8.16 Variation of drag coefficient with period parameter as determined by Keulegan and Carpenter (1958).

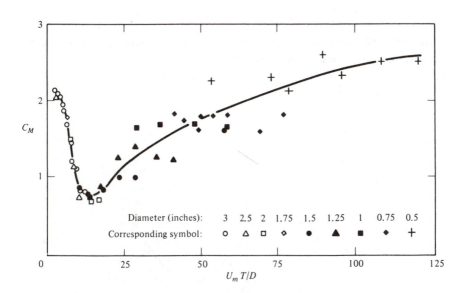

Figure 8.17 Inertia coefficient variation with period parameter as determined by Keulegan and Carpenter (1958).

$$\sin (\sigma t)_m = -\frac{F_I}{2F_D} \tag{8.68}$$

which, when substituted into Eq. (8.66), and recalling that $\cos^2 (\sigma t)_m$ $= 1 - \sin^2 (\sigma t)_m$, gives us

$$F_{T_m} = F_D + \frac{F_I^2}{4F_D} \tag{8.69}$$

The need for a second root is apparent upon examination of Eq. (8.68) and recognizing that if $F_I/2F_D > 1$, the first root is no longer possible. The second root to Eq. (8.67) is $\cos (\sigma t)_m = 0$, which was discarded by dividing this equation by $\cos (\sigma t)_m$. If $\cos (\sigma t)_m = 0$, $\sin (\sigma t)_m = -1$ and the maximum total force is

$$F_{T_m} = F_I \tag{8.70}$$

The interpretation of this second root can be seen by examining Figure 8.18. Because of the inflection of $\cos^2 \sigma t$ at $\sigma t = -\pi/2$, if $F_I > 2F_D$, the inertia force term decreases with increasing σt more rapidly than the term involving $\cos^2 \sigma t$ increases. Hence the maximum total force is *pure* inertia.

It is of interest to verify that the cause of the second root is the nature of the quadratic drag term. For example, if the drag force component were linear,

$$F_T = F_D \cos \sigma t - F_I \sin \sigma t \tag{8.71}$$

then, using the same procedures as before, there is only one root and the maximum total force is always given by

$$(F_T)_m = \sqrt{F_D^2 + F_I^2} \tag{8.72}$$

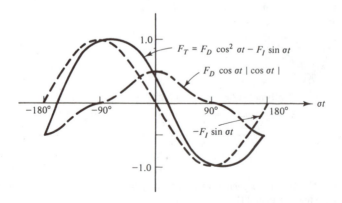

Figure 8.18 Illustration of force component combination for the case of $|F_I| = 2|F_D|$.

8.4 INERTIA FORCE PREDOMINANT CASE

As noted previously, if the structure is large relative to the length of the water particle excursion, flow separation will not occur, the drag force component is negligible, and the flow field can be treated by the classical methods of potential flow. There are sufficient numbers of structures in this class to be of practical interest and the applications include both wave forces and impulsive loading (i.e., due to earthquake motions). Two approaches have been developed and will be reviewed below. The first is an analytical approach and can be applied only for limited geometries. The second is a numerical method which is applicable for arbitrary geometries.

8.4.1 Rigorous and Approximate Analytical Methods for Wave Loading on Large Objects

MacCamy–Fuchs diffraction theory. As waves impinge on a vertical pile, they are reflected, or scattered, as in the case of a vertical wall, but in many directions. The scattering of acoustic and electromagnetic waves by a circular cylinder has long been known and understood. MacCamy and Fuchs (1954) applied the known theory to water waves. For linear wave theory, their results are exact, and can be used to predict C_M for a pile for which $D/H \gg 1$. The velocity potential for the incident wave can be written as

$$\phi_I = \frac{-gH}{2\sigma} \frac{\cosh k(h + z)}{\cosh kh} \cos (kx - \sigma t) \tag{8.73}$$

$$= \text{Re} \left\{ -\frac{gH}{2\sigma} \frac{\cosh k(h + z)}{\cosh kh} e^{i(kx-\sigma t)} \right\} \tag{8.74}$$

where Re means the real part of the now complex expression. From complex variables, $i = \sqrt{-1}$, and $e^{\pm i(kx-\sigma t)} = \cos (kx - \sigma t) \pm i \, \sin (kx - \sigma t)$. If the problem is expressed in terms of polar coordinates where r and θ are in the horizontal plane and z vertical, the incident wave may be written as

$$\phi_I = -\frac{gH}{2\sigma} \frac{\cosh k(h + z)}{\cosh kh} \left[J_0(kr) + \sum_{m=1}^{\infty} 2i^m \cos m\theta \, J_m(kr) \right] e^{-i\sigma t} \tag{8.75}$$

which satisfies the Laplace equation in polar form, and also the linearized form of the kinematic and dynamic free surface boundary conditions.

As this wave impinges on the pile, a reflected wave (which also satisfies the Laplace equation) radiates away and is assumed to have the following symmetric (about θ) form,

$$\phi_R = \sum_{m=0}^{\infty} A_m \cos m\theta [J_m(kr) + iY_m(kr)]e^{-i\sigma t} \frac{\cosh k(h+z)}{\cosh kh} \tag{8.76}$$

Equation (8.76) satisfies the Laplace equation and, for large kr, this solution has a periodic form which propagates away from the pile, ensuring that the assumed form satisfies the radiation boundary condition. Superimposing the incident and reflected waves gives the total flow field. The only remaining boundary condition is the no-flow condition at the cylinder, $-\partial(\Phi_I + \Phi_R)/\partial r = 0$ at $r = a$. Satisfying this condition determines the values of the terms in the infinite series A_m ($m = 0, 1, \ldots, \infty$). The final velocity potential is

$$\Phi_{I+R} = \text{Re}\left\{ \frac{gH \cosh k(h+z)}{2\sigma \cosh kh} e^{-i\sigma t} \right.$$

$$\left\{ \left[J_0(kr) - \frac{J_1'(ka)}{J_0'(ka) - iY_0'(ka)}(J_0(kr) + iY_0(kr)) \right] \right. \tag{8.77}$$

$$\left. + 2\sum_{m=1}^{\infty} i^m \left[J_m(kr) - \frac{J_m'(ka)}{J_m'(ka) - iY_m'(ka)}(J_m(kr) + iY_m(kr)) \right] \right\} \cos m\theta \right\}$$

where the primes denote derivatives of the Bessel functions with respect to their arguments.

Using the unsteady form of the Bernoulli equation to obtain the pressure, the force per unit length on the pile may be obtained.

$$dF_I = \frac{2\rho gH}{k} \frac{\cosh k(h+z)}{\cosh kh} G\left(\frac{D}{L}\right) \cos(\sigma t - \alpha) \tag{8.78}$$

where

$$\tan\alpha = \frac{J_1'(ka)}{Y_1'(ka)}; \qquad G\left(\frac{D}{L}\right) = \frac{1}{\sqrt{J_1'(ka)^2 + Y_1'(ka)^2}} \tag{8.79}$$

Comparing this to the general formula for inertial force,

$$dF_I = C_M \rho V \frac{\partial u}{\partial t} \tag{8.80}$$

where $V = \pi D^2/4$ and $\partial u/\partial t$ is calculated at the center of the pile, we find that $C_M = 4G(D/L)/\pi^3(D/L)^2$. A plot of C_M and α versus D/L is shown in Figure 8.19. Note that C_M and α reduce to 2.0 and 0, respectively, for small values of D/L, as predicted from potential flow theory for a cylinder in an oscillating flow.

Large rectangular objects.   In the MacCamy–Fuchs diffraction theory, the scattering of waves by the pile was included in the inertial force expression, thus allowing the determination of C_M for that case. If, however, the interaction between a structure and waves is not known, approximate

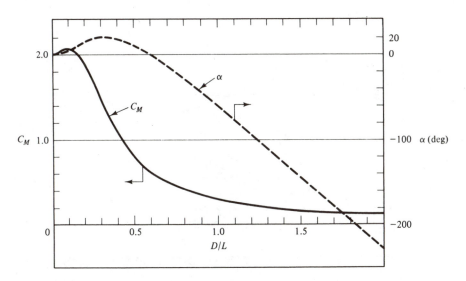

Figure 8.19   Variation of inertia coefficient C_M and phase angle α of maximum force with parameter D/L.

techniques at least allow the determination of the inertia force due to the pressure gradient. Experiments would be needed to determine the added mass, k_m.

Example 8.2

A large rectangular object with dimensions l_1, l_2, and l_3 in the x, y, and z directions is located somewhere within the water column. Calculate the horizontal inertial force on the structure due to a wave propagating in the x direction.

Solution. As before, we would like to integrate the dynamic pressure around the object. Figures 8.20 and 8.21 depict the object and S refers to the distance between the mean water level and the bottom of the object. The dynamic pressure induced by the waves in the absence of the structure is

$$p(x, z, t) = \rho g \eta K_p(z) = \frac{\rho g H}{2} \frac{\cosh k(h + z)}{\cosh kh} \cos (kx - \sigma t) \qquad (8.81)$$

In the configurations shown there is no variation of pressure in the y direction, as the waves are assumed to be long-crested and propagating in the x direction. In this example, the object will be considered to be totally submerged (Fig. 8.21).

Consider first the approximate wave-induced pressure on the face that is in the x direction, located at $x = x_1$, face (1). The total pressure force on this face, P_1, is

$$P_1 = \int_{-S}^{-S+l_3} l_2 p(x, z, t)\, dz = \frac{l_2 \rho g H \cos (kx_1 - \sigma t)}{2 \cosh kh} \int_{-S}^{-S+l_3} \cosh k(h + z)\, dz \qquad (8.82)$$

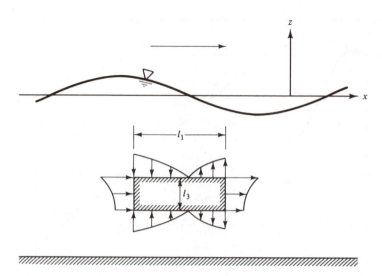

Figure 8.20   Dynamic wave pressures on rectangular object.

or

$$P_1 = \frac{l_2 \rho g H \cos(kx_1 - \sigma t)}{2 \cosh kh} \left[\frac{1}{k} (\sinh k(h - S + l_3) - \sinh k(h - S)) \right] \quad (8.83)$$

Using a trigonometric identity, we get

$$P_1 = \frac{l_2 l_3 \rho g H \cos k(x_1 - \sigma t)}{2 \cosh kh} \cosh k\left(h - S + \frac{l_3}{2} \right) \frac{\sinh\left(\frac{1}{2} k l_3 \right)}{\frac{1}{2} k l_3} \quad (8.84)$$

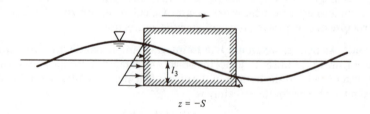

Figure 8.21   Wave forces on fixed rectangular object within free surface.

On the face at $x = x_1 + l_1$, the opposing force is

$$P_2 = \int_{-S}^{-S+l_3} l_2 p(x_1 + l_1, z, t)\, dz \tag{8.85}$$

$$= \frac{l_2 l_3 \rho g H}{2 \cosh kh} \cos\left[k(x_1 + l_1) - \sigma t\right] \cosh k\left(h - S + \frac{l_3}{2}\right) \frac{\sinh\left(\frac{1}{2} k l_3\right)}{\frac{1}{2} k l_3}$$

The net force in the x direction is then $P_1 - P_2$, as defined as F_x:

$$F_x = P_1 - P_2 \tag{8.86}$$

$$= \frac{l_1 l_2 l_3 \rho g H k \cosh k(h - S + l_3/2)}{2 \cosh kh} \frac{\sinh (k l_3/2)}{k l_3/2} \frac{\sin (k l_1/2)}{k l_1/2} \sin\left(k\left(x_1 + \frac{l_1}{2}\right) - \sigma t\right)$$

This can be rewritten in a more familiar form,

$$F_x = \rho V \frac{\sinh (k l_3/2)}{k l_3/2} \frac{\sin (k l_1/2)}{k l_1/2} \frac{\partial u}{\partial t} \tag{8.87}$$

where $\partial u/\partial t$ is evaluated at the center of the rectangular object.

In the limit as the size of the object becomes small, the term

$$\left. \frac{\sinh (k l_3/2)}{k l_3/2} \frac{\sin (k l_1/2)}{k l_1/2} \right|_{l_1, l_3 \to 0} \to 1 \tag{8.88}$$

as expected from the buoyancy analogy.[4] Remember, however, that the interaction of the structure with the waves was not accounted for, and thus the added mass is not included in this derivation. Therefore, the actual C_M should be larger than the terms above. The vertical force can be calculated in a similar manner (Dean and Dalrymple, 1972), yielding

$$F_z = \rho V \frac{\sinh (k l_3/2)}{k l_3/2} \frac{\sin (k l_1/2)}{k l_1/2} \frac{\partial w}{\partial t} \tag{8.89}$$

where again $\partial w/\partial t$ is evaluated at the center of the object. If the tank is situated on the bottom, such that the wave-induced pressure is not transmitted to the bottom of the tank, F_z is different.

$$F_z = -\rho V \frac{\coth (k l_3)}{k l_3} \frac{\sin (k l_1/2)}{k l_1/2} \frac{\partial w}{\partial t} \tag{8.90}$$

where $\partial w/\partial t$ is now evaluated at the center of the top of the object. The interested reader is referred to model tank experiments of Versowski and Herbich (1974) and to Chakrabarti (1973) for a verification of these formulae. Chakrabarti (1973) has developed the inertia force equations for other objects, valid for linear theory, such as a half-cylinder on the bottom and a hemisphere.

Wave forces on and motions of a floating body.   There are many naval architecture and marine engineering problems which are of importance to the ocean engineer. In this section a very approximate treatment will be

---

[4]The functions $(\sin w)/w$ and $(\sinh w)/w$ are shown in Figure 8.24.

presented for forces on and motions of floating bodies; the reader is referred to more extensive developments for additional detail and depth.

Generally, unless a large floating body is propelled, the size and/or streamlining are such that the dominant forces are related to the water particle accelerations rather than the drag forces which are velocity related. If we first consider an unrestrained floating object that is small compared to the wave length, since this object displaces its own weight of water, it is clear that the forces on the object are exactly those that would have occurred on the displaced fluid and hence the motions of the object will be the same as would have occurred for the displaced fluid. This result also applies to the case of a small neutrally buoyant object at some mean elevation within the water column. For objects that are restrained or large such that the kinematics change significantly over the object dimension, the situation becomes more complex as the object affects the waves. In the following section, a simplified case is considered for a.rectangular object either fixed or freely floating in the free surface. The treatment is similar to, but more general than the analysis for the submerged rectangular object.

Consider the case of a rectangular object fixed in the free surface as shown in Figure 8.22. The waves advance at an arbitrary angle θ, measured

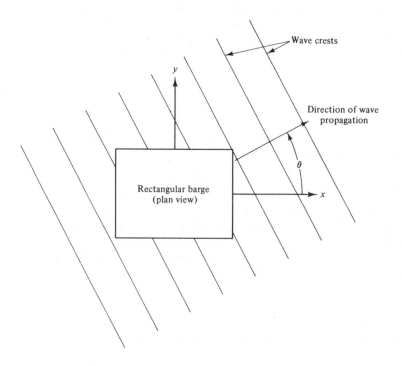

Figure 8.22 Waves propagating with angle α past a rectangular barge of draft d.

counterclockwise from the x axis. Representing the water surface displacement as

$$\eta = \frac{H}{2} \cos (k_x x + k_y y - \sigma t) \qquad (8.91)$$

where $k_x = k \cos \theta$ and $k_y = k \sin \theta$ and $\sigma^2 = gk \tanh kh$, the "undisturbed" pressure field due to this wave is given by

$$p = \rho g \frac{H}{2} \frac{\cosh k(h + z)}{\cosh kh} \cos (k_x x + k_y y - \sigma t) \qquad (8.92)$$

The forces due to this pressure field will be examined as a dominant contributor; however, it should be recognized that there is considerable wave reflection from the object and that this effect could contribute significantly to the wave forces.

The computation of forces will be illustrated in some detail for the surge (x) mode of motion. The force is given by

$$F_x = \int_{-l_y/2}^{l_y/2} \int_{-d}^{0} p\left(-\frac{l_x}{2}, y, z\right) dz \, dy - \int_{-l_y/2}^{l_y/2} \int_{-d}^{0} p\left(\frac{l_x}{2}, y, z\right) dz \, dy$$

$$(8.93)$$

in which d is the draft of the object. Inserting Eq. (8.92) for the pressure and carrying out the integration yields

$$F_x = \frac{-4\rho g(H/2)}{k_y} \frac{(\sinh kh - \sinh k(h - d))}{k \cosh kh} \sin \frac{k_x l_x}{2} \sin \frac{k_y l_y}{2} \sin \sigma t \quad (8.94)$$

which can be rendered dimensionless by normalizing with respect to the displaced weight:

$$\frac{F_x}{\rho g \, dl_x l_y} = -\frac{H}{2} k_x \frac{\sinh kh - \sinh k(h - d)}{kd \cosh kh} \frac{\sin k_x l_x/2}{k_x l_x/2} \frac{\sin k_y l_y/2}{k_y l_y/2} \sin \sigma t \qquad (8.95)$$

The interpretation of the equation above is interesting. Considering long waves, Eq. (8.95) reduces to

$$\frac{F_x}{\rho g \, dl_x l_y} = -\frac{H}{2} k_x \sin \sigma t = \left. \frac{\partial \eta}{\partial x} \right|_{x=0, y=0} \qquad (8.96)$$

As noted, comparison with Eq. (8.91) will show that this represents the instantaneous slope of the water surface in the x direction evaluated at the center of the platform. In other words, for very long waves, the horizontal wave force component is simply that due to the body tending to "slide" down the sloping surface (see Figure 8.23). Also, it is clear that if the wave propagation direction is 90°, then $k_x = 0$ and $F_x = 0$.

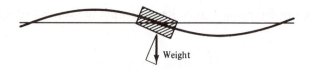

Weight

Figure 8.23 Wave forces on a "small" floating object are equivalent to the weight of the object acting down a surface slope.

For the more complete equation, the sin w/w terms are always less than unity (unless the argument is zero) and represent the reduction due to the finite length of the object; that is, the effective slope over the length of the object is less than the maximum slope (see Figure 8.24 for a plot of sin w/w).

The sway force (y direction) can be written down by inspection from Eq. (8.95), and is

$$\frac{F_y}{\rho g\, dl_x l_y} = -\frac{H}{2} k_y \frac{(\sinh kh - \sinh k(h-d))}{kd \cosh kh} \frac{\sin k_x l_x/2}{k_x l_x/2} \frac{\sin k_y l_y/2}{k_y l_y/2} \sin \sigma t \qquad (8.97)$$

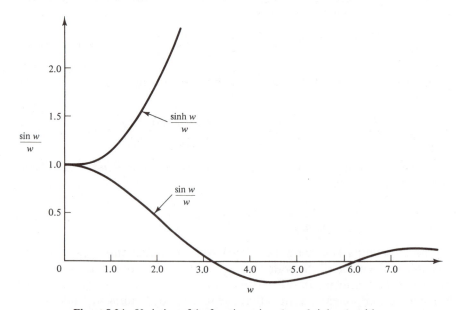

Figure 8.24 Variation of the functions sin w/w and sinh w/w with w.

The computation of the moments is somewhat more complicated than the forces. The pitch moment (about the y axis) will first be developed; the roll moment (about the x axis) could be written down by inspection.

The pitch moment about the center of gravity consists of a primary contribution due to pressure on the (large) horizontal bottom surface and a smaller contribution from the two ends of the object.

Referring to Figure 8.25 the pitch moment about the center of gravity M_{α_p} is

$$M_{\alpha_p} = I_1 + I_2 + I_3 \tag{8.98}$$

where

$$I_1 = \int_{-l_y/2}^{l_y/2} \int_{-d}^{0} (z - z_1) p\left(-\frac{l_x}{2}, y, z\right) dz\, dy \tag{8.99}$$

$$I_2 = -\int_{-l_y/2}^{l_y/2} \int_{-d}^{0} (z - z_1) p\left(\frac{l_x}{2}, y, z\right) dz\, dy \tag{8.100}$$

$$I_3 = -\int_{-l_y/2}^{l_y/2} \int_{-l_x/2}^{l_x/2} p(x, y, -d)\, x\, dx\, dy \tag{8.101}$$

in which z_1 represents the distance of the center of gravity above the mean water line, the first two integrals, I_1 and I_2, represent the contributions from the two ends, and the third integral is the moment due to the pressures acting on the bottom of the barge. The resulting expression for pitch moment is

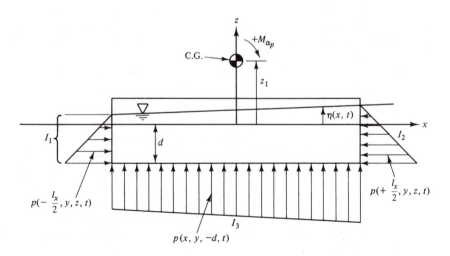

Figure 8.25 Definition sketch for pressures acting on a fixed rectangular barge to cause pitch moments.

$$M_{\alpha_r} = \frac{\gamma H}{2} \, \mathcal{S}\!\left(\frac{k_y l_y}{2}\right)\frac{l_y l_x}{k_x}\frac{\cosh k(h-d)}{\cosh kh} \tag{8.102}$$

$$\left\{-\cos\frac{k_x l_x}{2} + \mathcal{S}\!\left(\frac{k_x l_x}{2}\right)\left[1 - \frac{\cos^2 \theta}{\cosh k(h-d)}G\right]\right\}$$

in which

$$\mathcal{S}(x) \equiv \frac{\sin x}{x} \tag{8.103}$$

and

$$G \equiv kd \sinh k(h-d) - [\cosh kh - \cosh k(h-d)] \tag{8.104}$$

$$+ kz_1 [\sinh kh - \sinh k(h-d)]$$

The roll and yaw moment would be obtained similarly; however, the expressions will not be presented here.

8.4.2 Numerical Methods for Wave Loading on Large Objects of Arbitrary Shapes

For problems of this class, Garrison and a number of colleagues (1971, 1972, 1973, and 1974) have utilized numerical approaches in which the surface of the structure of interest is represented as a number of surface elements with an oscillating source located at the center of each of these elements. These sources, when combined with the incident wave field, satisfy the appropriate boundary conditions. In the following sections, the method will be outlined briefly and representative results presented; the reader is referred to the original papers for greater detail.

Although the boundary value problem will not be specified in detail, it is noted that it consists of the usual no-flow boundary conditions on the seafloor and the structure. For purposes of illustration, and since the problem is considered to be linear, it may be discussed in two parts. First, consider the object to be "transparent" to the flow which is due only to the incident wave field. Velocity components would occur normal to the surface of the structure. Denote this velocity as $V_{nI}(S)$, that is, the normal velocity through the structure due to the *incident* wave field. The objective then is to determine a second velocity potential which satisfies the Laplace equation, all of the boundary conditions, and which yields a velocity $V_{nG}(S)$ which is due to the Green's function and exactly cancels the normal velocity on the structure due to the incident velocity field, that is,

$$V_{nG}(S) = -V_{nI}(S) \tag{8.105}$$

Green's functions, G, are developed which satisfy the Laplace equation and the bottom and free surface boundary conditions, and are denoted by

$$G(\mathbf{x}, \xi) \tag{8.106}$$

in which the generalized field vector coordinate is represented as $\mathbf{x}$ and the surface coordinate as the vector ξ. The forms of the Green's functions may be found for various problems in Garrison et al. (1971, 1972, 1973, and 1974). The velocity potential at any location, $\mathbf{x}$, is given by

$$\Phi_G(\mathbf{x}) = \frac{1}{4\pi} \int \int_S f(\xi) G(\mathbf{x}; \xi)\, dS \tag{8.107}$$

in which $f(\xi)$ represents the proper weighting of all contributions on the surface S; this factor is determined in accordance with Eq. (8.105), that is

$$\frac{1}{4\pi} \int \int_S f(\xi) \frac{\partial G}{\partial n}(\mathbf{x};\xi)\, dS = V_{nI}(S) \tag{8.108}$$

The solution to this equation is carried out numerically by partitioning the surface into N area elements and expressing the integral as a matrix with N elements such that

$$f_j \alpha_{ij} = V_{nI_i} \tag{8.109}$$

and

$$\alpha_{ij} = \frac{1}{4\pi} \int \int_{\Delta S_i} \frac{\partial G}{\partial n}(\mathbf{x}_i; \xi)\, dS \tag{8.110}$$

The coefficient matrix is first calculated from Eq. (8.110) and then Eq. (8.109) is inverted to find the weighting factor matrix f_j.

Examples.   Garrison and Stacey (1977) have presented calculations of wave forces on a number of large offshore structures, including several for which exact solutions were available and other more complex structures for which wave tank experiments were conducted. Figure 8.26 shows a vertical cylindrical caisson for which calculations were carried out. The exact and approximate results are presented in Figure 8.27 for a caisson with a height-to-radius ratio of unity.

As a second example, consider the case of a CONDEEP structure

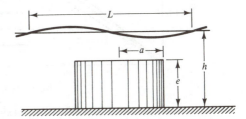

Figure 8.26 Fixed vertical caisson.
(From Garrison and Stacey, 1977.)

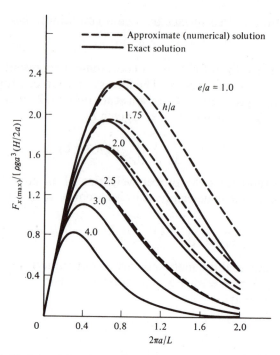

Figure 8.27 Horizontal inertia coefficients for vertical caisson for $e/a = 1.0$. (From Garrison and Stacey, 1977.)

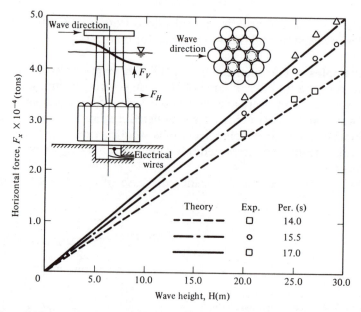

Figure 8.28 Measured and calculated maximum horizontal forces on 20-m CONDEEP platform. (From Garrison and Stacey, 1977.)

consisting of 19 cylinders extending upward approximately 47 m from the bottom; above this level three tapered cylinders extend up through the water surface and support the platform deck. In the idealization of this structure, Garrison and Stacey represented the lower caisson structure and portions of the three support columns by a distribution of sources; the upper portions of the three support columns were represented by the Morison equation [Eq. (8.32)]. Comparisons of calculated and measured maximum horizontal and vertical forces and moments are presented in Figures 8.28, 8.29, and 8.30.

8.4.3 Analytical Methods for Impulsive Loading on a Large Circular Cylinder

The forces imposed on a structure due to its motions can be determined in some cases from the solution of the wavemaker (Chapter 6) problem. For example, Jacobsen (1949) has presented the solution for the case of a vertical right circular cylinder oscillating in a direction perpendicular to its axis; Garrison and Berklite (1973) have also presented this solution with some corrections to Jacobsen's solution. There is no incident wave field and the boundary condition on the cylinder is expressed as

$$u(a, \theta) = U \cos \theta \cos \sigma t \qquad (8.111)$$

in which a is the cylinder radius and θ is the azimuth relative to the line of oscillation; the remainder of this "wavemaker" boundary value problem is as previously formulated. The solution is somewhat similar to that for the MacCamy–Fuchs problem and occurs as Bessel functions; the reader is referred to Garrison and Berklite or Dalrymple and Dean (1972) for the details. Although the solution is developed for a simple harmonic oscillation of the cylinder, it is possible, due to the linearity of the problem, to employ linear superposition and represent arbitrary time displacements such as those caused by earthquake motions of the seabed.

8.4.4. Forces Due to Impulsive Motions of Large Structures of Arbitrary Shape

The methodology employed by Garrison and Berklite (1973) for this more difficult problem is quite similar to that described previously for the case of large objects of arbitrary shape in which the use of Green's functions was outlined. The only differences are that there is no incident wave field and the normal velocity on the surface of the structure is now specified in accordance with the motions of the structure rather than specified as zero as for the case of a motionless structure. The linearity of the equations governing the problem allows each of the six motion components[5] to be solved

---

[5]Heave, pitch, roll, yaw, surge, and sway.

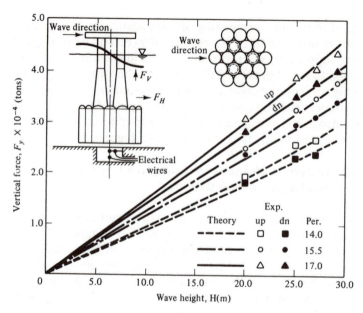

Figure 8.29 Measured and calculated maximum vertical forces on 20-m CONDEEP platform. (From Garrison and Stacey, 1977.)

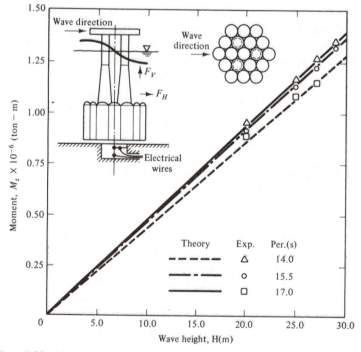

Figure 8.30 Measured and calculated maximum moments on 20-m CONDEEP platform. (From Garrison and Stacey, 1977.)

separately and then combined later. For any given structure, the added mass coefficients were found by Garrison and Berklite to be frequency dependent, and at very high and low frequencies it was possible to simplify the combined free surface boundary condition for periodic motion.

$$-\sigma^2\phi + g\frac{\partial\phi}{\partial z} = 0 \tag{8.112}$$

to

$$\phi = 0, \quad \sigma \text{ large} \tag{8.113}$$

or

$$\frac{\partial\phi}{\partial z} = 0, \quad \sigma \text{ small} \tag{8.114}$$

and

$$p = \rho\phi \tag{8.115}$$

The low-frequency limit corresponds to the case of a "rigid lid" boundary; that is, the motion is so slow that there is very little displacement at the free surface and the high-frequency limit corresponds to the case of standing waves located near the structure, with very little generation of waves propagating away from the structure.

It is of interest to note that the solutions for the limiting cases represented by Eqs. (8.113) and (8.114) do not represent wave-like behavior, but rather cases of antisymmetrical flow about the free surface and uniform flow as idealized in Figures 8.31 and 8.32, respectively.

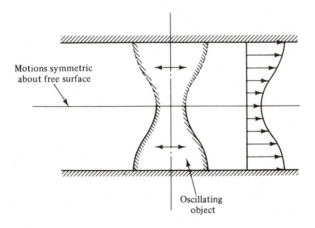

Figure 8.31 Interpretation of free surface boundary condition $\frac{\partial\phi}{\partial z} = 0$ (for low-frequency motions).

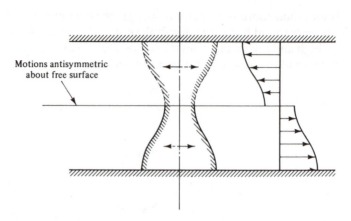

Figure 8.32 Interpretation of free surface boundary condition $\phi = 0$ (for high-frequency motions).

Example 8.3

Consider the case of a vertical circular cylinder oscillating along the x axis. The added mass for the cylinder is presented in Figure 8.33 for the case of a rigid boundary $(\partial\phi/\partial z = 0)$ and $(\phi = 0)$. It is seen that for the case of a rigid boundary, the added mass coefficient is unity as expected and that for the solution corresponding to a boundary condition, $\phi = 0$, the added mass approaches unity as h/a becomes large.

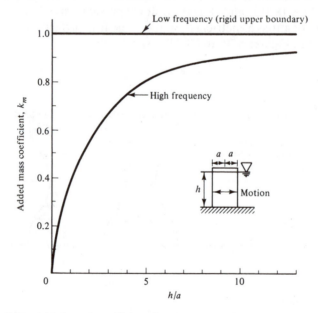

Figure 8.33 Added mass coefficient k_m versus ratio h/a for oscillating right circular cylinder. (Adapted from Garrison and Berklite, 1973.)

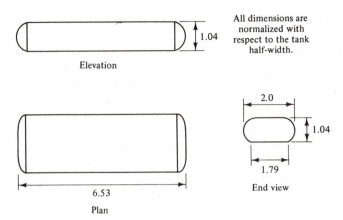

Figure 8.34 Dimensions of oil storage tank analyzed by Garrison and Berklite (1973).

A second example presented by Garrison and Berklite is that of an oil storage tank located on the seafloor as shown in Figure 8.34. The added mass and lever arm are presented in Figure 8.35.

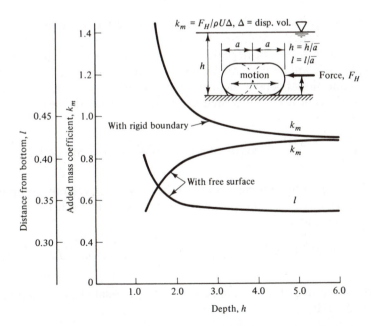

Figure 8.35 Added mass k_m and lever arm l for an oil storage tank on the seafloor. (From Garrison and Berklite, 1973.)

8.5 SPECTRAL APPROACH TO WAVE FORCE PREDICTION

The Morison equation for wave forces on a structural member is nonlinear in the water particle velocity u appearing in the drag force component [cf. Eq. (8.32)]. A second possible source of nonlinearity for the case of a surface-piercing piling is due to the variation in total immersed water depth due to the fluctuation of the free surface. However, in cases where the inertia force is dominant, the drag force component is negligible and the equation is now linear in the water particle acceleration $\dot{u}$. In addition, since the maximum acceleration occurs for the wave phase corresponding to zero water surface displacement, $\eta = 0$, there is no contribution at this phase from the second possible source of nonlinearity.

In view of the discussion above, for the case of inertia dominance, the local and total force are approximately linear in the wave height H, and the spectral methods described in Chapter 7 apply directly. From Eq. (8.35) the relationship can be expressed as

$$(F_I)_{max} = G(\sigma)H \tag{8.116}$$

in which for the total force on a structure,

$$G(\sigma) = C_M \frac{\rho \pi D^2}{8k} \sigma^2 \tag{8.117}$$

Borgman (1965a), (1965b), and (1967) has investigated the application of spectral methods to the problem of wave forces for the case in which the drag force components are not negligible. Only the most simple result of Borgman's approach will be presented here; the reader is referred to the original papers for additional detail. The incremental wave force dF on an elemental length ds of vertical piling located a distance s above the seafloor can be expressed as

$$dF = \left(\frac{C_D \rho D}{2} u_m^2 \cos \sigma t \, |\cos \sigma t| - \frac{C_M \rho \pi D^2}{4} u_m \sigma \sin \sigma t \right) ds \tag{8.118}$$

in which u_m represents the maximum of the horizontal velocity component. To apply linear spectral approaches, it is necessary to linearize the above equation. An intuitive form is

$$dF = \left(\frac{C_{DI} \rho D}{2} u_m \cos \sigma t - \frac{C_M \rho \pi D^2}{4} u_m \sigma \sin \sigma t \right) ds \tag{8.119}$$

Borgman shows that the force spectrum $S_{dF}(\sigma)$ is related to the sea surface spectrum $S_\eta(\sigma)$ by

$$S_{dF}(\sigma) = |X_{dF}(\sigma)|^2 S_\eta(\sigma) \tag{8.120}$$

in which

$$|X_{dF}(\sigma)|^2 = \left\{ \left[\frac{C_D\rho D}{2} \sqrt{\frac{8}{\pi}} \, U_{\text{rms}} X_u(\sigma, s) \right]^2 + \left[\frac{C_M\rho\pi D^2}{4} X_u(\sigma\,s)\sigma \right]^2 \right\} \qquad (8.121)$$

where

$$X_u(\sigma, s) \equiv \frac{u_m}{|\eta|} = \sigma\,\frac{\cosh ks}{\sinh kh} \qquad (8.122)$$

and the linearized drag coefficient C_{D_L} is defined in terms of the actual drag coefficient C_D and the root-mean-square velocity U_{rms} at the level s by

$$C_{D_L} = C_D \sqrt{\frac{8}{\pi}} \, U_{\text{rms}} \qquad (8.123)$$

That is, C_{D_L} has dimensions of velocity and U_{rms} is defined by

$$U_{\text{rms}} = \sqrt{\int_0^\infty |X_u(\sigma, s)|^2 S_\eta(\sigma)\,d\sigma} \qquad (8.124)$$

For the case of total wave forces over the entire water depth, the integration of Eq. (8.120) is carried out only up to the mean free surface, $z = 0$, and the result is

$$S_F(\sigma) = \left\{ \left[\left(\frac{C_D\rho D}{2} \right)^2 \sigma \frac{8}{\pi} G_1(\sigma) \right] + \left[\left(\frac{C_M\rho\pi D^2}{4} \right)^2 G_2(\sigma) \right] \right\} S_\eta(\sigma) \qquad (8.125)$$

in which

$$G_1(\sigma) = \frac{\displaystyle\int_0^h U_{\text{rms}}(S) \cosh ks\,ds}{\sinh kh} \qquad (8.126)$$

$$G_2(\sigma) = \frac{\sigma^2}{k} \qquad (8.127)$$

Borgman (1967) has extended this method to the computation of moments and to multilegged platforms.

REFERENCES

BORGMAN, L. E., "A Statistical Theory for Hydrodynamic Forces on Objects," Hydraulic Engineering Laboratory Rep. HEL 9-6, University of California at Berkeley, Oct. 1965a.

BORGMAN, L. E., "The Spectral Density for Ocean Wave Forces," Hydraulic Engineering Laboratory Rep. HEL 9-8, University of California at Berkeley, Dec. 1965b.

BORGMAN, L. E., "Spectral Analysis of Ocean Wave Forces on Piling," *J. Waterways Harbors Div., ASCE*, Vol. 93, No. WW2, May 1967.

CHAKRABARTI, S., "Wave Forces on Submerged Objects of Symmetry," *J. Waterways, Harbors Coastal Eng. Div., ASCE*, Vol. 99, No. WW2, May 1973.

DALRYMPLE, R. A., and R. G. DEAN, "The Spiral Wavemaker for Littoral Drift Studies," *Proc. 13th Conf. Coastal Eng., ASCE*, 1972.

DEAN, R. G., "Stream Function Representation of Nonlinear Ocean Waves," *J. Geophys. Res.*, Vol. 70, No. 18, 1965.

DEAN, R. G., and P. M. AAGAARD, "Wave Forces, Data Analysis and Engineering Calculation Method," *J. Petrol. Technol.*, Mar. 1970.

DEAN, R. G., and R. A. DALRYMPLE, Discussion of J. B. Herbich and G. E. Shank, "Forces Due to Waves on Submerged Structures," *J. Waterways, Harbors Coastal Eng. Div., ASCE*, Vol. 98, No. WW1, Feb. 1972.

GARRISON, C. J., "Dynamic Response of Floating Bodies," *Proc. Offshore Technol. Conf.*, 1974, Paper 2067.

GARRISON, C. J., and R. B. BERKLITE, "Impulsive Hydrodynamics of Submerged Rigid Bodies," *J. Eng. Mech. Div., ASCE*, Vol. EM1, Feb. 1973.

GARRISON, C. J., and P. Y. CHOW, "Wave Forces on Submerged Bodies," *J. Waterways, Harbors Coastal Eng. Div., ASCE*, Vol. 98, No. WW3, Aug. 1972.

GARRISON, C. J., and V. S. RAO, "Interaction of Waves with Submerged Objects," *J. Waterways, Harbors Coastal Eng. Div., ASCE*, Vol. 97, No. WW2, 1971.

GARRISON, C. J., and R. STACEY, "Wave Loads on North Sea Gravity Platforms: A Comparison of Theory and Experiment," *Proc. Offshore Technol. Conf.*, 1977, Paper 2794.

GARRISON, C. J., A. TØRUM, C. IVERSON, S. LEWSETH, and C. C. EBBESMEYER, "Wave Forces on Large Objects—A Comparison between Theory and Model Tests," *Proc. Offshore Technol. Conf.*, 1974, Paper 2137.

GOLDSTEIN, S., *Modern Developments in Fluid Dynamics*, Vol. 2, Oxford University Press, London, 1938.

JACOBSEN, L. S., "Impulsive Hydrodynamics of Fluid inside a Cylindrical Tank and of Fluid Surrounding a Circular Pier," *Bull. Seismol. Soc. Am.*, Vol. 39, No. 3, 1949.

KEULEGAN, G. H., and L. H. CARPENTER, "Forces on Cylinders and Plates in an Oscillating Fluid," *J. Res. Nat. Bur. Stand.*, Vol. 60, No. 5, May 1958.

LAMB, H., *Hydrodynamics*, Dover, New York, 1945.

MACCAMY, R. C., and R. A. FUCHS, "Wave Forces on Piles: A Diffraction Theory," Tech. Memo 69, Beach Erosion Board, 1954.

MORISON, J. R., M. P. O'BRIEN, J. W. JOHNSON, and S. A. SCHAAF, "The Force Exerted by Surface Waves on Piles," *Petrol. Trans., AIME*, Vol. 189, 1950.

SARPKAYA, T., "Vortex Shedding and Resistance in Harmonic Flow about Smooth and Rough Circular Cylinders at High Reynolds Numbers," Rep. NPS-59SL76021, U.S. Naval Postgraduate School, Feb. 1976.

SARPKAYA, T., and C. J. GARRISON, "Vortex Formation and Resistance in Unsteady Flow," *Trans. ASME*, Mar. 1963.

SCHLICHTING, H., *Boundary Layer Theory*, McGraw-Hill, New York, 1968.

VERSOWSKI, P. E., and J. B. HERBICH, "Wave Forces on Submerged Model Structures," *Proc. Offshore Technol. Conf.*, OTC 2042, Vol. II, 1974.

WILSON, B. W., and R. O. REID, Discussion of "Wave Force Coefficients for Offshore Pipelines," *J. Waterways Harbors Div., ASCE*, Vol. 89, No. WW1, 1963.

WRIGHT, J. C., and T. YAMAMOTO, "Waves Forces on Cylinders near Plane Boundaries," *J. Waterways, Ports, Coastal Ocean Div., ASCE*, Vol. 105, No. WW1, Feb. 1979.

PROBLEMS

8.1 The triangular cross section shown below is being considered for underwater petroleum storage. The "tank" would be in shallow water, so the waves may be regarded as long.

(a) Develop a relationship for the horizontal wave force on the tank. Express your answer in dimensionless form, normalizing by the displaced water weight.

(b) At what position of the wave profile would the horizontal wave force be a maximum?

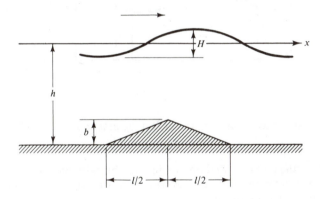

8.2 What is the maximum uplift force on the slab below due to a wave of 10-m height and 12-s period? Assume that the presence of the slab does not interfere with the wave motion. At what phase of the motion will be maximum uplift occur?

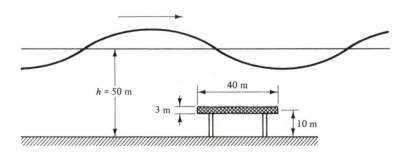

8.3   Given the following wave conditions:

$$H = 15.6 \text{ ft}$$

$$T = 14 \text{ s}$$

$$h = 20 \text{ ft}$$

(a) Consider the case of a single piling supporting a small observation deck. From corrosion considerations, the thickness of the tubular piling is 1 in. Assuming that the drag moment predominates, develop an equation for the stress σ in the outer fiber of the base of the piling as a function of the diameter D.

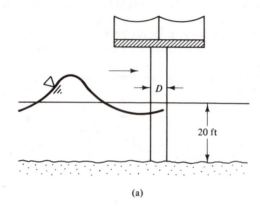

(a)

(b) What is the required diameter D if the maximum allowable σ is

$$\sigma_{max} = 20,000 \text{ psi?}$$

(c) For the diameter determined in part (b), calculate the maximum inertia moment component and express as a percentage of the maximum drag moment component.

(d) Allowing a freeboard for the lower deck elevation of 10 ft, at what elevation would this be?

Equations for Calculating Stress

$$\sigma = \frac{M}{S} = \text{stress on outer fiber}$$

$$S = \text{section modulus} = \frac{\pi}{32D}(D^4 - D_1^4)$$

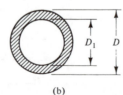

(b)

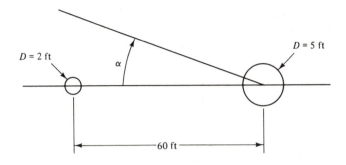

8.4 Given

$$H = 15.6 \text{ ft}$$

$$T = 14 \text{ s}$$

$$h = 20 \text{ ft}$$

(a) Calculate and tabulate the maximum total wave force on the two vertical cylinders shown above as a function of wave approach direction α for $\alpha =$ 0°, 30°, 60°, and 90°.

(b) What is the ratio of maximum inertia to drag force for the larger cylinder?

(c) What would be the total overturning moment for the direction α_{max} of maximum force?

8.5 Determine the inertia coefficients (horizontal and vertical) for a pipeline exactly half buried in the bottom.

8.6 Based on the experimental results presented in Figure 8.4 and using the C_D versus $\mathbb{R}$ relationship presented in Figure 8.5, develop a relationship of the separation angle θ_s versus Reynolds number $\mathbb{R}$. Use the approach of Eq. (8.11) and assume that p_{wake} can be taken as $p(a, \theta_s)$. Compare and comment on your results with those in Figure 8.4.

8.7 Referring to Eq. (8.13), and accounting for the effect of separation, develop a fairly simple equation for the added mass coefficient versus separation angle θ_s. (*Hint:* Use the same considerations suggested in Problem 8.6.)

8.8 Discuss the reasons for the decrease in C_M with increasing D/L, using the results of the MacCamy–Fuchs theory.

8.9 Consider the case of waves propagating past and aligned with the major axis of a barge.

(a) If the dominant forces on the barge are due to being "immersed" in the wave pressure field, develop an equation for the surge displacement $x_B(t)$ of the barge.

(b) Demonstrate that $x_B(t)$ is exactly the same as the *average* horizontal displacement of the water particles displaced by the barge.

8.10 A circular cylinder of diameter D and length l is held fixed in a horizontal plane at an elevation s above the bottom in a total water depth h as shown below. Considering only the inertia force component and a linear wave of height H and period T to the propagating in the x direction, develop expressions for the time-varying components of forces in the x and y directions, $F_x(t)$ and $F_y(t)$, and the moment $M_\alpha(t)$ about the z axis.

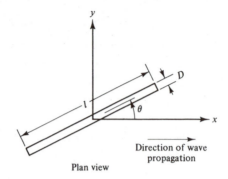

Plan view

8.11 Simplify Eqs. (8.35) and (8.38) for the cases of shallow and deep water. Discuss the variation of drag and inertia force and moment components with wave period for these two regions. Also evaluate the lever arms implied by the results.

8.12 A circular cylinder is immersed in an idealized flow of free stream velocity U. The cylinder is instrumented with strain gages to measure the force at the two locations shown. Develop an expression for the force per unit cylinder length measured by each of the two sets of strain gages. Interpret the sign of the force.

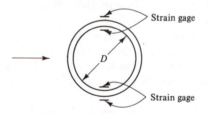

8.13 Consider a cylinder with axis horizontal located one-quarter wave length below the mean free surface with a wave of height H and period T propagating with crests parallel to the cylinder axis. The water depth is h.
  (a) Develop an expression for the time-varying magnitude and direction of the total wave force (i.e., drag plus inertia) acting on the cylinder.
  (b) Discuss your results for the limiting case of shallow and deep water.
  (c) Specifically for the case of deep water, discuss and interpret the time variation of the magnitude and direction of the total wave force.

9

Waves Over Real Seabeds

Dedication
JOSEPH VALENTIN BOUSSINESQ

Joseph Valentin Boussinesq (1842–1929) laid the foundations of hydro-dynamics, together with Cauchy, Poisson, and St.-Venant. His work in waves is largely remembered for the solitary wave theory that bears his name and the Boussinesq approximation, which facilitates the study of stratified flow.

Boussinesq was born in St.-André-de-Sangonis, France, and earned his baccalaureate from a seminary in Montpellier. Despite his informal education in the sciences, he produced a paper on capillarity in 1865 and presented it to the Academie des Sciences. From 1866 to 1872 he taught at the Colleges of Agdé, Le Vigan, and Gap. His doctoral work on the spreading of heat in 1867 won him the attention of Barre de St.-Venant.

In 1873 he became a professor at Lille and subsequently assumed the chair of physical and experimental mechanics in Paris.

Boussinesq's scientific work ranged over many fields of classical physics: light and heat, ether, fluid forces on bodies, waves, hydraulics, vortex motions, and elasticity. He also studied philosophical and religious matters such as determinism and free will.

9.1 INTRODUCTION

Historically, the mathematical treatment of water wave theory by various investigators has been carried out with the assumption of a rigid, impermeable horizontal seabed. In nature, of course, the actual bottom varies drastically from locales in the Gulf of Mexico where the muds behave as viscous fluids, to rippled porous sand beds, to rough rocky bottoms. The degree of bed rigidity (as measured by the shear modulus, say), the porosity, and the roughness all influence the water waves to varying degrees. This interaction

261

with the bed results in wave damping and a local change in wave kinematics. Significant wave damping can occur if the bed is very soft, or if the waves propagate a long distance; in either case, shoaling formulas developed earlier are no longer strictly valid.

If the presence of the wave over the bed causes significant bed deformation and stresses, the possibility exists of soil failure and significant forces on buried pipelines and on bottom-mounted structures.

9.2 WAVES OVER SMOOTH, RIGID, IMPERMEABLE BOTTOMS

9.2.1 Laminar Boundary Layer

The equations governing the water waves in a viscous fluid are the Navier–Stokes equation [Eqs. (2.39a) and (2.39c)], shown here in linearized form.

$$\frac{\partial u}{\partial t} = -\frac{1}{\rho}\frac{\partial p}{\partial x} + \nu\left(\frac{\partial^2 u}{\partial x^2} + \frac{\partial^2 u}{\partial z^2}\right) \tag{9.1}$$

$$\frac{\partial w}{\partial t} = -\frac{1}{\rho}\frac{\partial p}{\partial z} + \nu\left(\frac{\partial^2 w}{\partial x^2} + \frac{\partial^2 w}{\partial z^2}\right) - g \tag{9.2}$$

where $\nu \ (=\mu/\rho)$ is the kinematic viscosity.

It is useful to examine the relative sizes of the various terms in these equations; this can be done best by putting them in dimensionless form. Therefore, knowing a priori for waves that a length scale is the inverse of the wave number and a time scale is the inverse of the wave frequency, we can write

$$x = \frac{x'}{k}, \quad z = \frac{z'}{k}, \quad t = \frac{t'}{\sigma}, \quad u = a\sigma u', \quad p = \rho gap'$$

where a is the wave amplitude and the primed variables are dimensionless. Substituting into the equations for the x direction, we get

$$\frac{\partial u'}{\partial t'} = -\frac{gk}{\sigma^2}\frac{\partial p'}{\partial x'} + \frac{\nu k^2}{\sigma}\left(\frac{\partial^2 u'}{\partial x'^2} + \frac{\partial^2 u'}{\partial z'^2}\right) \tag{9.3}$$

The two dimensionless quantities that result are of different orders of magnitude. The first, the inverse of the square of a Froude number $(C/\sqrt{gk^{-1}})$ is of order unity [written as $O(1)$], from the dispersion relationship, while the second term, $\nu k^2/\sigma$, is the inverse of a Reynolds number and $O(10^{-7} \text{ to } 10^{-8})$ for normal ocean waves. Hence, in general, this term may be neglected—an a posteriori justification of something that was already done in Chapter 3.

Neglecting the frictional stresses implies that there is a slip boundary condition at the bottom, $z = -h$, as from Chapter 3 we know that the bottom velocity is nonzero. However, physically, there is no flow at the bottom, due to the presence of fluid viscosity; hence our argument above must be modified.

Consider that near the bottom there is a small region where u varies radically with elevation. The vertical length scale there must be different and thus, rescaling, we have $z = \delta z'$, where δ is the thickness of the region over which u changes rapidly. Again, the horizontal equation of motion is

$$\frac{\partial u'}{\partial t'} = -\frac{gk}{\sigma^2}\frac{\partial p'}{\partial x'} + \frac{vk^2}{\sigma}\frac{\partial^2 u'}{\partial x'^2} + \frac{v}{\sigma\delta^2}\frac{\partial^2 u'}{\partial z'^2} \tag{9.4}$$

The last term can become of $O(1)$ if $\delta \propto \sqrt{v/\sigma}$. The length scale δ is a convenient measure for the laminar boundary layer thickness and it is very small. For example, for a 5-s wave, $\delta \propto 1$ mm.

To summarize the scaling argument, very near the bottom, $O(\delta)$, viscous effects can become very important. It is therefore convenient to divide the flow field into two parts, an irrotational and a rotational component, or

$$u = u_p + u_r \tag{9.5}$$

where u_p satisfies the Euler equation,

$$\frac{\partial u_p}{\partial t} = -\frac{1}{\rho}\frac{\partial p}{\partial x} \tag{9.6}$$

and u_r satisfies the approximate rotational equation

$$\frac{\partial u_r}{\partial t} = v\frac{\partial^2 u_r}{\partial z^2} \tag{9.7}$$

The reader should verify the validity of this procedure using Eq. (9.1) and the principle of superposition. It is expected that u_r goes to zero away from the boundary.

For water waves, we know u_p from Chapter 3.

$$u_p = \frac{gak}{\sigma}\frac{\cosh k(h+z)}{\cosh kh}\cos(kx - \sigma t) \tag{9.8a}$$

or, in complex notation,

$$u_p = \frac{gak}{\sigma}\frac{\cosh k(h+z)}{\cosh kh}e^{i(kx-\sigma t)} \tag{9.8b}$$

where only the real part is used here and in the following complex-valued expressions. To find u_r, separation of variables is used, and keeping only the term that decays away from the bed, we find that[1]

---

[1]The u_r term is exactly the same expression as found by solving the problem of an oscillating bottom in a still fluid (Lamb, 1945, Sec. 345).

$$u_r = Ae^{\sqrt{-i\sigma/\nu}(z+h)}e^{i(kx-\sigma t)} = Ae^{-(1-i)\sqrt{\sigma/2\nu}(z+h)}e^{i(kx-\sigma t)} \tag{9.9}$$

The complex nature of the exponent of the $(z + h)$ term indicates that there is an exponential decay away from the bed modified by an oscillating term. The no-slip boundary condition at $z = -h$, $u = u_p + u_r = 0$, fixes A,

$$A = -\frac{gak}{\sigma}\frac{1}{\cosh kh} \tag{9.10}$$

The real part of the total horizontal velocity u is therefore

$$u = \frac{gak}{\sigma \cosh kh}[\cosh k(h + z) \cos (kx - \sigma t) \tag{9.11}$$

$$- e^{-\sqrt{\sigma/2\nu}(z+h)} \cos (kx - \sigma t + \sqrt{\frac{\sigma}{2\nu}}(z + h))]$$

which shows there is a phase shift of the viscous term with elevation. The horizontal velocity profile near the bed is shown in Figure 9.1, for a given wave, with $k\delta = 0.01$ and $\delta = \sqrt{\nu/2\sigma}$.

The vertical velocity in the bottom boundary layer is most conveniently found from the continuity equation,

$$w = -\int_0^{h+z} \frac{\partial u}{\partial x} ds = -\frac{gak}{\sigma \cosh kh}\left[\sinh k(h + z)e^{i(\psi+\pi/2)}\right. \tag{9.12}$$

$$\left. - k\sqrt{\frac{\nu}{\sigma}}(e^{-(1-i)\sqrt{\sigma/2\nu}(z+h)} - 1)e^{i(\psi+3\pi/4)}\right]$$

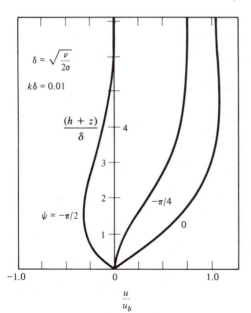

$$\delta = \sqrt{\frac{\nu}{2\sigma}}$$

$$k\delta = 0.01$$

$$\frac{(h + z)}{\delta}$$

$$\psi = -\pi/2$$

Figure 9.1 Normalized velocity profiles for various phase positions ψ in a laminar boundary layer. For $x = 0$, the velocity profiles depict the fluid motion in the boundary layer as the crest arrives.

where $\psi = kx - \sigma t$ and s is the elevation above the bottom. The vertical velocity consists of two terms near the bottom. The first is the wave-induced term and the second is the boundary layer correction term, which, incidentally, is much smaller than u_r.

The instantaneous shear stress exerted on the bed may be obtained from the Newtonian shear stress term

$$\tau_{xz} = \rho v \left(\frac{\partial u}{\partial z} + \frac{\partial w}{\partial x} \right) \Bigg|_{z=-h} \tag{9.13}$$

of which only the first term is large,

$$\tau_{xz} \approx \rho v \left(\frac{\partial u_r}{\partial z} \right) \Bigg|_{z=-h}$$

or

$$\tau_{xz} \simeq \rho \sqrt{\frac{v}{\sigma}} \frac{gak}{\cosh kh} \cos \left(kx - \sigma t - \frac{\pi}{4} \right) \tag{9.14}$$

The bed shear stress is thus harmonic in time and lags the free surface displacement by 45°. The mean bed shear stress is zero.

A conventional form for a shear stress in an oscillatory flow is

$$\tau_{xz} = \frac{\rho f}{8} u_b |u_b| \tag{9.15}$$

where u_b is the bottom velocity given by potential flow outside of the boundary layer (i.e., the potential flow value) and f is a friction factor. In terms of the maximum value $(\tau_{xz})_{max}$, we use

$$(u_b)_{max} = \frac{gak}{\sigma \cosh kh} = \zeta_b \sigma \tag{9.16}$$

where ζ_b is the maximum of the (inviscid) horizontal excursion of the water particle at the bottom. Relating the conventional form of the shear stress to the previously derived form,

$$\rho \sqrt{\frac{v}{\sigma}} \frac{gak}{\cosh kh} = \frac{\rho f (u_b)_{max}^2}{8} \tag{9.17}$$

or, after some manipulation,

$$f = \frac{8}{\mathbb{R}_b^{1/2}} \tag{9.18}$$

where $\mathbb{R}_b$ is the Reynolds number defined as

$$\mathbb{R}_b = \frac{u_b \zeta_b}{v} \tag{9.19}$$

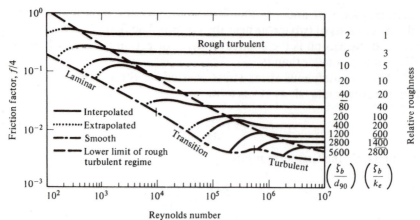

Figure 9.2   Stanton diagram for friction factor under waves as a function of $\mathbb{R}_b$ and ζ_b/k_e. The line labeled "Laminar" denotes Eq. (9.18). (From Kamphuis, 1975.)

The friction factor is plotted versus $\mathbb{R}_b$ in Figure 9.2. For smooth bottoms, the expression is valid for $\mathbb{R}_b$ up to 10^4.

Due to the presence of the shear stress, there is work done by the waves against the shear stress within the fluid. The mean rate of energy dissipation per unit time is given by

$$\epsilon_D = \rho v \overline{\int_0^h 2\left(\frac{\partial u}{\partial x}\right)^2 + \left(\frac{\partial w}{\partial x} + \frac{\partial u}{\partial z}\right)^2 + 2\left(\frac{\partial w}{\partial z}\right)^2} \, ds \qquad (9.20)$$

where the overbar denotes the time average over a wave period and $s = h + z$. The largest term in this expression is

$$\epsilon_D = \rho v \overline{\int_0^h \left(\frac{\partial u_r}{\partial z}\right)^2} \, ds$$

$$= \rho v \frac{g^2 a^2 k^2}{\sigma^2 \cosh^2 kh} \sqrt{\frac{\sigma}{2v}} \frac{1}{4} = \frac{\rho v g a^2 k}{2 \sinh 2kh} \sqrt{\frac{\sigma}{2v}} = \frac{vk\sqrt{\sigma/2v}\, E}{\sinh 2kh} \qquad (9.21)$$

If in the conservation of energy equation we set

$$\frac{dE}{dt} = -\epsilon_D \qquad (9.22)$$

where $E = \frac{1}{2}\rho g a^2$ and $a = a_0 e^{-\alpha bt}$, which is the assumed damping law for the wave amplitude, where $a = a_0$ at $t = 0$, we have for a damping coefficient,

$$\alpha_b = \frac{gk^2}{4\sigma \cosh^2 kh} \sqrt{\frac{2v}{\sigma}} = \frac{vk\sqrt{\sigma/2v}}{2 \sinh 2kh} \qquad (9.23)$$

Clearly, since the boundary layer thickness $\delta \ (= \sqrt{v/2\sigma})$ is in general small, the damping is also small. For a 5-s wave in 5 m of water, with a 1-mm-thick boundary layer, $\alpha_b = 4.8 \times 10^{-5}$ s^{-1}, or for a wave to decay to $e^{-1} = 0.368$ requires 1×10^4 s or a propagation distance equal to 126 km. (This is only considering the bottom effect.)

Example 9.1

Determine the amount of damping that will occur after a wave propagates a distance l in water of constant depth h.

Solution.   Using Eq. (9.21), we get

$$\frac{dE}{dt} = C_g \frac{dE}{dx} = -\epsilon_D = -\frac{vk}{\sinh 2kh} \sqrt{\frac{\sigma}{2v}} E \tag{9.24}$$

Now, since h and k are not functions of x, we can write this as

$$\frac{dE}{E} = -\frac{vk}{C_g \sinh 2kh} \sqrt{\frac{\sigma}{2v}} dx \tag{9.25}$$

Integrating yields

$$E = E_0 \exp\left(-\frac{vk^2}{n\sigma \sinh 2kh} \sqrt{\frac{\sigma}{2v}} x\right) \tag{9.26}$$

where the boundary condition of $E = E_0$ at $x = 0$ was used. The wave amplitude at $x = l$ where $l = 100$ km will be

$$a = a_0 \exp\left(-\frac{k^2 \sqrt{v/2\sigma l}}{2kh + \sinh 2kh}\right) \tag{9.27}$$

In the irrotational part of the wave motion, the loss of energy can be calculated in the same manner:

$$\epsilon_D = \rho v \overline{\int_{-h}^{0} \left\{ 2\left(\frac{\partial u}{\partial x}\right)^2 + \left(\frac{\partial w}{\partial x} + \frac{\partial u}{\partial z}\right)^2 + 2\left(\frac{\partial w}{\partial z}\right)^2 \right\} dz} \tag{9.28}$$

Integrating

$$\epsilon_D = 2\rho v a^2 g k^2 \tag{9.29}$$

where

$$a = a_0 e^{-\alpha_i t}$$

For this internal damping, $\alpha_i = 2vk^2$.

If we compare the two damping rates, we find that in deep water the latter damping is greater, as the bottom does not affect the waves, whereas in shallow water

$$\frac{(\epsilon_D)_{\text{bottom}}}{(\epsilon_D)_{\text{core}}} = \frac{1}{8}\left(\frac{1}{kh}\right)\frac{1}{k\delta} \tag{9.30}$$

Since in shallow water, $kh < \pi/10$ and $k\delta$ is much smaller, the bottom damping is much more significant.

At the free surface, there exists another boundary layer which contributes a small damping (Phillips, 1966),

$$\alpha_f = \frac{vk\sqrt{\sigma/2v}}{2\tanh kh} \tag{9.31}$$

This is always much smaller than the interior and the bottom boundary layer damping:

$$\frac{\alpha_f}{\alpha_i} = \frac{k\delta}{n}\tanh kh \tag{9.32}$$

$$\frac{\alpha_f}{\alpha_b} = \frac{k^2\delta^2}{\cosh^2 kh} \tag{9.33}$$

9.2.2 Turbulent Boundary Layers

When waves become large or the bottom is rough, the boundary layer is turbulent. In fact, for most cases in nature, a turbulent boundary layer exists. This implies (in analogy to steady flow over flat plates) that the boundary layer is thicker, the shear stress on the bottom is larger, and it depends on the square of the bottom velocity rather than linearly.

Experimental work by Jonsson (1966), Kamphuis (1975), and Jonsson and Carlsen (1976) as well as theoretical work by Kajiura (1968) has provided insight into the nature of the turbulent boundary layer and its dependency on Reynolds number and the relative roughness of the bed, which is defined as k_e/ζ_b, where k_e is the equivalent sand grain size on the bed and ζ_b is the excursion of the wave-induced water particle motion at the bottom in the absence of the boundary layer. Kamphuis (1975) indicates, with some reservations due to accuracy, that k_e can be related to the distribution of sand sizes present on the bottom by

$$k_e = 2d_{90}$$

where d_{90} is the sand size for which 90% of the sand is finer. Using a Stanton-type diagram (as used for pipe friction factors), Kamphuis has plotted the friction factor f versus Reynolds number and relative roughness as shown in Figure 9.2. As in pipe flow, for rough turbulent flow, there is no effect of $\mathbb{R}_b$, and Kamphuis proposed that

$$f \simeq 0.1\left(\frac{k_e}{\zeta_b}\right)^{3/4} \qquad \text{for } k_e/\zeta_b > 0.02 \tag{9.34}$$

and

$$\frac{1}{2\sqrt{f}} + \ell n \frac{1}{2\sqrt{f}} = -0.35 - \frac{4}{3}\ell n \frac{k_e}{\zeta_b} \qquad \text{for } k_e/\zeta_b < 0.02 \tag{9.35}$$

These equations are valid when

$$\frac{k_e}{\zeta_b} \mathbb{R}_b \sqrt{\frac{f}{8}} \geq 200 \tag{9.36a}$$

or, more stringently,

$$\frac{k_e}{\zeta_b} \geq \frac{2200}{\mathbb{R}_b} \tag{9.36b}$$

which is the condition for rough turbulent flow, when $\mathbb{R}_b > 5 \times 10^4$.
The mean bottom shear stress due to the action of the waves is still zero:

$$\overline{\tau_{xz}} = \overline{\frac{\rho f}{8} u_b |u_b|}$$

$$= \overline{\frac{\rho f}{8}(u_{b_{max}})^2 \cos(kx - \sigma t) |\cos(kx - \sigma t)|} \tag{9.37}$$

$$= 0$$

The energy damping however is nonzero and determined by the relationship

$$\overline{\tau_{xy} u_b} = \epsilon_D$$

or

$$\epsilon_D = \overline{\frac{\rho f}{8}(u_{b_{max}})^3 \cos^2(kx - \sigma t) |\cos(kx - \sigma t)|} \tag{9.38}$$

Averaging over a wave period, we have

$$\epsilon_D = \frac{\rho f}{6\pi}(u_{b_{max}})^3 = \frac{\rho f}{6\pi}\left(\frac{a\sigma}{\sinh kh}\right)^3 \tag{9.39}$$

which clearly increases as the depth decreases.

The decay of the wave height with distance over a flat bottom can be obtained from the energy equation.

$$\frac{dEC_g}{dx} = -\epsilon_D$$

or $$\tag{9.40}$$

$$\frac{1}{2}\rho g C_g \frac{da^2}{dx} = -\frac{\rho f}{6\pi}\frac{\sigma^3}{\sinh^3 kh} a^3$$

Solving for the wave amplitude a by separation, we find that

$$a(x) = \frac{a_0}{1 + \frac{2f}{3\pi}\frac{k^2 a_0 x}{(2kh + \sinh 2kh)\sinh kh}} \tag{9.41}$$

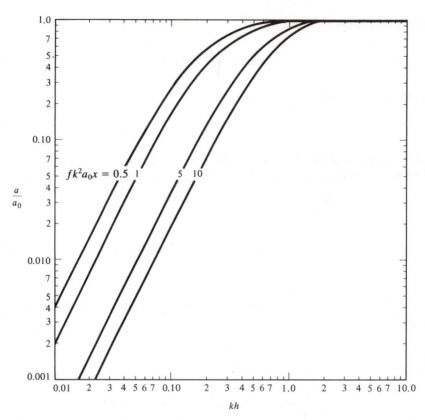

Figure 9.3 Damping of waves due to damping in a turbulent boundary layer.

This relationship is plotted in Figure 9.3. The amount of wave height decay clearly increases with friction factor as expected and depends on the water depth. In deep water a/a_0 goes to unity as the bottom friction becomes negligible, while the shallow water asymptote is

$$\frac{a}{a_0} = \frac{1}{1 + \dfrac{f}{6\pi} \dfrac{a_0 x}{h^2}} \tag{9.42}$$

The energy loss for a wave with a turbulent boundary layer can be compared to the laminar boundary layer case by relation to the two formulas (9.21) and (9.39):

$$\frac{(\epsilon_D)_{\text{laminar}}}{(\epsilon_D)_{\text{turbulent}}} = \frac{\dfrac{Evk\sqrt{\sigma/2v}}{\sinh 2kh}}{\dfrac{\rho f}{6\pi}\dfrac{a^3\sigma^3}{\sinh^3 kh}} = \frac{6\pi}{fa}\sqrt{\frac{v}{2\sigma}}\sinh kh \tag{9.43}$$

The smallest value of f is its laminar value f_L. Expressing f as βf_L for the turbulent case where β is greater than 1, the ratio is reduced to

$$\frac{(\epsilon_D)_{\text{laminar}}}{(\epsilon_D)_{\text{turbulent}}} = \frac{6\pi}{8\sqrt{2}\,\beta} \qquad (9.44)$$

or $\beta > 1.66$ for turbulent boundary layer to give greater damping. In general, this is the case, as can be deduced from the wave friction factor diagram.

Example 9.2

A wave of 5 m amplitude propagates a distance l with an average depth of 30 m. What is the final wave height? Given is $h = 30$ m, $T = 10$ s, $d_{90} = 0.3$ mm, and $l = 100$ km.

Solution.   From the dispersion relationship, $k = 0.0457$ m^{-1}. Next the friction factor must be determined.

$$\frac{k_e}{\zeta_b} = \frac{2(0.0003)}{5/\sinh 1.372} \qquad \text{as } \zeta_b = a/\sinh kh \qquad (9.45)$$

$$= 0.00022$$

and

$$\mathbb{R}_b = \frac{u_b \zeta_b}{\nu} = \frac{\zeta_b^2 \sigma}{\nu} = \frac{a^2 \sigma}{\nu \sinh^2 kh} = 3.3 \times 10^5 \qquad (9.46)$$

From Figure 9.2, $f = 0.004$.
    The quantity $fk^2 a_0 x = (0.004)\,(0.0457)^2\,(5)\,(100{,}000) = 4.18$ and

$$\frac{a}{a_0} = \cfrac{1}{1 + \cfrac{2}{3\pi}(fk^2 a_0 x)\cfrac{1}{(2kh + \sinh 2kh)\sinh kh}} = 0.956 \qquad (9.47)$$

or

$$a = 4.78 \text{ m}$$

This represents a 4% decrease in wave amplitude due to bottom frictional damping (over a smooth bottom).

9.3 WATER WAVES OVER A VISCOUS MUD BOTTOM

One representation of a soil bottom would be to characterize it as a viscous fluid. Examples of this type of bottom exist around the world, particularly near the mouths of large sediment-bearing rivers, such as in the Gulf of Mexico near Louisiana (Gade, 1958) and the coast of Surinam (Wells and Coleman, 1978). The mud bottom often damps out wave energy so rapidly that these areas can serve as a harbor of refuge for fishermen caught far away from home port by storms.
    The mathematical treatment follows by assuming a laminar flow of a highly viscous liquid overlain by an inviscid fluid. The surface water wave

Water

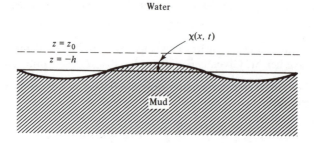

Figure 9.4 Schematic of waves over a mud bottom.

described by linear theory will drive an interfacial wave on the mud–water boundary that induces flows in the lower layer. These flows are rapidly damped by viscosity. Figure 9.4 shows a schematic of the waves and fluid regions.

9.3.1 Water Wave Region

In the overlying fluid, the Laplace equation and the linearized free surface boundary condition as discussed in Chapter 3 must be satisfied by the fluid motions. Further at the mud–water interface, continuity of pressure and vertical velocities must hold across the interface.

In the upper fluid region, denoted region 1, the velocity potential is assumed to be

$$\phi_1(x, z, t) = (A \cosh k(h + z) + B \sinh k(h + z))e^{i(kx - \sigma t)} \qquad (9.48)$$

The ϕ_1 is clearly periodic in space and time, and satisfies the Laplace equation (3.19). The LDFSBC (3.33b) yields

$$A \cosh kh + B \sinh kh = \frac{iga_0}{\sigma} \qquad (9.49)$$

while the LKFSBC (3.29) yields

$$A \sinh kh + B \cosh kh = \frac{i\sigma a_0}{k} \qquad (9.50)$$

The Bernoulli constant $C(t)$ has been taken to be zero, to ensure a zero spatial mean for $\eta(t)$, which has been assumed as the real part of $\eta(x, t) = a_0 e^{i(kx - \sigma t)}$.

With two equations and three unknowns, A, B, and k, we can solve for two of them.

$$A = \frac{ia_0 \cosh kh}{\sigma k} (gk - \sigma^2 \tanh kh) \tag{9.51}$$

$$B = \frac{ia_0 \cosh kh}{\sigma k} (\sigma^2 - gk \tanh kh) \tag{9.52}$$

Now, if we were solving the rigid bottom case, as in Chapter 3, we would finally specify that the vertical flow, at $z = -h$, was zero. This would require B to be zero, which implies that the terms within the parentheses must be zero. Hence the dispersion relationship, relating k to σ, results as before. However, in this case, since the bottom is not fixed and its location is unknown a priori, two interfacial boundary conditions are necessary to find an equivalent dispersion relationship. First, however, the fluid motion within the mud will be prescribed.

9.3.2 Mud Region

For convenience, we will assume that the mud region is infinite in depth (practically, this requires that it be at least as deep as $L/2$, where L is the wave length). Furthermore, a boundary layer approach will again be used; that is, the flow will be assumed inviscid except in the boundary layer regions (which, of course, can be very large). This is valid (Mei and Liu, 1973) as long as the kinematic viscosity v is very small. Therefore, the fluid mud region will be described by a solution to the Laplace equation, which is spatially and temporally periodic, since it is driven by the water wave. The potential function is then presumed to be of the following form, where d is unknown:

$$\phi_2(x, z, t) = de^{k(z+h)}e^{i(kx-\sigma t)} \tag{9.53}$$

A boundary layer correction for ϕ_2 is prescribed.

$$u_2 = fe^{(1-i)\sqrt{\sigma/2v}(z+h)}e^{i(kx-\sigma t)} \tag{9.54}$$

Recall from the laminar boundary layer treatment for waves that the vertical boundary layer velocity correction is very small.

The vertical velocity in each region must be the same as the motion at the interface (this is a kinematic boundary condition), so we have

$$\frac{\partial \chi}{\partial t} = -\frac{\partial \phi_1}{\partial z} = -\frac{\partial \phi_2}{\partial z} \qquad \text{on } z = -h + \chi(x, t) \tag{9.55}$$

where $\chi(x, t)$ is the vertical displacement of the interface, assumed to be

$$\chi(x, t) = m_0 e^{i(kx-\sigma t)} \tag{9.56}$$

Linearizing the kinematic boundary condition yields

$$\frac{\partial \chi}{\partial t} = -\frac{\partial \phi_1}{\partial z} = -\frac{\partial \phi_2}{\partial z} \qquad \text{on } z = -h \tag{9.57}$$

or

$$-i\sigma m_0 = -kB = -dk$$

Thus

$$d = B \tag{9.58}$$

and

$$m_0 = -\frac{ikB}{\sigma} \tag{9.59}$$

The continuity of pressure, which states that the pressure must be the same on both sides of the interface (since it is free and is assumed to have no surface tension, it cannot develop a force), can be written (in linear form) as

$$p_1 = p_2 \qquad \text{on } z = -h + \chi$$

or

$$\rho_1 \frac{\partial \phi_1}{\partial t} - \rho_1 g z = \rho_2 \frac{\partial \phi_2}{\partial t} - \rho_2 g z + (\rho_1 - \rho_2)gh \qquad \text{on } z = -h + \chi \tag{9.60}$$

Note that the last term on the right-hand side is necessary due to the two fluid densities present. Linearizing, we obtain

$$\rho_1 \frac{\partial \phi_1}{\partial t} - \rho_1 g \chi = \rho_2 \frac{\partial \phi_2}{\partial t} - \rho_2 g \chi \qquad \text{on } z = -h \tag{9.61}$$

Substituting for ϕ_1, ϕ_2, and z results in the following equation relating A to B:

$$A = \left[\frac{\rho_2}{\rho_1} \left(1 - \frac{gk}{\sigma^2} \right) + \frac{gk}{\sigma^2} \right] B \tag{9.62}$$

We have, however, already developed equations for A and B in terms of k [Eqs. (9.51) and (9.52)] and by substituting for A and B, we find the dispersion relationship, or

$$\left(\frac{\rho_2}{\rho_1} + \tanh kh \right) \sigma^4 - \frac{\rho_2}{\rho_1} gk (1 + \tanh kh)\sigma^2 + \left(\frac{\rho_2}{\rho_1} - 1 \right) (gk)^2 \tanh kh = 0 \tag{9.63}$$

This relationship, relating k to σ, can be factored as

$$(\sigma^2 - gk) \left[\sigma^2 \left(\frac{\rho_2}{\rho_1} + \tanh kh \right) - \left(\frac{\rho_2}{\rho_1} - 1 \right) gk \tanh kh \right] = 0 \tag{9.64}$$

Thus two possible roots exist for waves propagating in the positive x direction:

$$\sigma^2 = gk \tag{9.65}$$

and

$$\sigma^2 = \frac{gk\left(\dfrac{\rho_2}{\rho_1} - 1\right)\tanh kh}{\dfrac{\rho_2}{\rho_1} + \tanh kh} \tag{9.66}$$

These two dispersion relationships are plotted in Figure 9.5. The two possible wave modes can be distinguished by the ratio of the amplitudes of the surface wave and interfacial wave, which is for each case (Lamb, 1945)

$$\frac{a_0}{m_0} = e^{kh} \tag{9.67a}$$

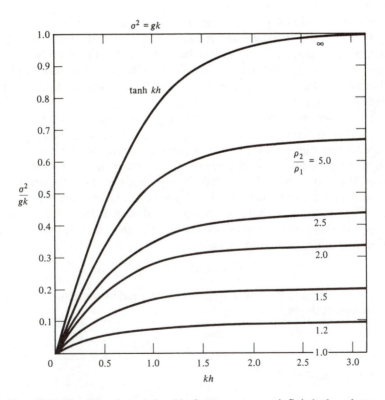

Figure 9.5 The dispersion relationship for waves over an infinitely deep denser lower fluid. Note that the deep water asymptotes are $\sigma^2/gk = (\rho_2/\rho_1 - 1)(\rho_2/\rho_1 + 1)$ for the model wave.

$$\frac{a_0}{m_0} = -\left(\frac{\rho_2}{\rho_1} - 1\right)e^{-kh} \tag{9.67b}$$

Therefore, the two cases are distinguished by which is larger, the surface or the interfacial amplitude, and whether the interface is in phase with or 180° out of phase with the free surface.

The first wave mode, $\sigma^2 = gk$, is interesting, as from Eq. (9.62), $A/B = 1$. The expressions for the two velocity potentials are

$$\phi_1 = \phi_2 = Ae^{k(h+z)}e^{i(kx-\sigma t)} \tag{9.68}$$

Thus the two regions, above and below the interface, are indistinguishable. The presence of a lower, more dense layer has no effect on the wave motion. This result, which is only true for the case of an infinitely deep mud layer, results from the fact that the interface is a constant pressure surface. Heuristically, we could remove the overlying water and the interfacial wave could propagate as a surface with the same (deep water) celerity. In shallow water, this would not be true as the interfacial wave no longer corresponds to a constant pressure surface.

For this mode of wave motion, there is no discontinuity of horizontal velocity across the interface and hence there is no boundary layer and no associated damping. (There would be damping if the mud were highly viscous, as damping would take place outside the boundary layers.) For the shallow water case, damping does occur and Dalrymple and Liu (1978) have treated this problem.

The other wave mode with the large out-of-phase interfacial wave creates an unusual effect in the upper layer. The free surface displacement can be viewed as a right-side-up wave, while the interfacial wave is an upside-down wave propagating at the same speed and in the same direction. In between the two, it could be intuitively expected that a quasi-bottom might exist, and in fact, one does. At the elevation z_0 in the upper layer where $w(x, z_0) = -\partial\phi/\partial z = 0$, there is no vertical flow and this then is the false bottom. For this elevation, it can be shown (Problem 9.1) that the dispersion relationship in Eq. (9.66) reduces to

$$\sigma^2 = gk \tanh k|z_0| \tag{9.69}$$

The damping in the lower layer is determined by matching the horizontal velocities at the interface.

$$-\frac{\partial\phi_1}{\partial x} = -\frac{\partial\phi_2}{\partial x} + u_2 \qquad \text{at } z = -h \tag{9.70}$$

yielding $f = ik(d - A)$ or

$$f = -\frac{a_0}{\sigma} e^{kh}(\sigma^2 - gk) \tag{9.71}$$

The damping in the boundary layer is found as

$$\epsilon_D = \rho_2 \nu_2 \int_{-\infty}^{-h} \left(\frac{\partial u_2}{\partial z}\right)^2 dz \qquad (9.72)$$

$$= \frac{\rho_2 \sqrt{\sigma \nu_2}}{4} \frac{a_0^2}{\sigma^2} e^{2kh} (\sigma^2 - gk)^2 \qquad (9.73)$$

Of the two possible wave modes discussed, the problem remains as to which mode is more "realistic." The quotation marks are used as both solutions are in fact realistic, but the means by which the waves are generated determines the mode. For example, for waves propagating into a muddy region, it is probable that the first mode ($\sigma^2 = gk$) is the most likely one, as the wave lengths associated with the second mode are very short, particularly for small values of ρ_2/ρ_1. However, as ρ_2/ρ_1 becomes large, it is possible that both modes are excited. If, on the other hand, the waves are generated at the interface by a displacement of the mud, it is more likely the second mode will be the only one present. This wave, which exists primarily at the interface, propagates very slowly, due to the fact that the restoring force which causes the wave to propagate is a result of the density differences between the two fluids.

Example 9.3

Determine the wave lengths of the two possible modes of wave propagation over an infinitely deep mud layer, with $\rho_2/\rho_1 = 1.2$. The overlying water column is 4.6 m in depth and the wave period is 8 s.

Solution. In Figure 9.5, the ordinate may be written as $\sigma^2 h/gkh$ for convenience. $\sigma^2 h/g$ is computed as 0.287 for this case. For mode 1 we have $\sigma^2 h/gkh = 1$ at $kh = 0.287$. This yields a wave length of 100 m. For the second mode, we have to use an iterative technique. If we guess $kh = 2.0$, from the figure we find for $kh = 2.0$ and $\rho_2/\rho_1 = 1.2$ that $\sigma^2 h/gkh \simeq 0.087$. Dividing this number into $\sigma^2 h/g$ yields kh; $kh = 3.30$. Therefore, an estimate of 2 for kh was too low. Now we estimate kh as 3.0, which yields $\sigma^2 h/g = 0.09$, or $kh = 3.19$. Iterating, we find that 3.16 is a good value. Therefore, $L = 9.1$ m. By comparison, the wave length of the wave over a rigid bottom at 4.6 m is 51 m.

9.4 WAVES OVER RIGID, POROUS BOTTOMS

Sandy seabeds can be characterized as a porous medium, thus permitting mathematical treatment. Since Darcy's experiments in the 1800s, investigators have treated soils as a continuum, with spatially averaged flows, rather than worrying about the flows in the tortuous channels between the sand grains. The solution of this problem will be similar to the preceding case. A governing equation will be developed for the flows in the bed; these flows will be matched to those induced by the waves in the fluid region, and the damping due to the forced flow in the granular medium will be calculated.

For a full saturated soil, which is assumed to be incompressible (as is the fluid), the conservation of mass leads to

$$\nabla \cdot \mathbf{u} = 0 \qquad (9.74)$$

where $\mathbf{u}$ is the discharge velocity or the average velocity across a given area of soil (including both the intercepted areas of the soil particles and the pores between them). Darcy's law relates the velocity to the pressure gradients in the fluid:

$$\mathbf{u} = -\frac{K}{\mu} \nabla p_s \qquad (9.75)$$

where K is a constant called the permeability, which is a characteristic of the soil, and μ is the dynamic viscosity of the fluid.[2] The governing equation for the fluid in the soil is obtained by substituting for $\mathbf{u}$ into the conservation of mass equation, Eq. (9.74), or

$$\nabla \cdot \left(-\frac{K}{\mu} \nabla p_s \right) = 0 \qquad (9.76)$$

or

$$\nabla^2 p_s = 0$$

Thus the pore pressure satisfies the Laplace equation, as does the velocity potential in the fluid. In order to match the two solutions, p_s and ϕ, the boundary conditions will be that the pressure be continuous across the soil–water interface, as are the vertical velocities.

The assumed progressive wave forms of ϕ and p_s are

$$\phi(x, z) = [A \cosh k(h + z) + B \sinh k(h + z)] \, e^{i(kx - \sigma t)} \qquad (9.77)$$

and

$$p_s(x, t) = D e^{k(h+z)} \, e^{i(kx - \sigma t)} \qquad (9.78)$$

The continuity of pressure across the interface requires that

$$p(x, -h) = p_s(x, -h) \qquad (9.79)$$

where the subscript s again denotes the soil region pressure. Rewriting, we have

$$\rho \left. \frac{\partial \phi}{\partial t} \right|_{z=-h} = p_s(x, -h) \qquad (9.80)$$

---

[2]This equation, which neglects the acceleration terms, assumes that the flow can be treated quasi-statically. An order-of-magnitude analysis bears this out for most sand beds.

or

$$-i\sigma\rho A = D$$

For the vertical velocities to be continuous,

$$-\frac{\partial\phi}{\partial z} = -\frac{K}{\mu}\frac{\partial p_s}{\partial z} \qquad \text{at } z = -h$$

or

$$B = \frac{KD}{\mu} \tag{9.81}$$

So far we have two equations for the three unknowns A, B, and D; now we use the linear free surface boundary conditions to relate them to the wave amplitude a and to obtain the dispersion relationship. The linear dynamic free surface boundary condition yields

$$\eta = \frac{1}{g}\frac{\partial\phi}{\partial t} = -\frac{i\sigma}{g}(A \cosh kh + B \sinh kh)e^{i(kx-\sigma t)} = ae^{i(kx-\sigma t)} \tag{9.82}$$

Substituting for A and B from above yields

$$D = \rho ga\left\{\cosh kh\left[1 - \left(\frac{i\rho\sigma K}{\mu}\right)\tanh kh\right]\right\}^{-1} \tag{9.83}$$

Application of the linear kinematic free surface boundary condition provides the dispersion relationship,

$$\frac{\partial\eta}{\partial t} = -\frac{\partial\phi}{\partial z} \qquad \text{on } z = 0 \tag{9.84}$$

$$i\sigma a = Ak \sinh kh + Bk \cosh kh$$

or, substituting for a, A, and B, in terms of D, results in

$$\sigma^2\left[1 - i\left(\frac{\sigma K}{v}\right)\tanh kh\right] = gk\left(\tanh kh - \frac{i\sigma K}{v}\right) \tag{9.85}$$

where $v = \mu/\rho$, the kinematic viscosity. Reordering gives

$$\sigma^2 - gk \tanh kh = -i\left(\frac{\sigma K}{v}\right)(gk - \sigma^2 \tanh kh) \tag{9.86}$$

This dispersion relationship is complex, yielding a complex k, which may be written as $k = k_r + ik_i$. The real part of k represents the real wave number, that is, it is related to the wavelength, while the imaginary component determines the spatial damping rate. This follows by examining the free surface profile,

$$\eta(x, t) = a_0 e^{i(kx-\sigma t)} = a_0 e^{i((k_r+ik_i)x-\sigma t)}$$

$$= a_0 e^{-k_i x} e^{i(k_r x-\sigma t)} \tag{9.87}$$

Thus there is exponential damping due to k_i being greater than zero.

The quantity $\sigma K/v$ in Eq. (9.86) is generally small. For sand, K ranges from about 10^{-9} to 10^{-12} m^2, while the kinematic viscosity is $O(10^{-6})$. Therefore, $\sigma K/v$ ranges from 10^{-2} to 10^{-5}, which is small.

Approximate solutions can be obtained from the dispersion relationship. In intermediate depth we can replace $\cosh kh = \cosh k_r h + ik_i h \sinh k_r h$, as a priori we expect $k_i h \ll 1$; similarly for $\sinh kh$. Substituting into Eq. (9.86) for the hyperbolic functions and k, we can separate it into real and imaginary parts.

Real: $\qquad (\sigma^2 - Rgk_i) - Rgk_i hk_r \tanh k_r h$

$$= gk_r \tanh k_r h - (gk_i + R\sigma^2)k_i h \qquad (9.88)$$

Imaginary: $\quad (\sigma^2 - Rgk_i)k_i h \tanh k_r h$

$$= gk_r(R + k_i h) + (gk_i + R\sigma^2) \tanh k_r h \qquad (9.89)$$

where $R = \sigma K/v$. Neglecting the small products of Rk_i and k_i^2 in the real expression gives

$$\sigma^2 \simeq gk_r \tanh k_r h \qquad (9.90)$$

while the second expression, after some algebra, yields

$$k_i \simeq \frac{2(\sigma K/v)k_r}{2k_r h + \sinh 2k_r h} \qquad (9.91)$$

as found by Reid and Kajiura (1957). This result is plotted in Figure 9.6.

In shallow water, $|kh| < \pi/10$, the dispersion relationship can be written as

$$\sigma^2 - gk^2 h = -iRgk\left(1 - \frac{\sigma^2 h}{g}\right) \qquad (9.92)$$

Substituting again $k = k_r + ik_i$ and separating into real and imaginary parts gives

Real: $\qquad \sigma^2 - g(k_r^2 - k_i^2)h = Rgk_i\left(1 - \frac{\sigma^2 h}{g}\right) \qquad (9.93)$

Imaginary: $\quad 2gk_r k_i h = -Rk_r g\left(1 - \frac{\sigma^2 h}{g}\right) \qquad (9.94)$

Solving for k_i and k_r gives us

$$k_i = \frac{(1 - \sigma^2 h/g)\,(\sigma K/v)}{2h} = \frac{(1 - k_0 h)\,(\sigma K/v)}{2h} \qquad (9.95)$$

$$k_r = \frac{\sigma}{\sqrt{gh}}\left[1 - \frac{R^2 g}{4h\sigma^2}\left(1 - \frac{\sigma^2 h}{g}\right)\right]^{1/2} \qquad (9.96)$$

These expressions are more accurate in shallow water than the previous expressions. The shallow water asymptote for k_i is $(1/2h)(K\sigma/v)$.

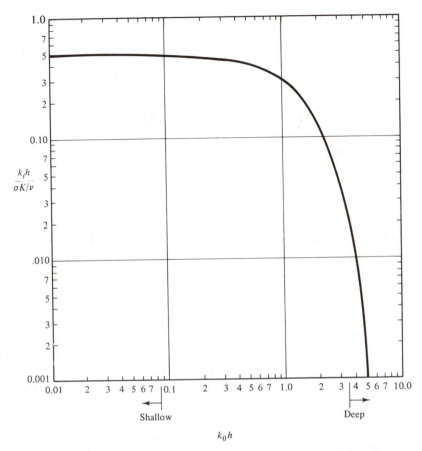

Figure 9.6 Dimensionless damping coefficient versus depth.

Liu (1973) included a laminar boundary layer at the fluid–soil interface, so as to eliminate the discontinuity in the horizontal velocity, and developed an approximate expression for the combined damping due to the porous media and the laminar boundary layer. These can be shown to $O(\sigma K/v)$ to be the sum of the damping rates due to the porous media, Eq. (9.91), and that due to the laminar boundary layer, Eq. (9.27):

$$k_i = \frac{2k_r}{2k_r h + \sinh 2k_r h}\left(\frac{\sigma K}{v} + k_r \sqrt{\frac{v}{2\sigma}}\right) \tag{9.97}$$

The damping rate of energy per unit time and per unit area ϵ_D is related by k_i by the energy conservation equation,

$$\frac{\partial C_g E}{\partial x} = -\epsilon_D \tag{9.98}$$

or approximately for a constant depth and $E = \frac{1}{2} \rho g a_0^2 e^{-2k_i x}$,

$$\epsilon_D = \frac{\rho g^2 a^2 K k_r}{2\nu \cosh^2 k_r h} \qquad (9.99)$$

for the porous damping alone and

$$\epsilon_D = \frac{\rho g^2 a^2 k_r}{2 \cosh^2 k_r h} \left(\frac{K}{\nu} + \frac{k_r}{2\sigma} \sqrt{\frac{\nu}{2\sigma}} \right) \qquad (9.100)$$

including the laminar boundary layer.

REFERENCES

DALRYMPLE, R. A., and P. L.-F. LIU, "Waves over Soft Muds: A Two-Layer Fluid Model," *J. Phys. Ocean.*, Vol. 8, pp. 1121–1131, 1978.

GADE, H. G., "Effects of a Non-rigid Impermeable Bottom on Plane Surface Waves in Shallow Water," *J. Mar. Res.*, Vol. 16, pp. 61–82, 1958.

JONSSON, I. G., "Wave Boundary Layers," *Proc. 10th Conf. Coastal Eng., ASCE,* Tokyo, 1966, pp. 127–148.

JONSSON, I. G., and N. A. CARLSEN, "Experimental and Theoretical Investigations in an Oscillatory Rough Turbulent Boundary Layer," *J. Hydraulics Res.*, Vol. 14, 1976.

KAJIURA, K., "A Model of the Bottom Boundary Layer in Water Waves," *Bull. Earthquake Res. Inst.* (Japan), Vol. 46, pp. 75–123, 1968.

KAMPHUIS, J. W., "Friction Factor under Oscillatory Waves," *J. Waterways, Harbors Coastal Eng. Div., ASCE*, Vol. 101, pp. 135–144, 1975.

LAMB, H., *Hydrodynamics*, Dover, New York, 1945.

LIU, P. L.-F., "Damping of Water Waves over Porous Bed," *J. Hydraulics Div., ASCE,* Vol. 99, pp. 2263–2271, 1973.

MEI, C. C., and P. L.-F. LIU, "A Note on the Damping of Surface Gravity Waves in a Bounded Liquid," *J. Fluid Mech.*, Vol. 59, p. 279, 1973.

PHILLIPS, O. M., *The Dynamics of the Upper Ocean*, Cambridge University Press, Cambridge, 1966.

REID, R. O., and K. KAJIURA, "On the Damping of Gravity Waves over a Permeable Seabed," *Trans. Am. Geophys. Union*, Vol. 38, 1957.

WELLS, J. T., and J. M. COLEMAN, "Longshore Transport of Mud by Waves: Northeastern Coast of South America," in H. J. MacGillavry and D. J. Beets (eds.), 8th Caribbean Geol. Conf. (Willemstad), *Geol. Mijnbouw*, Vol. 57, pp. 353–359, 1978.

PROBLEMS

9.1 Show that the dispersion relationship given for waves propagating over a viscous mud can be expressed as $\sigma^2 = gk \tanh k |z_0|$ [Eq. (9.69)].

9.2 For Example 9.3, find the damping ϵ_D for both modes.

9.3 Develop and solve the boundary value problem for waves propagating over a porous layer of thickness d.

9.4 The dynamic bottom pressure under a wave can be written as

$$p(x, -h,t) = \frac{\rho g\, a \cos (kx - \sigma t)}{\cosh kh}$$

With this as the boundary condition at $z = -h$ for the pressure $p_s(x, z)$ in a porous medium, develop the expression for $p_s(x, z)$. Compare this solution to that obtained in the text. What are the physical differences?

9.5 Relate the laminar damping under a progressive wave with distance [Eq. (9.27)] to the damped long wave [Eq. (5.80)]. What is f in terms of R_b for the long wave? Why the difference from Eq. (9.18)?

10

Nonlinear Properties Derivable from Small-Amplitude Waves

Dedication

HERMANN LUDWIG FERDINAND VON HELMHOLTZ

Hermann Ludwig Ferdinand von Helmholtz (1821–1894) was born in Potsdam, southwest of Berlin. The dedication of this chapter to Helmholtz is in recognition of his extensive contributions to fluid dynamics and physics in general. While he did work in the area of waves, his major contribution to this text is the Helmholtz equation, which governs the motion of waves in harbors.

Helmholtz entered the Pepiniere Berlin University in 1838 to study medicine. During his formal education, Gustav Magnus and others influenced him to expand his interest to natural sciences. In 1842 he graduated, successfully defending his work on ganglia. From 1842 to 1845, simultaneous to Kelvin's activities, he investigated the mechanical equivalent of heat. In 1849 he took a professorship in physiology at Königsburg, where he developed an interest in the importance of electricity in the working of the human body and studied ophthalmology and color vision. In 1855 he moved to Bonn and in 1858 to another chair at Heidelberg. There he developed his theories on vortex motion, free streamline flows, and the viscosity of water. In 1871 he succeeded Magnus at the University of Berlin, where he built a physical sciences institute which educated many well-known scientists, such as Heinrich Hertz and Max Planck. Planck has been quoted as observing: "Wir hatten das Gefühl, dass er sich selber mindestens ebenso langweilte wie wir" ("We had the feeling that he himself was at least as bored as we were"). Clearly, he engendered a testimonial distinct from the one Lamb received from his students.

In 1883 Helmholtz became a Prussian noble in recognition of his scientific contributions. In 1888 he assumed the leadership of the Physical Technical Government Institute (Reichsanstalt) in Charlotten-burg, West Berlin.

Other areas of interest for Helmholtz included the physiology of optics, binocular vision, acoustics, and the physiology of the ear, sound (harmony), and electrodynamics.

10.1 INTRODUCTION

Wave energy and power, which were derived in Chapter 4, are nonlinear quantities obtained from the linear wave theory—nonlinear in the sense that they involve the wave height to the second power. In this chapter other nonlinear quantities will be sought which have a bearing on coastal and ocean design. These quantities, which are time averaged, are correct to second order in ak, yet have their origin strictly in *linear* theory. In Chapter 11 a further and more complete study of nonlinear waves is undertaken.

10.2 MASS TRANSPORT AND MOMENTUM FLUX

If a small neutrally buoyant float is placed in a wave tank and its trajectory traced as waves pass by, a small mean motion in the direction of the waves can be observed. The closer to the water surface, the greater the tendency for this net motion. This motion of the float, which is indicative of the mean fluid motion, is a nonlinear effect, as the trajectory of the water particles from linear theory are predicted to be closed ellipses (see Chapter 4).

There are two approaches for examining this mass transport: the Eulerian frame, using a fixed point to measure the mean flux of mass, or the Lagrangian frame, which involves moving with the water particles.

10.2.1 Eulerian Mass Transport

Examining the horizontal velocity at any point below the water surface and averaging over a wave period shows that

$$\overline{u}(x, z) = \frac{1}{T} \int_0^T u(x, z)\, dt = 0 \tag{10.1}$$

However, in the region between the trough and the wave crest, the horizontal velocity must be obtained by the Taylor series. For example, for the surface velocities we have, approximately,[1]

[1]Neglecting some contributions from second-order theory.

$$u(x, \eta) = u(x, 0) + \eta \frac{\partial u}{\partial z}\bigg|_{z=0} \tag{10.2}$$

$$= \frac{gak}{\sigma} \frac{\cosh k(h + z)}{\cosh kh}\bigg|_{z=0} \cos (kx - \sigma t) + \frac{ga^2k^2}{\sigma} \tanh kh \cos^2 (kx - \sigma t)$$

$$= \frac{gak}{\sigma} \cos (kx - \sigma t) + a^2 k \sigma \cos^2 (kx - \sigma t)$$

The surface velocity is periodic, yet faster at the wave crest than at the wave trough, as the second term is always positive at these two phase positions. This asymmetry of velocity indicates that more fluid moves in the wave direction under the wave crest than in the trough region. This is, in fact, true. If we average $u(x, \eta)$ over a wave period (an operation denoted by an overbar), there is a mean transport of water[2]

$$\overline{u(x, \eta)} \equiv \frac{1}{T} \int_0^T u(x, \eta) \, dt = \frac{a^2 k \sigma}{2} = \frac{(ka)^2 C}{2} \tag{10.3}$$

To obtain the total mean flux, or flow of mass, we perform the following integration, where M is defined as the mass transport

$$M \equiv \overline{\int_{-h}^{\eta} \rho u \, dz} = \overline{\int_{-h}^{0} \rho u \, dz} + \overline{\eta \rho u} = \frac{E}{C} \tag{10.4}$$

a result first presented by Starr (1947). Note that the first term in Eq. (10.4) is zero; again, there is no mean flow except due to the contribution of the region bounded vertically by η. The depth-averaged time-mean velocity, due to mass transport, is

$$U = \frac{M}{\rho h} \tag{10.5}$$

10.2.2 Lagrangian Mass Transport

The Eulerian velocity discussed above is obtained by examining the velocity at a fixed point. A Lagrangian velocity is one obtained by moving with a particle as it changes location. The velocity of a particular water particle with a mean position of (x_1, z_1) is $u(x_1 + \zeta, z_1 + \xi)$, where ζ and ξ are locations on the trajectory of the particle. An approximation to the instantaneous velocity is

$$u_L(x_1 + \zeta, z_1 + \xi) = u(x_1, z_1) + \frac{\partial u}{\partial x} \zeta + \frac{\partial u}{\partial z} \xi \tag{10.6}$$

---

[2]Clearly, $\overline{u}(x, \eta)$ is much less than the phase speed of the wave, C.

Using the values of the trajectory obtained in Chapter 4 [Eqs. (4.9) and (4.10)] evaluated at (x_1, z_1)], u_L can be written as

$$u_L = \frac{gak}{\sigma} \frac{\cosh k(h+z)}{\cosh kh} \cos (kx - \sigma t) \qquad (10.7)$$

$$+ \frac{a^2\sigma k}{\sinh^2 kh}[\cosh^2 k(h+z) \sin^2 (kx - \sigma t) + \sinh^2 k(h+z) \cos^2 (kx - \sigma t)]$$

The mean value of u_L is

$$\bar{u}_L (x_1 + \zeta, z_1 + \zeta) = \frac{a^2\sigma k \cosh 2k(h+z)}{2 \sinh^2 kh} = \frac{ga^2k^2 \cosh 2k(h+z)}{\sigma \sinh 2kh} \qquad (10.8)$$

This mean Lagrangian velocity indicates that the water particles drift in the direction of the waves and move more rapidly at the surface than at the bottom.

Integrating over the water column to obtain the total transport and multiplying by the density of the fluid yields, as before,

$$M = \int_{-h}^{0} \rho \bar{u}_L \, dz = \frac{\rho ga^2k}{2\sigma} = \frac{E}{C} \qquad (10.9)$$

10.3 MEAN WATER LEVEL

The Bernoulli equation at the free surface, Eq. (3.13), is

$$\frac{(\partial\phi/\partial x)^2 + (\partial\phi/\partial z)^2}{2} - \frac{\partial\phi}{\partial t} + gz = C(t) \qquad \text{on } z = \eta \qquad (10.10)$$

Expanding to the free surface by the Taylor series yields to *first* order in η after time averaging (which is denoted by the overbar),

$$\frac{\overline{(\partial\phi/\partial x)^2 + (\partial\phi/\partial z)^2}}{2} + g\bar{\eta} - \overline{\eta \frac{\partial^2\phi}{\partial t \partial z}} = \overline{C(t)} \qquad (10.11)$$

where $\bar{\eta}$ is a mean displacement in water level from $z = 0$. Substituting for η and ϕ from the linear *progressive* wave theory, we have

$$\bar{\eta} = -\frac{a^2k}{2 \sinh 2kh} + \frac{\overline{C(t)}}{g} \equiv -f(x) + \frac{\overline{C(t)}}{g} \qquad (10.12)$$

There are several choices for $C(t)$ here, depending on the problem. If the problem is one of waves propagating from deep to shallow water, a customary boundary condition is $\bar{\eta}$ is zero in deep water, which fixes $\overline{C(t)} = 0$ everywhere. Thus $\bar{\eta}$ is always negative, becoming more so as the wave enters shallow water until breaking commences. This is called the setdown. Alternatively, we can force the x axis ($z = 0$) to be the mean water level at some fixed

x_1 by setting $\overline{C(t)} = f(x_1)g$ in Eq. (10.12), where f is now a constant. As another example, in an enclosed tank where the amount of water in the tank must be conserved, a continuity argument must be invoked for $\overline{C(t)}$. If the tank is of length l, then

$$\frac{1}{l} \int_0^l \overline{\eta}(x) \, dx = 0 \tag{10.13a}$$

or, from Eq. (10.12),

$$\frac{1}{l} \int_0^l f(x) \, dx = \frac{\overline{C(t)}}{g} \tag{10.13b}$$

The mean water level associated with standing waves is

$$\overline{\eta} = \frac{\overline{C(t)}}{g} + \frac{a^2 k}{4 \sinh 2kh} (\cosh 2kh \cos 2kx - 1)$$

This is left as an exercise for the reader (Problem 10.3).

10.4 MEAN PRESSURE

The mean pressure under a wave can be most easily obtained by time-averaging the Bernoulli equation:

$$p(z) = \rho \frac{\partial \phi}{\partial t} - \rho \frac{u^2 + w^2}{2} - \rho g z + C(t) \tag{10.14a}$$

or

$$\overline{p}(z) = -\rho \frac{\overline{u^2 + w^2}}{2} - \rho g z + \overline{C}(t) \tag{10.14b}$$

under a progressive wave. If $\overline{C}(t) = 0$, the case for shoaling progressive waves, then it is clear that the mean pressure is decreased from its hydrostatic value. As (u, w) decrease with depth into the water, the mean pressure approaches hydrostatic with depth. Substituting into the equation above yields

$$\overline{p}(z) = -\frac{\rho g a^2 k \cosh 2k(h + z)}{2 \sinh 2kh} - \rho g z \tag{10.15}$$

Alternatively, if the coordinate system is located at the mean water level such that $\overline{C}(t) = f(x_1)g$ and $\overline{\eta} = 0$, it can be shown that

$$\overline{p}(z^*) = -\overline{\rho w^2} - \rho g z^* \tag{10.16}$$

where z^* differs by $\overline{\eta}$ from the z of the other coordinate system (see Figure 10.1).

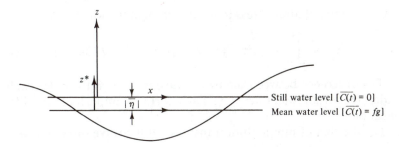

Figure 10.1 The two vertical reference systems and associated Bernoulli constants.

Under a standing wave of amplitude a,

$$\overline{p}(z) = -\frac{\rho g a^2 k}{4 \sinh 2kh}[\cosh 2k(h+z) - \cos 2kx] - \rho g z \qquad (10.17)$$

and at the bottom,

$$\overline{p}(-h) = \frac{\rho g a^2 k}{4 \sinh 2kh}(\cos 2kx - 1) + \rho g h \qquad (10.18)$$

10.5 MOMENTUM FLUX

At a point above the trough level, there is a mean momentum flux as well as mass flux. The mean vertically averaged momentum flux correct to second order in ka is

$$\overline{\int_{-h}^{\eta}(\rho u)u\, dz} = MC_g \qquad (10.19)$$

where C_g is the group velocity, the speed at which the wave energy propagates.

The flux of momentum in the direction of the wave past a section and the pressure force per unit width is defined as

$$I_x = MC_g + \overline{\int_{-h}^{\eta}p(z)\, dz} \qquad (10.20)$$

From Newton's second law, this quantity is unchanged between any two sections unless forces are applied. Evaluating the last integral yields the expression

$$I_x = MC_g + \tfrac{1}{2}\rho g h^2 \qquad (10.21)$$

I_x can be rewritten as

$$I_x = S_{xx} + \tfrac{1}{2}\rho g(h + \overline{\eta})^2 \qquad (10.22)$$

where S_{xx} is the radiation stress in the direction of the waves.

$$S_{xx} \equiv \overline{\int_{-h}^{\eta} p(z) \, dz - \tfrac{1}{2} \rho g (h + \eta)^2} + MC_g = E(2n - \tfrac{1}{2}) \qquad (10.23)$$

The difference between the two forms for I is that the latter explicitly includes the mean water level $\bar{\eta}$. Each form is important for different applications.

For the flux of momentum transverse to the wave direction, we have

$$\overline{\int_{-h}^{\eta} (\rho v) v \, dz} = 0 \qquad (10.24)$$

The sum of momentum flux and pressure force in the transverse direction is

$$I_y = \overline{\int_{-h}^{\eta} p(z) \, dz} = \tfrac{1}{2} \rho g h^2$$

or

$$I_y = S_{yy} + \tfrac{1}{2} \rho g (h + \bar{\eta})^2$$

where

$$S_{yy} \equiv \overline{\int_{-h}^{\eta} p(z) \, dz - \tfrac{1}{2} \rho g (h + \bar{\eta})^2}$$

$$= -\rho g h \bar{\eta} \text{ to } O(ka)^2$$

$$= E(n - \tfrac{1}{2})$$

If a progressive wave is propagating at some angle θ to the x axis, then S_{xx} and S_{yy} are modified to the following forms:

$$S_{xx} = E[n (\cos^2 \theta + 1) - \tfrac{1}{2}] \qquad (10.25)$$

$$S_{yy} = E[n (\sin^2 \theta + 1) - \tfrac{1}{2}] \qquad (10.26)$$

in which n is the ratio of group velocity to wave celerity ($n = C_G/C$). In addition, for this case there is an additional term representing the flux in the x direction of the y component of momentum, denoted S_{xy}:

$$S_{xy} = \overline{\int_{-h}^{\eta} \rho u v \, dz} \approx \overline{\int_{-h}^{0} \rho(u v) \, dz} \qquad (10.27)$$

and employing linear wave theory, it can be shown that

$$S_{xy} = \frac{E}{2} n \sin 2\theta \qquad (10.28)$$

It is of interest to note that, if the bathymetry is composed of straight and parallel contours and if no energy dissipation or additions occur, there is no change in S_{xy} from deep to shallow water.

For further information on radiation stresses and their uses, the reader is referred to Longuet-Higgins and Stewart (1964), Longuet-Higgins (1976), and Phillips (1966).

Example 10.1: Wave Setdown and Setup

As waves shoal and break on a beach, the momentum flux in the onshore direction is reduced and results in compensating forces on the water column. Consider a train of waves encountering the coast with normal incidence. For a short distance dx (Figure 10.2), a force balance can be developed

$$I_1 = I_2 - R_x \tag{10.29a}$$

$$I - \frac{dI}{dx}\frac{dx}{2} = I + \frac{dI}{dx}\frac{dx}{2} - R_x \tag{10.29b}$$

or finally,

$$\frac{dI}{dx}\,dx = R_x \tag{10.29c}$$

using the Taylor series expansion, where I is evaluated at the center and R_x is the reaction force of the bottom in the $(-x)$ direction. Using the radiation stress approach,

$$\frac{dI_x}{dx} = \frac{d}{dx}[S_{xx} + \tfrac{1}{2}\rho g(h + \bar{\eta})^2]$$

$$= \frac{dS_{xx}}{dx} + \rho g(h + \eta)\frac{d(h + \bar{\eta})}{dx} \tag{10.30}$$

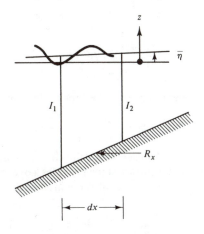

Figure 10.2 Schematic diagram for calculation of wave setup or setdown.

For a mildly sloping bottom, the reaction force R is due to the weight of the column of fluid and thus

$$R_x = \rho g(h + \bar{\eta}) \frac{dh}{dx} dx$$

Substituting yields

$$-\frac{1}{\rho g(h + \bar{\eta})} \frac{dS_{xx}}{dx} = \frac{d\bar{\eta}}{dx} \qquad (10.31)$$

There is therefore a change in mean water surface slope whenever there is a change in S_{xx}. The change in $\bar{\eta}$ offshore of the breaker line is described by Eq. (10.12), which describes a gradual reduction of the mean water level as the shoreline is approached. At $x = x_b$, the breaker line, the wave amplitude is $a = \kappa(h + \bar{\eta})/2$, where κ is the breaking index (Chapter 4), and $\bar{\eta}$ (in shallow water) is

$$\bar{\eta} = -\frac{a^2}{4h_b}$$

as given by Longuet-Higgins and Stewart (1964) or

$$\bar{\eta} = -\frac{\kappa^2 h_b}{16} \qquad (10.32)$$

The setdown therefore is less than 5% of the breaking depth for $\kappa = 0.8$.

Inside the surf zone, where $a(x) = \kappa(h + \bar{\eta})/2$, based on a spilling breaker model, the setup is found from the force balance, Eq. (10.31):

$$-\frac{1}{\rho g(h + \bar{\eta})} \frac{d}{dx} \left[\frac{1}{2} \rho g \frac{\kappa^2(h + \bar{\eta})^2}{4} \frac{3}{2} \right] = \frac{d\bar{\eta}}{dx}$$

Simplifying yields

$$\frac{d\bar{\eta}}{dx} \left(1 + \frac{3\kappa^2}{8} \right) = -\frac{3\kappa^2}{8} \frac{dh}{dx} \qquad (10.33)$$

Finally,

$$\bar{\eta} = -\frac{3\kappa^2/8}{1 + 3\kappa^2/8} h + C \qquad (10.34)$$

Evaluating the constant at $x = x_b$, the breaker line, where $\bar{\eta} = \bar{\eta}_b$, gives finally

$$\bar{\eta}(x) = \bar{\eta}_b + \frac{3\kappa^2/8}{1 + 3\kappa^2/8} [h_b - h(x)] \qquad (10.35)$$

The mean water surface displacement $\bar{\eta}$ thus increases linearly with depth as the shore is approached. This water surface slope provides a hydrostatic pressure gradient directed offshore to counter the change of wave momentum by breaking across the shoreline.

$$\bar{\eta}(0) = \bar{\eta}_b + \frac{3\kappa^2/8}{1 + 3\kappa^2/8} h_b \qquad (10.36)$$

or, for $\kappa = 0.8$, $\bar{\eta}(0)$ is about 15% of the breaker depth or about 19% of the breaking wave height.

Example 10.2: Applied Longshore Wave Thrust

For waves propagating obliquely into the surf zone, breaking will result in a reduction in wave energy and an associated decrease in S_{xy} [cf. Eq. (10.28)], which is manifested as an applied longshore wave thrust F_y on the surf zone. For straight and parallel bottom contours, thrust per unit area is given by

$$F_y = -\frac{\partial S_{xy}}{\partial x} \qquad (10.37)$$

Thus gradients of the momentum flux terms provide a useful framework for the driving forces in the nearshore zone. In the present case, the longshore wave thrust per unit area is resisted by shear stresses on the bottom and lateral faces of the water column (Longuet-Higgins, 1970).

10.6 SUMMARY

The results of linear wave theory may be used to calculate nonlinear mean quantities, correct to second order in ka. These quantities, such as mass transport and mean momentum flux, play a major role in coastal engineering. In fact, the mean momentum flux of the waves in the longshore direction, relative to a coastline, is related to the currents engendered at the coastline and the amounts of sediments transported along the coast. See, for an overview, the book by Komar (1976). In the open ocean, the mean momentum flux results in the drifting of objects, such as ships, ice flows, and oil slicks.

REFERENCES

KOMAR, P. D., *Beach Processes and Sedimentation*, Prentice-Hall, Englewood Cliffs, N.J. 1976.

LONGUET-HIGGINS, M. S., "Longshore Currents Generated by Obliquely Incident Sea Waves, 1," *J. Geophys. Res.*, Vol. 75, No. 33, 1970.

LONGUET-HIGGINS, M. S., "The Mean Forces Exerted by Waves on Floating or Submerged Bodies, with Applications to Sand Bars and Wave Power Machines," *Proc. Roy. Soc. A*, Vol. 106, June 1976.

LONGUET-HIGGINS, M. S., and R. W. STEWART, "Radiation Stresses in Water Waves: A Physical Discussion with Applications," *Deep-Sea Res.*, Vol. 2, 1964.

PHILLIPS, O. M., *The Dynamics of the Upper Ocean*, Cambridge University Press, Cambridge, 1966.

STARR, V. P., "A Momentum Integral for Surface Waves in Deep Water," *J. Mar. Res.*, Vol. 6, No. 2, 1947.

PROBLEMS

10.1 Determine the mean water level due to a wave train impinging on a perfectly reflecting vertical wall with an angle θ.

10.2 Calculate the mean water level associated with an edge wave,

$$\phi = \frac{ga}{\sigma} e^{-k(y\cos\beta - z\sin\beta)} \cos kx \sin \sigma t$$

where y is positive offshore, x is alongshore, and β is the bottom slope.

10.3 Show that the setdown under a standing wave system is

$$\bar{\eta}(x) = \frac{a^2 k}{4 \sinh 2kh} (\cosh 2kh \cos 2kx - 1)$$

10.4 Show by two different methods that for the origin of the vertical coordinate taken at the mean water line, the mean pressure for a progressive wave system is

$$\bar{p} = -\rho g z - \rho \overline{w^2}$$

One method is suggested in the paragraph following Eq. (10.15). A second method involves integration of the vertical equation of motion from an arbitrary depth z up to the free surface, the use of the Leibniz rule, and time averaging over a wave period.

10.5 For the case of straight and parallel bottom contours, combine energy conservation consideration with Snell's law to demonstrate that S_{xy} is the same from deep to shallow water.

10.6 Verify Eqs. (10.25) and (10.26) for the radiation stresses developed by a wave train traveling at an angle θ to the x axis. Use $\phi(x, y, z, t)$ as developed in Chapter 4.

11

Nonlinear Waves

Dedication
SIR GEORGE GABRIEL STOKES

Sir George Gabriel Stokes (1819–1903) was born in Skreen, Ireland. He entered Bristol College at 16 and matriculated at Pembroke College, Cambridge, in 1837. He became a Fellow of Pembroke College in 1841 and in 1849 received the Lucasian Professorship of Mathematics at Cambridge—the same professorship held by Airy from 1826. To bolster his teaching salary he also taught at the Government School of Mines.

Stokes's contributions range from optics, acoustics, and hydrodynamics to viscous fluid problems (a unit of viscosity is named for him) and to the proof that the wave of maximum height has a crest angle of 120°. He also did a great deal of work related to the concept of ether which was hypothesized to exist between the planets and stars.

In 1842 he solved three-dimensional flow problems by introducing an axisymmetric stream function. In 1849 he developed the dynamical theory of diffraction using Bessels series and Fourier integral theory, and in 1852 he received the Rumford Medal of the Royal Society for the discovery of the nature of natural fluorescence.

His inclusion in this chapter derives from the development of Stokes waves, large-amplitude waves that he conceived through a nonlinear wave theory. This theory, although usually extended to higher orders of accuracy than he was able to achieve, remains in use today.

In 1845 Stokes produced a number of papers on viscous flow. He was unaware that the French scientists Navier, Poisson, and St.-Venant had treated these problems, and he independently derived the now-called Navier–Stokes equations.

Stokes received a number of awards and prizes as well as numerous honorary doctorates for his work, a process that culminated in 1889 when he became a Baronet.

11.1 INTRODUCTION

The water waves that have been discussed thus far have been small-amplitude waves, which satisfied linearized forms of the kinematic and dynamic free surface boundary conditions. We have seen that the linear wave theory has been useful in many respects, even when the requirements of linear theory, small $kH/2$, have been violated. In this chapter, extension of the linear theory to a second-order Stokes (1847) theory and then an "any"-order theory will be developed. The desire is to develop a water wave theory to best satisfy the mathematical formulation of the water wave theory. In shallow water a different expansion will then be explored, where the classical Stokes expansion is inefficient.

11.2 PERTURBATION APPROACH OF STOKES

Reviewing the periodic water wave boundary value problem for waves propagating in the $+x$ direction, we have linear and nonlinear boundary conditions applied to a linear governing differential equation.

11.2.1 Linear Equation and Boundary Conditions

$$\nabla^2\phi = 0 \quad \text{governing differential equation} \tag{11.1}$$

$$-\frac{\partial\phi}{\partial z} = 0 \quad \text{on } z = -h \quad \text{bottom boundary condition} \tag{11.2}$$

$$\phi(x, z, t) = \phi(x + L, z, t) \quad \text{lateral boundary condition} \tag{11.3}$$

$$\phi(x, z, t) = \phi(x, z, t + T) \quad \text{periodicity requirement} \tag{11.4}$$

11.2.2 Nonlinear Boundary Conditions

Dynamic free surface boundary condition (DFSBC):

$$\frac{p}{\rho} + \frac{(\partial\phi/\partial x)^2 + (\partial\phi/\partial z)^2}{2} - \frac{\partial\phi}{\partial t} + gz = C_B(t) \quad \text{on } z = \eta(x, t) \tag{11.5}$$

Kinematic free surface boundary condition (KFSBC):

$$-\frac{\partial\phi}{\partial z} = \frac{\partial\eta}{\partial t} - \frac{\partial\phi}{\partial x}\frac{\partial\eta}{\partial x} \quad \text{on } z = \eta(x, t) \tag{11.6}$$

It is convenient at this juncture to put the governing equations and the related boundary conditions into dimensionless forms. We define the following dimensionless variables, developed in terms of g, a, and k, which are gravity, the wave amplitude, and the wave number, respectively.

$$X = kx$$

$$Z = kz$$

$$\Pi = \frac{\eta}{a}$$

$$\Phi = \frac{k\phi}{a\sqrt{gk}}$$

$$T = \sqrt{gk}\, t$$

$$Q = \frac{k}{g} C_B(t)$$

$$\omega = \frac{\sigma}{\sqrt{gk}}$$

$$P = \frac{kp}{\rho g}$$

The governing equation is thus

$$\frac{\partial^2 \Phi}{\partial X^2} + \frac{\partial^2 \Phi}{\partial Z^2} = 0 \tag{11.7}$$

The periodicity and lateral boundary conditions remain the same in dimensionless form; however, the free surface boundary conditions are modified to be

$$P + (ka)^2 \left[\frac{(\partial\Phi/\partial X)^2 + (\partial\Phi/\partial Z)^2}{2} \right] - (ka)\frac{\partial\Phi}{\partial T} + Z = Q(t) \quad \text{on } Z = ka\Pi \tag{11.8}$$

where P will be taken as zero on the free surface. (Note that if $ka = 0$, then $Z = 0$; there are no waves and therefore only a trivial solution exists.) The KFSBC becomes

$$\frac{\partial\Pi}{\partial T} - (ka)\frac{\partial\Phi}{\partial X}\frac{\partial\Pi}{\partial X} = -\frac{\partial\Phi}{\partial Z} \quad \text{on } Z = ka\Pi \tag{11.9}$$

In our previous derivation of small-amplitude wave theory, we expanded the nonlinear conditions about $Z = 0$, the mean water level, and then neglected products of very small quantities, such as $(\partial\Phi/\partial X)^2$. This clearly was neglecting terms of order $(ka)^2$ when compared to ka.

In the perturbation approach, we will assume that the solution will depend on the presumed small quantity ka, which we will define as ϵ. The linear solution will not depend on ϵ, while the second order will, the third order will depend on ϵ^2, and so on. Therefore, we will decompose all quantities into a power series in ϵ, which is presumed to be less than unity.

$$\Pi = \Pi_1 + \epsilon\Pi_2 + \epsilon^2\Pi_3 + \cdots$$

$$\Phi = \Phi_1 + \epsilon\Phi_2 + \epsilon^2\Phi_3 + \cdots$$

$$Q(t) = \epsilon Q_1(T) + \epsilon^2 Q_2(T) + \epsilon^3 Q_3(T) + \cdots \tag{11.10}$$

$$\omega = \omega_1 + \epsilon\omega_2 + \epsilon^2\omega_3 + \cdots$$

Again, as we a priori do not know the location of the free surface $Z = (ka)\Pi(X, T)$, we will resort to expanding the nonlinear free surface boundary conditions about $Z = 0$ in terms of $\epsilon\Pi$, retaining the higher-order terms up to ϵ^2, denoted as $O(\epsilon^2)$. Using the Taylor series we have

$$\left\{ \frac{1}{2}\epsilon^2\left[\left(\frac{\partial\Phi}{\partial X}\right)^2 + \left(\frac{\partial\Phi}{\partial Z}\right)^2\right] - \epsilon\frac{\partial\Phi}{\partial T} + Z \right\} + \epsilon\Pi\frac{\partial}{\partial Z}\left\{\frac{1}{2}\epsilon^2\left[\left(\frac{\partial\Phi}{\partial X}\right)^2\right.\right. \tag{11.11}$$

$$\left.\left. + \left(\frac{\partial\Phi}{\partial Z}\right)^2\right] - \epsilon\frac{\partial\Phi}{\partial T} + Z \right\} - \frac{\epsilon^3\Pi^2}{2}\frac{\partial^3\Phi}{\partial Z^2\,\partial T} = Q(t) \qquad \text{on } Z = 0$$

and

$$\left(-\frac{\partial\Phi}{\partial Z} - \frac{\partial\Pi}{\partial T} + \epsilon\frac{\partial\Phi}{\partial X}\frac{\partial\Pi}{\partial X}\right) + \epsilon\Pi\frac{\partial}{\partial Z}\left(-\frac{\partial\Phi}{\partial Z} + \epsilon\frac{\partial\Phi}{\partial X}\frac{\partial\Pi}{\partial X}\right) \tag{11.12}$$

$$-\frac{\epsilon^2\Pi^2}{2}\frac{\partial^3\Phi}{\partial Z^3} = 0 \qquad \text{on } Z = 0$$

where we have accounted for the fact that Π and $Q(T)$ are not functions of elevation.

Substituting the perturbation expansions, Eqs. (11.10), into the linear conditions, Eqs. (11.1) to (11.4), we have, retaining only terms of first order in ϵ (the others being much smaller)

$$\nabla^2\Phi_1 + \epsilon\nabla^2\Phi_2 + \cdots = 0 \tag{11.13}$$

$$-\frac{\partial\Phi_1}{\partial Z} - \epsilon\frac{\partial\Phi_2}{\partial Z} + \cdots = 0 \qquad \text{at } Z = -kh$$

$$\Phi_1(X, Z, T) + \epsilon\Phi_2(X, Z, T) + \cdots = \Phi_1(X + L, Z, T) + \epsilon\Phi_2(X + L, Z, T)$$

$$\Phi_1(X, Z, T) + \epsilon\Phi_2(X, Z, T) + \cdots = \Phi_1(X, Z, T + T_p) + \epsilon\Phi_2(X, Z, T + T_p)$$

where T_p is the dimensionless wave period, $2\pi/\omega$. At the free surface, we obtain for the DFSBC and KFSBC, respectively:

$$\frac{\epsilon}{2}\left[\left(\frac{\partial \Phi_1}{\partial X}\right)^2 + \left(\frac{\partial \Phi_1}{\partial Z}\right)^2\right] - \frac{\partial \Phi_1}{\partial T} - \epsilon\frac{\partial \Phi_2}{\partial T} + \Pi_1 + \epsilon\Pi_2 - \epsilon\Pi_1\frac{\partial^2 \Phi_1}{\partial T\,\partial Z}\cdots \qquad (11.14)$$

$$= Q_1(T) + \epsilon Q_2(T)\cdots \qquad \text{on } Z = 0$$

$$-\frac{\partial \Phi_1}{\partial Z} - \epsilon\frac{\partial \Phi_2}{\partial Z} - \frac{\partial \Pi_1}{\partial T} - \epsilon\frac{\partial \Pi_2}{\partial T} + \epsilon\frac{\partial \Phi_1}{\partial X}\frac{\partial \Pi_1}{\partial X} - \epsilon\Pi_1\frac{\partial^2 \Phi_1}{\partial Z^2}\cdots \qquad (11.15)$$

$$= 0 \qquad \text{on } Z = 0$$

The original nonlinear boundary value problem has now been reformulated into an infinite set of linear equations of ascending orders. To visualize the manner in which the linear equations are obtained, consider the following general form of the perturbed equations:

$$A_1 + \epsilon A_2 + \epsilon^2 A_3 \cdots = B_1 + \epsilon B_2 + \epsilon^2 B_3 \cdots \qquad (11.16)$$

The required condition that the equality holds for *arbitrary* ϵ is that the coefficients of like powers of ϵ must be equal. Therefore,

$$A_1 = B_1$$

$$A_2 = B_2$$

$$A_3 = B_3, \quad \text{etc.}$$

This procedure will now be used to separate the equations by order.

11.2.3 First-Order Perturbation Equations

If we gather together all the terms that do not depend on ϵ, the linear equations result.

$$\nabla^2 \Phi_1 = 0$$

$$-\frac{\partial \Phi_1}{\partial Z} = 0 \qquad \text{on } Z = -kh$$

$$-\frac{\partial \Phi_1}{\partial T} = -\Pi_1 + Q_1(T) \qquad \text{on } Z = 0 \qquad (11.17)$$

$$\frac{\partial \Pi_1}{\partial T} = -\frac{\partial \Phi_1}{\partial Z} \qquad \text{on } Z = 0$$

$$\Phi_1(X, Z, T) = \Phi_1(X + 2\pi, Z, T)$$

$$\Phi_1(X, Z, T) = \Phi_1(X, Z, T + T_p)$$

These are the equations that were used in Chapter 3.

The solutions are, in dimensionless form,

$$\Phi_1 = -\frac{\cosh{(kh + Z)}}{\omega \cosh{kh}} \sin{(X - \omega T)}$$

$$\Pi = \cos{(X - \omega T)} \tag{11.18}$$

$$\omega_1^2 = \tanh{kh}$$

$$Q_1(T) = 0$$

which in dimensional form are Eqs. (3.42), (3.43), and (3.34).

11.2.4 Second-Order Perturbation Equation

To the order of ϵ,

$$\nabla^2 \Phi_2 = 0$$

$$-\frac{\partial \Phi_2}{\partial Z} = 0 \qquad \text{on } Z = -kh$$

$$-\frac{\partial \Phi_2}{\partial Z} - \frac{\partial \Pi_2}{\partial T} = -\frac{\partial \Phi_1}{\partial X}\frac{\partial \Pi_1}{\partial X} + \Pi_1 \frac{\partial^2 \Phi_1}{\partial Z^2} \qquad \text{on } Z = 0 \tag{11.19}$$

$$\Pi_2 - \frac{\partial \Phi_2}{\partial T} - Q_2(T) = -\frac{1}{2}\left[\left(\frac{\partial \Phi_1}{\partial X}\right)^2 + \left(\frac{\partial \Phi_1}{\partial Z}\right)^2\right] + \Pi_1 \frac{\partial^2 \Phi_1}{\partial Z \partial T} = 0 \qquad \text{on } Z = 0$$

$$\Phi_2(X, Z, T) = \Phi_2(X + 2\pi, Z, T)$$

$$\Phi_2(X, Z, T) = \Phi_2(X, Z, T + T_p)$$

Note that all the equations and conditions are linear in the variables of interest, $\Phi_2(X, Z, T)$ and $\Pi_2(X, T)$, but the free surface boundary conditions have inhomogeneous terms that depend on the first-order solution. Since the first-order solution is known, the terms on the right-hand side are known also.

To solve for the second-order solution it is convenient to use the combined free surface boundary condition, which is found by eliminating Π_2 from the free surface conditions,

$$\frac{\partial^2 \Phi_2}{\partial T^2} + \frac{\partial \Phi_2}{\partial Z} + \frac{\partial Q_2(T)}{\partial T} = \frac{\partial \Phi_1}{\partial X}\frac{\partial \Pi_1}{\partial X} - \frac{\partial \Pi_1}{\partial T}\frac{\partial^2 \Phi_1}{\partial Z \partial T} \tag{11.20}$$

$$- \Pi_1 \frac{\partial}{\partial Z}\left(\frac{\partial^2 \Phi_1}{\partial T^2} + \frac{\partial \Phi_1}{\partial Z}\right) + \frac{1}{2}\frac{\partial}{\partial T}\left[\left(\frac{\partial \Phi_1}{\partial X}\right)^2 + \left(\frac{\partial \Phi_1}{\partial Z}\right)^2\right] \qquad \text{on } Z = 0$$

For convenience, the right-hand side of this expression will be defined as D.

Substituting for Φ_1 and Π_1 from Eqs. (11.18) into the expression for D and using trigonometric identities, it is possible to express D simply as

$$D = \frac{3\omega_1}{\sinh 2kh} \sin 2(X - \omega T) \tag{11.21}$$

As a trial solution for $\Phi_2(X, Z, T)$, the following form is taken:

$$\Phi_2(X, Z, T) = a_2 \cosh 2(kh + Z) \sin 2(X - \omega T) \tag{11.22}$$

which satisfies the Laplace equation and the bottom boundary condition. Examining the second-order combined free surface boundary condition, Eq. (11.20), it is clear that $\partial Q_2(T)/\partial T = 0$, as it cannot depend on $\sin 2(X - \omega T)$ as do all the other terms (being *only* a function of time), and thus the inequality could not otherwise be satisfied. Therefore, $Q_2(t) = $ constant, Q_2. Substituting Φ_2 into the combined condition yields a_2.

$$a_2 = -\frac{3}{8} \frac{\omega}{\sinh^4 kh} \tag{11.23}$$

Therefore,

$$\Phi_2(X, Z, T) = -\frac{3}{8} \frac{\omega \cosh 2(kh + Z)}{\sinh^4 kh} \sin 2(X - \omega T) \tag{11.24}$$

To determine the corresponding free surface elevation, $\Pi_2(X, T)$, the second-order dynamic free surface boundary condition is used,

$$\Pi_2 = \frac{\partial \Phi_2}{\partial T} + Q_2 - \frac{1}{2}\left[\left(\frac{\partial \Phi_1}{\partial X}\right)^2 + \left(\frac{\partial \Phi_1}{\partial Z}\right)^2\right] + \Pi_1 \frac{\partial^2 \Phi_1}{\partial Z \partial T} \quad \text{on } Z = 0 \tag{11.25}$$

Substituting for Φ_2 and Φ_1 yields, in dimensional form,

$$\eta_2(x, t) = \frac{3}{16} \frac{H_1^2 \sigma^2 \cosh 2kh}{g \sinh^4 kh} \cos 2(kx - \sigma t)$$

$$+ Q_2 - \frac{H_1^2}{16} \frac{\sigma^2}{g \sinh^2 kh} [\cosh 2kh + \cos 2(kx - \sigma t)] \tag{11.26}$$

$$+ \frac{H_1^2 \sigma^2}{8g} [1 + \cos 2(kx - \sigma t)]$$

where H_1 is the first-order wave height ($H_1 = 2a$).

There are two options that can be applied to this equation in order to proceed. First, as in Chapter 10, we can specify the Bernoulli constant to be zero, corresponding to no setdown in deep water and then separating η into a mean $\bar{\eta}$ and a fluctuating $\hat{\eta}$ term.

$$\eta_2 = \bar{\eta} + \hat{\eta}_2 \tag{11.27}$$

and from Eq. (11.26),

$$\bar{\eta} = -\frac{H_1^2 \sigma^2}{16\, g \sinh^2 kh} = -\frac{H_1^2 k}{8 \sinh 2kh} \tag{11.28}$$

as in Chapter 10, and

$$\hat{\eta}_2 = \frac{ka^2}{4} \frac{\cosh kh}{\sinh^3 kh}(2 + \cosh 2kh)\cos 2(kx - \sigma t) \tag{11.29}$$

The second alternative is to specify h as the mean water level depth and then η has a zero mean. Then the Bernoulli constant is

$$Q = \frac{H_1^2 \sigma^2}{16 \sinh^2 kh} \tag{11.30}$$

and the fluctuating part of η_2, as before is given by Eq. (11.29). The resulting second-order wave profile is much more peaked at the wave crest and flatter at the wave troughs than the previous sinusoidal wave form. This is shown in Figure 11.1.

The velocity potential and water surface displacement, to second order then, in dimensional form are

$$\phi = \epsilon\phi_1 + \epsilon^2\phi_2$$

$$= -\frac{H_1 g}{2\sigma} \frac{\cosh k(h + z)}{\cosh kh}\sin(kx - \sigma t) \tag{11.31}$$

$$- \frac{3}{32} H_1^2 \sigma \frac{\cosh 2k(h + z)}{\sinh^4 kh}\sin 2(kx - \sigma t)$$

and $\quad \eta = \epsilon\eta_1 + \epsilon^2\eta_2 \tag{11.32}$

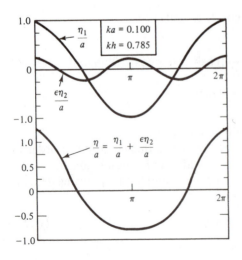

Figure 11.1 A second-order stokes water surface profile as composed of η_1 and $\epsilon\eta_2$ contributions where $\epsilon = ka$.

$$\eta = \frac{H_1}{2} \cos{(kx - \sigma t)} + \frac{H_1^2 k}{16} \frac{\cosh{kh}}{\sinh^3{kh}} (2 + \cosh{2kh}) \cos{2(kx - \sigma t)}$$

The dispersion equation relating σ to k remains the same,

$$\sigma^2 = gk \tanh{kh} \tag{11.33}$$

However, it is noted that a correction occurs to the dispersion equation at the third order.

Convergence.   A measure of the validity of the Stokes expansion procedure is whether or not the series for ϕ converges. This can be checked for the second-order theory by examining the ratio of the second-order term to the first-order term, which must be less than 1 in order for the series for ϕ, Eq. (11.10), to converge.[1]

$$R = \frac{\epsilon \phi_2}{\phi_1} = \frac{3}{8} \frac{ka \cosh{2kh}}{\cosh{kh} \sinh^3{kh}} \ll 1 \tag{11.34}$$

In deep water, defined as $kh > \pi$, the asymptotic forms of hyperbolic functions can be substituted to reduce R to

$$R = 3e^{-2kh}ka \tag{11.35}$$

R is thus very small in deep water, particularly since ka has been assumed small previously. The highest value in deep water would occur for $kh = \pi$, $ka = \pi/7$, occurring for the wave of maximum steepness,

or

$$R = \frac{3\pi}{7} e^{-2\pi} = 0.0025 \tag{11.36}$$

In shallow water, $kh < \pi/10$, the hyperbolic functions can again be replaced by the asymptotic values,

$$R = \frac{3}{8} \frac{ka}{k^3 h^3} = \frac{3}{64\pi^2} \left(\frac{L^2 H}{h^3} \right) < 1 \tag{11.37}$$

The relative depth kh thus becomes an important parameter in shallow water. In fact, $ka < 8(kh)^3/3$; this is a severe restriction on wave height, as this can be written as $a/h < (8/3)(kh)^2$, where kh is small. The maximum that the ratio a/h can obtain is $a/h = 8\pi^2/300$ for $kh = \pi/10$, or the maximum wave amplitude is about one-fourth of the water depth. (In shallower water, this ratio must decrease.) However, as mentioned in Chapter 4, the wave amplitude for breaking is almost 0.4 the water depth. Therefore, for high waves in

---

[1]Properly for the power series for ϕ in terms of ϵ to converge, the *ratio test* requires that ratio of the $n + 1$ term divided by the nth term be less than unity as $n \to \infty$.

shallow water, the Stokes expansion is not very good, at least when carried out to only the second order.

The term in parentheses in Eq. (11.37) is called the Ursell parameter (Ursell, 1953), which, for second-order Stokes theory to be valid, has a magnitude

$$\frac{L^2 H}{h^3} << \frac{64 \pi^2}{3} \tag{11.38}$$

The value of the Ursell parameter actually should be less than indicated above, due to the fact that in shallow water the theoretical wave form will develop an anomalous bump in the trough for large waves due to the largeness of the second-order term. To investigate this, the free surface equation will be examined at the trough and the second derivative will be obtained. From the calculus, a negative second derivative indicates a concave downward curvature, or, for this application, a secondary crest or bump.

$$\eta_T = \frac{H}{2} \cos (kx - \sigma t) + \frac{H^2 k}{16} \frac{\cosh kh}{\sinh^3 kh} (2 + \cosh 2kh) \cos 2(kx - \sigma t) \tag{11.39}$$

and

$$\frac{\partial^2 \eta_T}{\partial x^2} = \frac{H}{2} k^2 - \frac{H^2 k^3}{4} \frac{\cosh kh}{\sinh^3 kh} (2 + \cosh 2kh) \qquad \text{for } kx - \sigma t = \pi \tag{11.40}$$

Setting the second equation to zero and solving for ka yields

$$ka = \frac{\sinh^3 kh}{\cosh kh (2 + \cosh 2kh)} \tag{11.41}$$

This is the maximum value of ka for which there is no bump in the trough. In deep water, the maximum permissible ka from this equation is $\frac{1}{2}$, which is greater than the limiting steepness value of $\pi/7$; therefore, in deep water a secondary crest will not occur in the wave profile, while in shallow water, the maximum value of ka is

$$ka < \frac{(kh)^3}{3} \tag{11.42}$$

In comparing this rate to that for R, determined previously, this latter condition is eight times more stringent. In fact, the Ursell parameter reduces to

$$\frac{L^2 H}{h^3} < \frac{8 \pi^2}{3} \tag{11.43}$$

Therefore, for shallow water, the requirement that the wave be single crested should be used as the criterion for the maximum height wave. This idea has been used for fifth-order Stokes waves by Ebbesmeyer (1974).

Kinematics.   The velocities under the second-order wave are, in dimensional form,

$$u = -\frac{\partial \phi}{\partial x} = \frac{H}{2} \frac{gk}{\sigma} \frac{\cosh k(h+z)}{\cosh kh} \cos (kx - \sigma t)$$

$$+ \frac{3}{16} \frac{H^2 \sigma k \cosh 2k(h+z)}{\sinh^4 kh} \cos 2(kx - \sigma t)$$

$$(11.44)$$

$$w = -\frac{\partial \phi}{\partial z} = \frac{H}{2} \frac{gk}{\sigma} \frac{\sinh k(h+z)}{\cosh kh} \sin (kx - \sigma t)$$

$$+ \frac{3}{16} \frac{H^2 \sigma k \sinh 2k(h+z)}{\sinh^4 kh} \sin 2(kx - \sigma t)$$

The presence of the second-order term increases the velocities, but in a manner that varies along the wave due to the $2(kx - \sigma t)$ phase function. For the horizontal velocity the velocities are greater under the crest but are reduced under the trough when compared to linear wave theory.

The total horizontal acceleration is, to second order,

$$\frac{Du}{Dt} = \frac{H}{2} gk \frac{\cosh k(h+z)}{\cosh kh} \sin (kx - \sigma t) - \frac{H^2}{4} gk^2 \frac{\sin 2(kx - \sigma t)}{\sinh 2kh} \qquad (11.45)$$

$$+ \frac{3}{8} \frac{H^2 \sigma^2 k}{\sinh^4 kh} \cosh 2k(h+z) \sin 2(kx - \sigma t)$$

The total vertical acceleration is found similarly (see Problem 11.1).

11.3 THE STREAM FUNCTION WAVE THEORY

Should the reader have followed through the details of the second-order wave theory, it would have been quite arduous. Clearly, higher-order Stokian wave theories [third order, Borgman and Chappelear (1958); fifth order, Skjelbreia and Hendrickson (1961)] become quite difficult. Expanding to even higher orders becomes extremely formidable. For this reason, it was desirable to have wave theories that could be developed on the computer to any order. The first such theory was developed by Chappelear (1961) involving the use of the velocity potential. Dean (1965) used the stream function to develop the stream function wave theory, which was computationally simpler than Chappelear's technique.

Cokelet (1977) has extended the method originally developed by Schwartz (1974) to allow a very accurate calculation of the characteristics of water waves, including heights ranging up to near breaking. The procedure involves expressing the complex potential solution in a Fourier series and represents the Fourier coefficients as series in terms of a perturbation parameter. An interesting result is that the wave speed, wave energy, and wave momentum all exhibit maxima at wave heights slightly smaller than the breaking height.

At present, the Cokelet method appears to yield the most accurate results for nearly breaking waves; however, the differences from the numerical theories (Chappelear and Dean) are generally small and the Cokelet approach is not known to have been applied to design.

11.3.1 Formulation and Solution

In Chapter 3 the linear form of the stream function for water waves was given as

$$\psi(x, z, t) = -\frac{H}{2}\frac{g}{\sigma}\frac{\sinh k(h+z)}{\cosh kh}\cos(kx - \sigma t) \qquad (11.46)$$

or if the coordinate system is moved with celerity of the wave, C, thereby rendering the system steady, as

$$\psi(x, z) = Cz - \frac{Hg}{2\sigma}\frac{\sinh k(h+z)}{\cosh kh}\cos kx \qquad (11.47)$$

The advantage of moving the coordinate system with speed C is that the problem is rendered steady, thus reducing the number of terms in the boundary conditions.

The boundary value problem for progressive water waves is, in stream function form,

$$\nabla^2\psi = 0, \quad \text{throughout the fluid} \qquad (11.48a)$$

$$\frac{1}{2}\left[\left(\frac{\partial\psi}{\partial z}\right)^2 + \left(\frac{\partial\psi}{\partial x}\right)^2\right] + g\eta = Q_B, \quad \begin{array}{l}\text{a constant, on } z = \eta(x),\\ \text{the DFSBC}\end{array} \qquad (11.48b)$$

$$\frac{\partial\psi}{\partial x} = -\frac{\partial\psi}{\partial z}\frac{\partial\eta}{\partial x}, \quad \text{on } z = \eta(x), \text{ the KFSBC} \qquad (11.48c)$$

Using the stream function, the latter condition is true by definition; that is, the free surface, wherever it is, is a streamline. This condition, therefore, is satisfied exactly.

$$\frac{\partial\psi}{\partial x} = 0 \quad \text{on } z = -h, \quad \text{BBC} \qquad (11.48d)$$

$$\psi(x, z) = \psi(x + L, z), \quad \text{lateral boundary condition} \qquad (11.48e)$$

Now, from analogy to the second-order wave theory, we might assume that the Nth-order stream function might look like

$$\psi(x, z) = Cz + \sum_{n=1}^{N} X(n) \sinh \{nk(h + z)\} \cos nkx \qquad (11.49)$$

and

$$u = -\frac{\partial \psi}{\partial z}$$

$$w = \frac{\partial \psi}{\partial x} \qquad (11.50)$$

Note that for the linear theory, we must have the coefficient

$$X(1) = -\frac{Hg}{\sigma} \frac{1}{\sinh kh}$$

The only condition not satisfied by this assumed form is the dynamic free surface boundary condition. The $X(N)$ are, therefore, chosen to satisfy this condition. On the computer, this condition is satisfied at I discrete points along the wave profile, each point being denoted by i. The DFSBC is thus evaluated at each i point along the profile, giving Q_{Bi}. According to the DFSBC, all the Q_{Bi} must be equal to Q_B, where Q_B is a constant.

$$Q_{Bi} = \frac{\left(\frac{\partial \psi}{\partial z}\right)_i^2 + \left(\frac{\partial \psi}{\partial x}\right)_i^2}{2} + g\eta_i = Q_B \qquad (11.51)$$

However, to get the Q_{Bi}, the $X(n)$'s ($n = 1, 2,..., N$) must be known to calculate $\partial \psi/\partial z$, $\partial \psi/\partial x$, and η. The procedure then must be an iterative one; values of $X(n)$ are used to determine the Q_{Bi}, the Q_{Bi} are then used to get new $X(n)$, and so on, until the boundary condition is satisfied.

The measure of the satisfaction of the boundary condition will be defined as E_1, which is the mean squared error to the boundary condition

$$E_1 = \frac{2}{L} \int_0^{L/2} (Q_{Bi} - Q_B)^2 \, dx \qquad (11.52)$$

where

$$Q_B = \frac{2}{L} \int_0^{L/2} Q_{Bi} \, dx$$

For an exact solution, E_1 must be zero.

As occurred with the second-order analytical solution for which Q_B is different from zero, $\eta(x)$ must have a zero mean, that is,

$$(2/L) \int_0^{L/2} \eta(x) \, dx = 0$$

Further, for design purposes, it is desirable to be able to prescribe the wave height a priori. These last two conditions can be considered as constraints to the condition that E_1 be zero, or at least very small. To solve for the $X(n)$'s, E_1 must be minimized, subject to the constraints. Note that there are two additional unknowns, due to the necessity of also determining the wavelength L and the value of the free surface streamline $\psi(x, \eta)$, which is a constant. Using the method of Lagrange multipliers (Hildebrand, 1965), we minimize the objective function O_f:

$$O_f = E_1 + \frac{2\lambda_1}{L} \int_0^{L/2} \eta(x)\, dx + \lambda_2\left[\eta(0) - \eta\left(\frac{L}{2}\right) - H \right] \tag{11.53}$$

where λ_1 and λ_2 are Lagrange multipliers. The objective function is nonlinear and in order to facilitate the solution, it is expanded by a truncated Taylor series:

$$O_f^{j+1} = O_f^j + \sum_{n=1}^{N+2} \frac{\partial O_f^j}{\partial X(n)} \Delta X^j(n) \tag{11.54}$$

where $\Delta X(n)^j$ is a small correction to $X(n)$:

$$X^{j+1}(n) = X^j(n) + \Delta X^j(n) \tag{11.55}$$

and the superscript j indicates the number of iterations that have been made. Minimizing the expanded objective function with respect to all the $X(n)_j$ plus λ_1 and λ_2 yields a series of linear equations for the $\Delta X^j(n)$ for fixed j.

Solving the equations for $\Delta X(n)$ in matrix form yields the solution for iteration, $j + 1$. This process is repeated for several iterations until O_f^{j+1} is acceptably small. This technique is simply a Newton–Raphson procedure, but applied to a set of nonlinear equations (see, e.g., Gerald, 1978).

The stream function wave theory has been used to generate 40 representations of nonlinear waves by Dean (1974) and the results tabulated in dimensionless form. Using these tables, most designs using nonlinear wave theory can be carried out without the use of a computer.

Chaplin (1980) has developed an improved approach to that of Dean (1965) for calculating the stream function coefficients, although it is not clear that his method is an improvement over that of Dalrymple (1974), which is presented above. Chaplin formulates the problem in dimensionless form with h, H, and T as the independent parameters and the dimensionless surface displacements as the unknowns. The method, which is more complex, but yields greater accuracy, particularly for nearly-breaking waves, commences by determining a set of orthonormal functions representing the terms in the series given by Eq. (11.49). These functions then allow a more direct solution of the stream function coefficients which satisfy the dynamic free surface boundary condition [Eq. (11.48b)]. The method has the advantage that, in contrast to that originally developed by Dean, a maximum in wave length (or celerity) is represented at wave heights slightly smaller than breaking. Chaplin carried out comparisons of a number of parameters and concluded that for waves up to 75% of the breaking height the errors in the tables of Dean were less than 1% except in extremely shallow water. For waves of 90% of the breaking height the errors were less than 5% in most cases.

Extension of the theory to waves on vertically sheared currents has been done by Dalrymple (1974) and Dalrymple and Cox (1976), and for irregular measured water surfaces by Dean (1965). The latter procedure involves determining the best-fit stream function to a given water surface profile.

11.4 FINITE-AMPLITUDE WAVES IN SHALLOW WATER

In the Stokes perturbation procedure, the perturbation parameter was ka, the wave steepness. In very shallow water the Stokes wave profile [Eq. (11.32)] becomes (using shallow asymptotic expansions for the hyperbolic functions)

$$\eta(x, t) = a \cos (kx - \sigma t) + \frac{3ka^2}{4(kh)^3} \cos 2(kx - \sigma t) \qquad (11.56)$$

The second term is a function of wave amplitude and length, as well as the water depth, being proportional to the Ursell number or $(a/h)(L^2/h^2)$, which will be defined as the ratio α/β, where $\alpha \equiv a/h$, $\beta \equiv h^2/L^2$. In fact, the Stokian wave profile for higher orders in shallow water is an expansion using the ratio α/β as the perturbation parameter. This implies that α/β must be much less than unity or $\alpha \ll \beta$. In shallow water, this requires quite a short wavelength or a small-amplitude wave, as discussed previously. It would be desirable for design purposes to have a perturbation expansion in shallow water which would at least allow α and β to be of the same magnitude. This can be achieved with a different perturbation procedure than that used previously.

First, the shallow water wave will be assumed to be propagating without change in form; thus, by moving with the wave celerity C, the motion becomes stationary, and a stream function approach becomes convenient, as in the preceding section.

The free surface boundary conditions are

$$\left(\frac{\partial \psi}{\partial x}\right)^2 + \left(\frac{\partial \psi}{\partial z}\right)^2 + 2g(h + \eta) = Q \qquad \text{on } z = h + \eta \qquad (11.57a)$$

and

$$\psi = Ch \qquad \text{on } z = h + \eta \qquad (11.57b)$$

In this context, the coordinate system is taken to be on the bottom and Q is the Bernoulli constant. At the bottom,

$$\psi = 0 \qquad \text{on } z = 0 \qquad (11.57c)$$

This condition ensures that there is no flow through the horizontal bottom, as

$$w = \frac{\partial \psi}{\partial x} = 0 \qquad \text{on } z = 0$$

For a wave propagating on a quiescent fluid, $Q = C^2 + 2gh$, which is determined from the dynamic free surface boundary by moving far upstream of the wave, where the wave motion is negligible.

It is again convenient to express the equation in nondimensional form prior to the perturbation procedure. In contrast to the Stokes expansion, however, the x, z coordinates will be nondimensionalized differently, recognizing the fact that there will be larger gradients in the vertical direction than the horizontal.

$$X = \frac{x}{L}$$

$$Z = \frac{z}{h}$$

$$\Pi = \frac{\eta}{a}$$

$$\Psi = \frac{\psi}{h\sqrt{ga}} \tag{11.58}$$

The governing Laplace equation, in terms of the nondimensional variables, is written as

$$\beta \frac{\partial^2 \Psi}{\partial X^2} + \frac{\partial^2 \Psi}{\partial Z^2} = 0 \tag{11.59a}$$

where, again, $\beta \equiv (h/L)^2$.

The two free surface conditions are

$$\Psi = \frac{C}{\sqrt{ga}} \qquad \text{on } Z = 1 + \alpha\Pi \tag{11.59b}$$

where

$$\alpha \equiv \frac{a}{h} \tag{11.59c}$$

and

$$\beta\left(\frac{\partial \Psi}{\partial X}\right)^2 + \left(\frac{\partial \Psi}{\partial Z}\right)^2 + \frac{2}{\alpha}(1 + \alpha\Pi) = \frac{Q}{ga} \qquad \text{on } Z = 1 + \alpha\Pi \tag{11.59d}$$

By differentiating[2] with respect to X, we can eliminate the constants to obtain the form we will use:

$$\beta\frac{\partial\Psi}{\partial X}\frac{\partial^2\Psi}{\partial X^2} + \alpha\beta\frac{\partial\Psi}{\partial X}\frac{\partial^2\Psi}{\partial X\partial Z}\frac{\partial\Pi}{\partial X} + \frac{\partial\Psi}{\partial Z}\frac{\partial^2\Psi}{\partial X\partial Z} + \alpha\frac{\partial\Psi}{\partial Z}\frac{\partial^2\Psi}{\partial Z^2}\frac{\partial\Pi}{\partial X} + \frac{\partial\Pi}{\partial X} \tag{11.60}$$

$$= 0 \quad \text{on } Z = 1 + \alpha\Pi$$

Using a Frobenius power series solution technique, we will assume a solution in terms of a series in Z (see, e.g., Wylie, 1960):

$$\Psi(X, Z) = \sum_{n=0}^{\infty} Z^n f_n(X) \tag{11.61}$$

To satisfy the bottom boundary condition, f_0 must be zero. Substituting the assumed solution into the dimensionless Laplace equation and grouping terms yields

$$2f_2 Z^0 + \left(6f_3 + \beta\frac{d^2 f_1}{dX^2}\right)Z^1 + \left(12f_4 + \beta\frac{d^2 f_2}{dX^2}\right)Z^2 \tag{11.62}$$

$$+ \left(20f_5 + \beta\frac{d^2 f_3}{dX^2}\right)Z^3 + \cdots = 0$$

For this equation to be satisfied for any Z, the coefficients of the Z^n terms must be zero. Therefore,

$$f_2 = 0$$

$$f_3 = -\frac{\beta}{6}\frac{d^2 f_1}{dX^2}$$

$$f_4 = 0 \tag{11.63}$$

$$f_5 = -\frac{\beta}{20}\frac{d^2 f_3}{dX^2} = \frac{\beta^2}{120}\frac{d^4 f_1}{dX^4}$$

$$f_6 = 0$$

and so on. Therefore, the series may be written

$$\Psi = Zf_1 - \frac{\beta}{\cdot 6}Z^3\frac{d^2 f_1}{dX^2} + \frac{\beta^2 Z^5}{120}\frac{d^4 f_1}{dX^4} + \cdots$$

or

$$\Psi = \sum_{n=0}^{\infty} (-1)^n\frac{\beta^n Z^{(2n+1)}}{(2n+1)!}\frac{d^{2n} f_1}{dX^{2n}} \tag{11.64}$$

---

[2]Since Z at the free surface is a function of X, the total derivative is used.

Clearly, we now have a series in terms of β, the relative depth parameter. The objective is to determine the functional form of f_1 in order that Ψ satisfies the two free surface boundary conditions. Substituting the expansion for Ψ into the kinematic and dynamic free surface boundary conditions yields

$$(1 + \alpha\Pi)f_1 - \frac{1}{6}\beta(1 + \alpha\Pi)^3 \frac{d^2 f_1}{dX^2} + O(\beta^2) = \frac{C}{\sqrt{ga}} \qquad (11.65)$$

$$\frac{\beta}{2}(1 + \alpha\Pi)^2 \frac{df_1}{dX}\frac{d^2 f_1}{dX^2} + f_1 \frac{df_1}{dX} - \frac{\beta}{2}(1 + \alpha\Pi)^2 f_1 \frac{df_1^3}{dX^3} + \frac{d\Pi}{dX} + O(\beta^2) = 0 \qquad (11.66)$$

First, examining the zeroth-order solution for Ψ in β, that is, the solution depending on β^0, it is clear that the horizontal velocity, $U = -\partial\Psi/\partial Z$, is uniform over depth, as f_1 is not a function of Z. In this case, the kinematic boundary condition reduces to

$$(1 + \alpha\Pi)f_1 = \frac{C}{\sqrt{ga}}$$

or

$$f_1 = \frac{C}{\sqrt{ga}}(1 + \alpha\Pi)^{-1} \qquad (11.67)$$

Substituting into Eq. (11.66) will yield, to order β^0, an expression for C:

$$-\frac{C^2 \alpha(d\Pi/dX)}{ga(1 + \alpha\Pi)^3} + \frac{d\Pi}{dX} = 0 \qquad (11.68)$$

or

$$\left[-\frac{C^2}{ga}\alpha(1 + \alpha\Pi)^{-3} + 1\right]\frac{d\Pi}{dX} = 0 \qquad (11.69)$$

For this last equation to be true everywhere, the term within the parentheses must be zero. Therefore,

$$C^2 = gh(1 + \alpha\Pi)^3 \qquad (11.70a)$$

or[3]

$$C \simeq \sqrt{gh}\left(1 + \frac{3\,\alpha\Pi}{2}\right) \qquad \text{to order } (\alpha^2, \beta^0) \qquad (11.70b)$$

To the first approximation in α, we have the usual shallow water wave celerity, which depends solely on the mean water depth, $C = \sqrt{gh}$. The wave

---

[3]Recall the binominal series approximation:

$$(1 + \epsilon)^n \simeq 1 + n\epsilon + \frac{n(n-1)}{2!}\epsilon^2$$

form Π can be arbitrarily chosen for this case. The next approximation, to $O(\alpha^2, \beta^0)$, provides a correction term, $\frac{3}{2}\alpha\Pi$, which indicates that the larger the local water surface displacement, the faster the local wave speed. This result was first due to Airy (1845). The difficulty is that we originally postulated that we were moving the coordinate system with wave speed C, which was assumed to be constant for the wave. Clearly, this is not the case, so we expect the wave to deform as it propagates, with the higher portions of the wave profile moving faster than the lower portions, so that, in fact, the wave profile continually steepens in front until our assumption of α being small is violated. Physically, the wave eventually breaks in the form of a bore. Theoretically, we must find a better solution, one that yields a constant celerity.

To a higher order, $O(\alpha^2, \alpha\beta)$, the solution is assumed to be

$$f_1 = \frac{C}{\sqrt{ga}} (1 + \alpha\Pi)^{-1} + \beta A \tag{11.71}$$

where A is an unknown function of x. Substituting into the kinematic free surface boundary condition and retaining terms $O(\beta)$ yields the following equation for A in terms of Π and its derivatives:

$$A = -\frac{\alpha}{6} \frac{C}{\sqrt{ga}} \left[\frac{d^2\Pi}{dX^2} - \frac{2\alpha}{1 + \alpha\Pi} \left(\frac{d\Pi}{dX} \right)^2 \right] \tag{11.72}$$

Substituting f_1 and A into the dynamic free surface boundary condition yields a very complicated expression, which, however, to $O(\alpha^2, \alpha\beta)$ reduces to this nonlinear equation:

$$\frac{d\Pi}{dX} \left[1 - \frac{C^2\alpha}{ga} (1 - 3\alpha\Pi) \right] + \frac{\alpha\beta C^2}{3ga} \frac{d^3\Pi}{dX^3} = 0 \tag{11.73}$$

or

$$\frac{1}{3} \frac{d^3\Pi}{dX^3} + \frac{3\alpha\Pi}{\beta} \frac{d\Pi}{dX} + \frac{d\Pi}{dX} \left(\frac{ga}{\beta C^2\alpha} - \frac{1}{\beta} \right) = 0 \tag{11.74}$$

This equation is the steady-state form of the Korteweg–DeVries (1895) equation. The solution to the linearized form of this equation,[4] which is of $O(\alpha, \alpha\beta)$, is

$$\Pi = \cos 2\pi X$$

with the following equation for the wave celerity:

$$C = \sqrt{\frac{gh}{1 + 4\pi^2\beta/3}} \approx \sqrt{gh} \left(1 - \frac{2\pi^2\beta}{3} \right) \tag{11.76}$$

[4]That is, neglecting the second term.

The effect of including the parameter β is to reduce the celerity, just the opposite of the parameter α. In fact, by introducing β, the relative depth (i.e., making it different from zero, the infinite wave length case), we have developed a wave moving at constant C, a wave that does not form a bore as did the solution when $\beta = 0$. This wave is equivalent to the small-amplitude wave theory we have developed in the first eight chapters. In fact, it is easy to show that the celerity given above is equal to the first two terms in the shallow water expansion of Eq. (3.35).

11.4.1 The Solitary Wave

We will now seek a solution containing both α and β such that their influence results in nonlinear waves of permanent form. The equation above can be solved without the necessity of linearization. The procedure is to integrate once with respect to X:

$$\frac{1}{3}\frac{d^2\Pi}{dX^2} + \frac{3}{2}\frac{\alpha}{\beta}\Pi^2 + \Pi\left(\frac{ga}{\beta C^2\alpha} - \frac{1}{\beta}\right) + D = 0 \tag{11.77}$$

Multiply by $d\Pi/dX$ and integrate again to yield

$$\frac{1}{6}\left(\frac{d\Pi}{dX}\right)^2 + \frac{\alpha}{2\beta}\Pi^3 + \frac{\Pi^2}{2}\left(\frac{ga}{\beta C^2\alpha} - \frac{1}{\beta}\right) + D\Pi + E = 0 \tag{11.78}$$

where D and E are constants of integration. If we solve this equation for the case of a single wave which has no influence at infinity, then $\Pi = d\Pi/dX = 0$ at $X = \infty$. Clearly, D and E must be zero, from Eqs. (11.77) and (11.78). The remaining equation is, therefore,

$$\left(\frac{d\Pi}{dX}\right)^2 = \frac{3\Pi^2}{\beta}\left(1 - \frac{ga}{C^2\alpha}\right)\left(1 - \frac{\alpha}{1 - ga/C^2\alpha}\Pi\right) \tag{11.79}$$

For the wave form to be symmetric about the X axis, $d\Pi/dX$ must go to zero at $\Pi = 1$, the wave crest. Thus

$$\alpha = 1 - \frac{ga}{C^2\alpha} \tag{11.80}$$

or

$$C = \sqrt{\frac{gh}{1-\alpha}} = \sqrt{gh}\left(1 + \frac{\alpha}{2}\right) \qquad \text{to } O(\alpha^2, \alpha\beta) \tag{11.81}$$

and the equation becomes

$$\left(\frac{d\Pi}{dX}\right)^2 = 3\Pi^2\frac{\alpha}{\beta}(1 - \Pi) \tag{11.82}$$

The solution is

$$\Pi = \text{sech}^2 \sqrt{\frac{3}{4}\frac{\alpha}{\beta^3}}\, X \tag{11.83}$$

or in dimensional form

$$\eta = a\, \text{sech}^2 \sqrt{\frac{3}{4}\frac{a}{h^3}}\, x \tag{11.84}$$

This is called the solitary wave of Boussinesq (1872). Munk (1949) has advocated the use of superimposed solitary waves to describe waves in the surf zone. The solitary wave form is shown in Figure 11.2. The entire wave profile is positive for this wave; there is no η less than zero. The a therefore represents the height of the wave and h the depth at infinity. The volume of water contained in a solitary wave, V, over a distance $-l < x < l$, that is, the amount of water above the mean water level, is found by integrating the profile.

$$V = \int_{-l}^{l} \eta\, dx = \frac{2a\, \tanh\, \sqrt{\tfrac{3}{4}(a/h^3)}\,l}{\sqrt{\tfrac{3}{4}(a/h^3)}} \tag{11.85}$$

For l equal to infinity, the hyperbolic tangent is unity and

$$V_\infty = 4h\sqrt{\frac{ah}{3}} \tag{11.86}$$

Clearly, for engineering use, an infinitely long wave has no value; however, the effective length of the solitary wave is much less. For example, 95% of this volume is contained within the distance

$$l = \frac{2.12h}{\sqrt{a/h}} \tag{11.87}$$

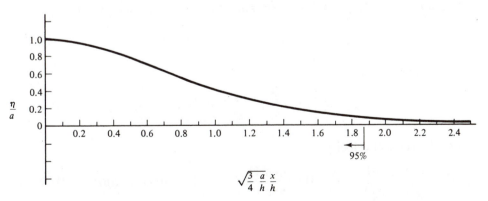

Figure 11.2 Dimensionless free surface profile of a solitary wave.

For example, if $a/h = 0.5$, then 95% of the volume is within a space of about six water depths.

The water particle velocities under the solitary wave are found by $U = -\partial\Psi/\partial Z$ and $W = \partial\Psi/\partial X$ from Eqs. (11.64), (11.71), and (11.72).

$$U = -f_1 + \frac{1}{2}\beta Z^2 \frac{d^2 f_1}{dX^2} + O(\beta^2)$$

$$= \sum_{n=0}^{\infty} (-1)^{n+1} \frac{\beta^n Z^{2n}}{2n!} \frac{d^{2n} f_1}{dX^{2n}} \tag{11.88}$$

$$W = Z\frac{df_1}{dX} - \frac{1}{6}\beta Z^3 \frac{d^3 f_1}{dX^3} + O(\beta^2)$$

$$= \sum_{n=0}^{\infty} (-1)^n \frac{\beta^n Z^{(2n+1)}}{(2n+1)!} \frac{d^{2n+1} f_1}{dX^{2n+1}} \tag{11.89}$$

Substituting for f_1 from Eqs. (11.71) and (11.72) for the horizontal velocity yields

$$U = \frac{C}{\sqrt{ga}}\left[-1 + \alpha\Pi - (\alpha\Pi)^2 + \alpha\beta\left(\frac{1}{6} - \frac{Z^2}{2}\right)\frac{d^2\Pi}{dX^2}\right] \qquad \text{to } O(\alpha^2, \alpha\beta)$$

or

$$U = \frac{C}{\sqrt{ga}}\left\{-1 + \left[\alpha + 3\alpha^2\left(\frac{1}{6} - \frac{Z^2}{2}\right)\right]\Pi - \alpha^2\left(\frac{7}{4} - \frac{9}{4}Z^2\right)\Pi^2\right\} \tag{11.90}$$

where Π is given as a function of position by Eq. (11.83). The first term in brackets, that is, the minus 1, is to account for the speed of translation of the coordinate system. For a fixed observer, this term would be neglected. The remainder of the expression for U consists of terms proportional to Π; therefore, away from the crest the velocity becomes small. Under the crest of a solitary wave, $\Pi = 1$ and the expression for U is greatly simplified:

$$U_C = \frac{C}{\sqrt{ga}}\left\{-1 + \alpha\left[1 + \frac{\alpha}{4}(3Z^2 - 5)\right]\right\} \tag{11.91}$$

or in dimensional form,

$$u_C = C\left\{-1 + \frac{a}{h}\left[1 + \frac{a}{4h}\left(3\left(\frac{z}{h}\right)^2 - 5\right)\right]\right\} \tag{11.92}$$

For the vertical velocity

$$W = \frac{C}{\sqrt{ga}}\left[Z\alpha\frac{d\Pi}{dX}(2\alpha\Pi - 1) - \frac{\alpha\beta}{6}\frac{d^3\Pi}{dX^3}(Z - Z^3)\right] \qquad \text{to } O(\alpha^2, \alpha\beta) \tag{11.93}$$

or

$$W = -\frac{C}{\sqrt{ga}} \left\{ \alpha Z \frac{d\Pi}{dX} \left[1 + \frac{\alpha}{2}(1 - 7\Pi - Z(1 - 3\Pi)) \right] \right\} \qquad (11.94)$$

where Π is given by Eq. (11.83) and

$$\frac{d\Pi}{dX} = -2\Pi \sqrt{\frac{3}{4}\frac{\alpha}{\beta}} \tanh \sqrt{\frac{3}{4}\frac{\alpha}{\beta}} X \qquad (11.95)$$

In dimensional form

$$w = C \sqrt{\frac{3a}{h}\frac{z}{h}\frac{\eta}{h}} \tanh \left(\sqrt{\frac{3}{4}\frac{a}{h^3}} X \right) \left\{ 1 + \frac{a}{2h} \left[1 - 7\frac{\eta}{a} - \left(\frac{z}{h}\right)^2 \left(1 - \frac{3\eta}{a}\right) \right] \right\}$$

$$(11.96)$$

For applications of the solitary wave theory, the reader is referred to the extensive work of Munk (1949).

11.4.2 Cnoidal Wave Theory

In 1895, Korteweg and Devries (1895) developed a shallow water wave theory which allowed periodic waves to exist. These waves have the unique feature of reducing to the solitary wave theory at one limit and to a profile expressed in terms of cosines at the other limit, thus spanning the range between the linear and solitary theories. The wave profile is developed in terms of a Jacobian elliptic integral, $cn(u)$, and they called the theory "cnoidal" to be consonant with the sinusoidal, or Airy theory.

The development of the periodic theory follows the previous perturbation procedure for solitary waves with the exception that in Eq. (11.78) we cannot force the unknown constants D and E to be zero. If, however, for our cnoidal waves we force $\Pi = 0$ at $Z = 1$, defined as the wave trough, then $d\Pi/dX$ should be zero there also, as the wave form is periodic. Therefore, E must be zero and the integrated equation becomes

$$\frac{1}{6}\left(\frac{d\Pi}{dX}\right)^2 + \frac{\alpha}{2\beta}\Pi^3 - F\frac{\Pi^2}{2} + D\Pi = 0 \qquad (11.97)$$

where

$$F = \frac{1}{\beta}\left(1 - \frac{ga}{\alpha C^2}\right) \qquad (11.98)$$

At the wave crest, $\Pi = 1$ and again $d\Pi/dX = 0$; thus D can be readily found:

$$D = \tfrac{1}{2}\left(F - \frac{\alpha}{\beta} \right) \tag{11.99}$$

The equation is now

$$\left(\frac{d\Pi}{dX} \right)^2 = -\frac{3\alpha}{\beta}\Pi^3 + 3F\Pi^2 - 3\left(F - \frac{\alpha}{\beta} \right)\Pi \tag{11.100}$$

or

$$\left(\frac{d\Pi}{dX} \right)^2 = \frac{3\alpha}{\beta}\Pi(1 - \Pi)\,(\Pi + S) \tag{11.101}$$

where

$$S = 1 - F\frac{\beta}{\alpha} \tag{11.102}$$

The substitution $\Pi = \cos^2 \chi$ will be used to transform this equation into a more tractable form, involving χ. From the imposed conditions on Π at the crest and trough, the values of χ are seen to be 0 and $\pi/2$ for the crest and first trough, respectively. Substituting, we obtain

$$\frac{\partial \chi}{\sqrt{\dfrac{3\alpha}{4\beta}}\,\sqrt{1 + S - \sin^2 \chi}} = dX$$

or

$$X = \frac{1}{\sqrt{\dfrac{3}{4}\dfrac{\alpha}{\beta}(1 + S)}} \int^{\chi} \frac{ds}{\sqrt{1 - \dfrac{1}{1 + S}\sin^2 s}}$$

$$X = \frac{1}{\sqrt{\dfrac{3}{4}\dfrac{\alpha}{\beta}(1 + S)}} F\left(\frac{1}{\sqrt{1 + S}}, \chi \right) = \frac{1}{\sqrt{\dfrac{3}{4}\dfrac{\alpha}{\beta k^2}}} F(k, \chi) \tag{11.103}$$

where

$$k \equiv \frac{1}{\sqrt{1 + S}} \tag{11.104}$$

and where $F(k, \chi)$ is the notation for the elliptic integral of the first kind with modulus k and amplitude χ. The amplitude of χ is then given, from the theory of elliptic functions, as

$$\chi = \cos^{-1}\left[\operatorname{cn}\left(X \sqrt{\frac{3\alpha}{4\beta k^2}} \right) \right] \tag{11.105}$$

or

$$\Pi = \operatorname{cn}^2\left(X \sqrt{\frac{3\alpha}{4\beta k^2}} \right) \operatorname{mod}(k) \tag{11.106}$$

or in dimensional form,

$$\eta = a \operatorname{cn}^2\left(x \sqrt{\frac{3\alpha}{4h^3 k^2}} \right) = a \operatorname{cn}^2[F(k, \chi)] \tag{11.107}$$

To be consistent, the parameter a has been used in the definition for η; however, in this connection, a is the wave height, as in the solitary wave theory.

The Jacobian elliptic function cn is a periodic function with a period of $4K$, where K is the complete elliptic integral of the first kind, $K \equiv F(k, \pi/2)$, as shown in Figure 11.3. The function $cn^2 u$ is periodic with period $2K$. The wave length of the cnoidal wave is found by setting X equal to unity in the argument of $cn\ u$. Therefore,

$$\sqrt{\frac{3\alpha}{4\beta k^2}} 1 = 2K \tag{11.108}$$

or

$$\frac{aL^2}{h^3} \equiv U_r = \frac{16}{3} K^2 k^2 \tag{11.109}$$

The parameter k is uniquely related to wave amplitude a, the length L, and the water depth h. A graph of k versus the Ursell parameter U_r, $K(k)$, and $E(k)$, the complete elliptic integral of the second kind, is shown in Figure 11.4. For shallow water (Chapter 3) $h/L < 1/20$, and therefore the Ursell parameter has a minimum value of $U_r = 400(a/h)$. For nearly-breaking waves, a (the wave height) is about $0.8h$. This gives an Ursell value of 320 and a k value of

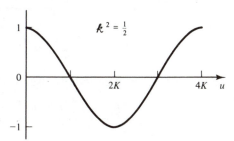

Figure 11.3   The Jacobian elliptic function, $cn\ u$.

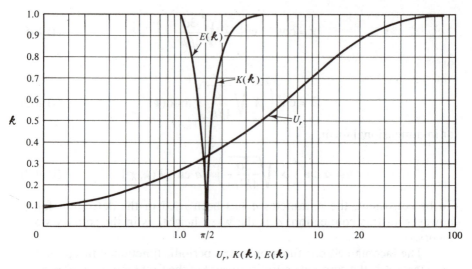

Figure 11.4 Complete elliptic integrals of the first and second kinds and the Ursell parameter as a function of the modulus k.

0.999999 or larger for shallow water. Various water surface profiles are shown in Figure 11.5 for various values of k.

The parameter h refers to the water depth at the wave trough. To determine the mean water depth, the wave profile is averaged and denoted $\overline{\Pi}$.

$$\overline{\Pi} = \int_0^1 \text{cn}^2(2KX)\,dX \tag{11.110}$$

or

$$\bar{\eta} = \left[1 + \frac{1}{k^2}\left(\frac{E}{K} - 1\right) \right]a \tag{11.111}$$

where E is the complete elliptic integral of the second kind. The total depth is then $(h + \bar{\eta})$.

The cnoidal wave celerity can be found using the definition for F, S, and k following Eqs. (11.98), (11.102), and (11.104). Solving for C, we have

$$C = \sqrt{\frac{gh}{1 + \dfrac{a}{h}\left(\dfrac{1}{k^2} - 2\right)}} \tag{11.112}$$

To find the related wave period, we use the definition of L $(C = L/T)$,

$$T = \frac{L}{C} = \sqrt{\frac{16}{3}\frac{K^2 k^2 h^2}{ga}\left[1 + \frac{a}{h}\left(\frac{1}{k^2} - 2\right)\right]} \tag{11.113}$$

from Eqs. (11.109) and (11.112).

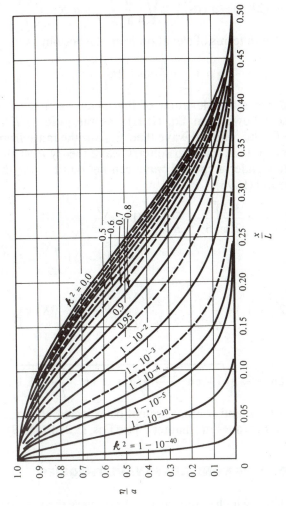

Figure 11.5 Surface profile of the cnoidal wave. (From Wiegel, 1960.)

Several interesting asymptotic features of the cnoidal wave should be pointed out. As $k \to 1$, the wave length becomes infinite, as $K(1) \to \infty$ and $cn^2(x) \to sech^2(x)$, the solitary wave.[5] On the other hand, as $k \to 0$, $cn(x) \to cos(x)$, and $K \to \pi/2$, and the wave form changes:

$$\eta = a\, cn^2(2KX) \to a\, cos^2\left(x\sqrt{\frac{3}{4}\frac{a}{h^3}\frac{1}{k^2}} \right) = a\, cos^2\left(\frac{\pi X}{L} \right) \tag{11.114}$$

which can be written in terms of elevation from the bottom as

$$z = h + \frac{a}{2} + \frac{a}{2}\, cos\left(2\pi\frac{X}{L} \right) \tag{11.115}$$

where $h + a/2$ denotes the elevation of the mean water level above the bottom. This also follows from Eq. (11.111), as the ratio of E/K goes to $(1 - k^2/2)$ for $K \to 0$. Thus cnoidal wave theory spans the range from sinusoidal or Airy theory in deep water to solitary wave theory in shallow water.

The velocities under a cnoidal wave can be found as for the solitary wave, Eqs. (11.90) and (11.96).

$$u = C\left\{ -1 + \frac{\eta}{h} - \left(\frac{\eta}{h}\right)^2 + \frac{3}{2}\frac{1}{k^2}\left[\frac{1}{6} - \left(\frac{z}{h}\right)^2 \right]\left[(1+k)\frac{a^2}{h^2} \right.\right. \tag{11.116}$$

$$\left.\left. + 2(2k-1)\left(\frac{\eta}{h}\right)^2 - 3k^2\left(\frac{\eta}{h}\right)^4 h^2 \right] \right\}$$

$$w = -C\left\{ \frac{d\eta}{dx}\frac{z}{h}\left(-2\frac{\eta}{h} + 1 \right) + \frac{3}{hk^2}\left[1 - \left(\frac{z}{h}\right)^2 \right]\left[(2k-1)\frac{a}{h} - 3Kk\eta \right] \right\} \tag{11.117}$$

where

$$\frac{\partial\eta}{\partial x} = -a\sqrt{\frac{3a}{h}\frac{1}{kh}}\, cn\,\frac{2Kx}{L}\sqrt{1 - cn^2\frac{2Kx}{L}}\sqrt{1 + k\left(cn^2\frac{2Kx}{L} - 1 \right)} \tag{11.118}$$

The leading terms for u and w are, as might be expected, the same as those developed for the long waves in Chapter 5 [see Eqs. (5.2) and (5.3)].

11.5 THE VALIDITY OF NONLINEAR WAVE THEORIES

It is important to know which of the various water wave theories to apply to a particular problem, where the wave characteristics and water depth are specified. For example, is the linear wave theory suitable or must cnoidal theory be used? In order to address these problems, the validity of the various

---

[5]Iwagaki (1968), using this asymptotic behavior, has developed the hyperbolic wave theory (valid for $K > 3$), which means that $k > 0.98$, which is a blend of solitary and cnoidal theory having the mathematical advantage of the solitary theory and some of the properties of the cnoidal theory.

theories must be known. This "validity" is composed of two parts: the mathematical validity and the physical validity. The first is the ability of any given wave theory to satisfy the mathematically posed boundary value problem. For example, all the theories in the book satisfy the bottom boundary condition exactly, but the cnoidal and solitary wave theories only approximately satisfy the Laplace equation within the fluid. All of the theories only satisfy the dynamic free surface boundary approximately, while the kinematic free surface boundary condition is satisfied (to the numerical accuracy of the computer) by the stream function theory. On the other hand, the physical validity refers to how well the prediction of the various theories agrees with actual measurements. This part of the validity has been difficult to obtain due to the problem of wave tank design and measurement requirements. The interested reader is referred to Dean (1974).

The analytical validity of many wave theories was examined by Dean (1970) (see also Dean, 1974). Figure 11.6 shows the results of the comparison of the theories, denoting the regions for which each theory provides the best fit to the dynamic free surface boundary condition. As would be expected, the

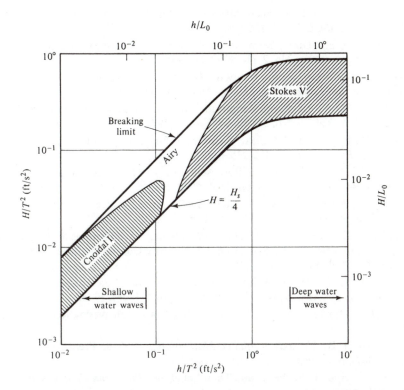

Figure 11.6   Periodic wave theories providing best fit to dynamic free surface boundary condition (analytical theories only).

cnoidal wave theory does well in shallow water, while in deep water, the Stokes V wave theory proved to be more applicable. Somewhat surprisingly the linear wave theory did well for the intermediate water depths. However, when high-order stream function wave theory is used, it provides the best fit of all the theories, even in shallow water (although quite high orders, such as twentieth order, are necessary).

REFERENCES

AIRY, G. B., "Tides and Waves," *Encyclopaedia Metropolitana*, 1845.

BORGMAN, L. E., and J. E. CHAPPELEAR, "The Use of the Stokes–Struik Approximation for Waves of Finite Height", *Proc. 6th Conf. Coastal Eng., ASCE*, Council on Wave Research, Berkeley, Calif., 1958.

BOUSSINESQ, J. "Théorie des ondes et des remous qui se propagent le long d'un canal rectangulaire horizontal, en communiquant au liquide contenu dans ce canal des vitesses sensiblement pareilles de la surface au fond," *J. Math. Pures Appl.*, Vol. 17, pp. 55–108, 1872.

CHAPLIN, J. R., "Developments of Stream Function Wave Theory, "*Proc. 17th Conf. Coastal Eng., ASCE*, Vol. 3, 1980, pp. 179–205.

CHAPPELEAR, J. E., "Direct Numerical Calculation of Nonlinear Ocean Waves," *J. Geophys. Res.*, Vol. 66, No. 2, pp. 501–508, 1961..

COKELET, E. D., "Steep Gravity Waves in Water of Arbitrary Uniform Depth," *Philos. Trans. Roy. Soc. Lond. A*, Vol. 286, pp. 183–230, 1977.

DALRYMPLE, R. A., "A Finite Amplitude Wave on a Linear Shear Current," *J. Geophys. Res.*, Vol. 79, No. 30, pp. 4498–4504, 1974.

DALRYMPLE, R. A., and J. C. COX, "Symmetric Finite Amplitude Rotational Water Waves," *J. Phys. Ocean.*, Vol. 6, No. 6, 1976.

DEAN, R. G., "Stream Function Representation of Nonlinear Ocean Waves," *J. Geophys. Res.*, Vol. 70, No. 18, pp. 4561–4572, 1965.

DEAN, R. G., "Relative Validity of Water Wave Theories," *J. Waterways Harbors Div., ASCE*, Vol. 96, No. WW1, pp. 105–119, Feb. 1970.

DEAN, R. G., "Evaluation and Development of Water Wave Theories for Engineering Application," Vols. 1 and 2, Spec. Rep. 1, U.S. Army, Coastal Engineering Research Center, Fort Belvoir, Va., 1974.

EBBESMEYER, C. C., "Fifth Order Stokes Wave Profiles," *J. Waterways, Harbors Coastal Eng. Div., ASCE*, Vol. 100, No. WW3, pp. 264–265, 1974.

GERALD, C. F., *Applied Numerical Analysis*, 2nd ed., Addison-Wesley, Reading, Mass., 1978.

HILDEBRAND, F. B., *Methods of Applied Mathematics*, 2nd ed., Prentice-Hall, Englewood Cliffs, N.J., 1965.

IWAGAKI, Y., "Hyperbolic Waves and Their Shoaling, "*Proc. 10th Conf. Coastal Eng., ASCE*, London, 1968.

KORTEWEG, D. J., and G. DE VRIES, "On the Change of Form of Long Waves Advancing in a Rectangular Channel, and on a New Type of Long Stationary Waves," *Philos. Mag.*, 5th Ser., Vol. 39, pp. 422–443, 1895.

MUNK, W. H., "The Solitary Wave Theory and Its Applications to Surf Problems," *Ann. N.Y. Acad. Sci.*, Vol. 51, pp. 376–424, 1949.

SCHWARTZ, L. W., "Computer Extension and Analytic Continuation of Stokes' Expansion for Gravity Waves," *J. Fluid Mech.*, Vol. 62, 1974.

SKJELBREIA, L., and J. A. HENDERSON, "Fifth Order Gravity Wave Theory," *Proc. 7th Conf. Coastal Eng., ASCE*, 1961, pp. 184–196.

STOKES, G. G., "On the Theory of Oscillatory Waves," *Trans. Camb. Philos. Soc.*, Vol. 8, pp. 441–455, 1847.

URSELL, F., "The Long-Wave Paradox in the Theory of Gravity Waves," *Proc. Camb. Philos. Soc.*, Vol. 49, 1953, pp. 685–694.

WIEGEL, R. L., "A Presentation of Cnoidal Wave Theory for Practical Application," *J. Fluid Mech.* Vol. 7, Pt. 2 (1960).

WIEGEL, R. L., *Oceanographical Engineering*, Prentice-Hall, Englewood Cliffs, N.J., 1964.

WYLIE, C. R., Jr., *Advanced Engineering Mathematics*, 2nd ed., McGraw-Hill, New York, 1960.

PROBLEMS

11.1 Verify that the total horizontal acceleration given in Eq. (11.45) for the Stokes wave theory is correct to second order. Determine the total vertical acceleration.

11.2 Develop the horizontal and vertical velocities, correct to $O(\alpha, \alpha\beta)$ for $\eta = a \cos kx$, Eq. (11.75). Compare with linear (Airy) theory.

11.3 For shallow water waves, develop the equation correct to $O(\alpha^2, \alpha\beta)$ for the pressure under the waves.

11.4 Determine the region of validity for the second-order Stokes theory. Which value of the Ursell parameter is more restrictive?

11.5 Calculate the pressure under Stokes waves, correct to second order.

11.6 What is the α^0, β^0 order solution of Eq. (11.74)? What is the physical significance of this flow?

11.7 Verify Eqs. (11.90) and (11.96).

11.8 Assuming equipartitioning of the energy and finding the potential energy, show that the total energy in a solitary wave per unit crest width is

$$E = \frac{8\rho g h^2 a}{3} \sqrt{\frac{a}{3h}}$$

12

A Series of Experiments for a Laboratory Course Component in Water Waves

12.1 INTRODUCTION

There are several important reasons to include a laboratory component as a portion of a course in water waves. First, since the field of water waves is evolving rapidly with new significant developments, the experience in laboratory techniques will develop a student's capability to test new analytic results and will provide a better basis for evaluating the validity of experimental results reported in the literature. Second, and probably of greater significance, is the confidence (hopefully) and perspective gained by the student in conducting measurements and assessing the associated theoretical results.

12.2 REQUIRED EQUIPMENT

Most of the equipment required for the experiments to be described is usually available with wave tank facilities.

12.2.1 Wave Tank

The size of the wave tank is not critical, but should be of a sufficient size that capillary waves are not significant and that a plane beach of small slope (say 1:15) can be placed in the tank and still allow room for measurements. It assists greatly if a portion of the tank is glass- or Lucite-walled. Also, a movable carriage mounted on level rails is useful for transporting the wave gage and possibly other equipment. The tank at the University of Delaware is approximately 24 m long, 1 m deep, and 0.5 m wide, although a smaller tank would be suitable. The experiments to be described will be based on a capability to generate monochromatic waves; however, the range of experiments would be greatly expanded with the availability of a spectral-generating capability.

12.2.2 Wave Gages and Recording Equipment

Laboratory wave gages and recording oscillographs are quite standard and will not be described in detail. Either capacitance or resistance gages are suitable. It is helpful to mount the wave gages on a point gage support to allow static calibrations to be carried out readily (see Figure 12.1). Generally, two wave gages are required with output on the same oscillograph and as noted previously, it is desirable if one of the gages is movable on a level surface.

12.2.3 Velocity Sensor

A small laboratory version of a biaxial electronic current meter is useful in conducting measurements of the water particle velocity field. If an equivalent current meter is not available, it is possible to measure water particle excursions visually.

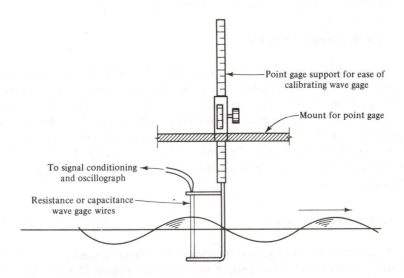

Figure 12.1   Wave gage mounted on graduated point gage support.

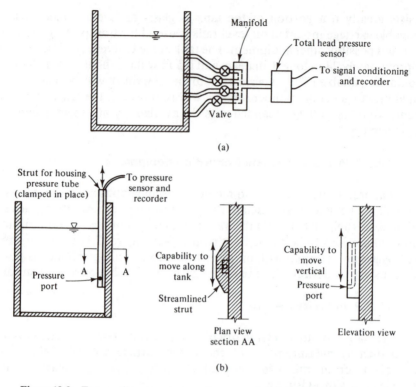

Figure 12.2   Two possible arrangements for measuring pressure field in waves: (a) permanent taps through Lucite wall of wave tank; (b) movable pressure port with pressure tubing housed in movable streamlined strut.

12.2.4 Pressure Sensor

A reasonably sensitive pressure sensor is desirable. A strain gage total head sensor with a range of 0.005 to 1 psi is very satisfactory. If the observational section of the wave tank is made of Lucite it may be possible to drill ports and connect these to a manifold as shown in Figure 12.2a. If the walls are glass or it is not desired to tap through the walls, a somewhat streamlined strut can be placed flush with the tank wall (see Figure 12.2b). With either system it is essential to be able to bleed any air from lines connecting the port to the sensor.

12.2.5 Wave Forces

A "portal-type" force gage is inexpensive to construct and useful since it responds to forces and is insensitive to moments. Figure 12.3 portrays the main features of a portal gage. The upper and lower plates are rigid relative to the side plates. The sensing is by four strain gages connected to a full bridge

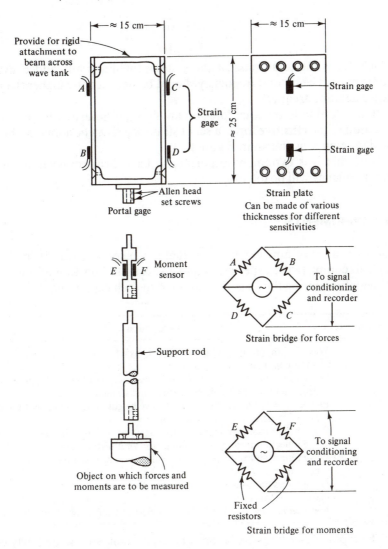

Figure 12.3   Force and moment sensors.

circuit as shown. For purposes of measuring a wide range of forces with good sensitivity, different sets of web plates can be constructed. The strain ϵ at the extremes (top and bottom) of the web plates can be shown to be

$$\epsilon = \frac{3}{2}\frac{Fl}{Ewt^2} \tag{12.1}$$

in which F is the applied force, l, w, and t are the plate length, width, and thickness, respectively, and E is the modulus of elasticity of the material. The natural frequency σ_n of the system is

$$\sigma_n = \sqrt{\frac{2}{l^3}\frac{Ewt^3}{M_T}}$$ (12.2)

in which M_T is the total mass of the system, including any added hydro-dynamic mass. The natural frequency should be significantly higher than the highest excitation frequency.

If, in addition to the total force on an object, it is desired to determine the location of the effective force, a set of strain gages can be added to the rod to yield moments, as shown in Figure 12.3.

Note that it is extremely important to have firm connections or the natural frequency will be too low.

12.3 EXPERIMENTS

Following is a list of nine experiments that can be carried out. It should be possible to complete the experiment and a substantial portion of the report documentation during the class time allotted to each experiment.

Experiment No.	Description
1	Wave length, profile, and group velocity as a function of wave period, water depth, and wave height
2	Wave profiles and particle trajectories as functions of wave height, water depth, and wave period; progressive and standing waves
3	Pressure variations as a function of wave height, water depth, and wave period; progressive and standing waves
4	Wave height transformation in shoaling water; wave breaking
5	Wave reflection from beach; comparison with Miche's theory
6	Wave reflection from a partial vertical barrier; comparison with approximate theory
7	Wave forces on cylinders and spheres
8	Plane wavemaker
9	Approximate wavemaker theory for a perfectly reflecting "beach"

The report describing the laboratory experiment should be fairly concise. A reasonable format for the reports is as follows:

1. Purpose—stating the objectives of the experiment.

2. Background and/or theory—describing the problem and presenting theoretical relationships to be tested.

3. Equipment description—this section can be quite brief, especially if the equipment has been used previously and is described in an earlier report.

4. Procedure—describing the experimental, data reduction, and/or analysis procedures.

5. Results (and conclusions)—presentation of results and possible reasons for any significant differences between theory and experiment. Can you suggest a procedure (experimental or analytical) that would verify or disprove your suggested reasons for any differences noted between theory and experimental results?

The report, excluding graphs and data sheets, should not exceed several pages. Items 1 through 4 and any graphs for item 5 can be a laboratory group effort; the conclusions and interpretation of results in item 5 should be an individual effort. The "group effort" portions of the report can be copies; however, each group member should turn in a complete report.

Each of the experiments above is described briefly in the following sections.

12.3.1 Experiment 1: Wave Length, Profile, and Group Velocity as a Function of Wave Period, Water Depth, and Wave Height

The purpose of this experiment is to compare measured wave profiles, wave lengths, and group velocities with the corresponding values as predicted by small-amplitude wave theory.

Small-amplitude wave theory.
Wave Profile η.   The wave profile η generated by a simple harmonic wavemaker is

$$\eta = \frac{H}{2} \cos \left(\frac{2\pi t}{T} - \frac{2\pi x}{L} \right)$$ (12.3)

where H, T, L, x, and t are the wave height, wave period, wave length, and distance and time coordinates, respectively.

Wave Length L.   The small-amplitude relationship for wave length L is

$$L = L_0 \tanh 2\pi \frac{h}{L}$$ (12.4)

where h is the water depth and L_0 is the "deep water" wave length expressed by

$$L_0 = \frac{gT^2}{2\pi}$$ (12.5)

The quantity L/L_0 is plotted against h/L_0 in Figure 3.9.

Group Velocity C_G.   The group velocity C_G is the speed at which the wave energy propagates and is also the speed of propagation of the leading edge of a train of waves. The group velocity can be expressed as

$$C_G = \frac{C_0}{2} \tanh \left(2\pi\frac{h}{L} \right) \left[1 + \frac{4\pi(h/L)}{\sinh 4\pi(h/L)} \right] \qquad (12.6)$$

where C_0 is the deep water celerity, that is

$$C_0 = \frac{gT}{2\pi} \qquad (12.7)$$

The ratio C_G/C_0 is also plotted against h/L_0 (Figure 3.9).

Measurements. The major piece of equipment for this experiment is the wave tank. Two capacitance wave gages connected to a two-channel oscillograph are used to sense and record the moving water surface.

For each of the runs, the water depth, wave height, and wave period should be observed.

Wave Length. The wave length can be established by first spacing the two wave gages approximately one wave length apart along the channel. A final spacing can be established by adjusting the position of one gage until the oscillograph traces are observed to be in phase.

Group Velocity. The group velocity is determined by spacing the two wave gages 5 to 10 m apart and then starting the wave generator. The "leading edge" or front of the wave train will travel at the group velocity. The group velocity can be calculated from the known separation distance between the two gages and the observed difference in "leading edge" arrival times at the two gages.

Wave Profile. It is desirable to obtain a reasonably high speed oscillograph record of one or two wave periods.

12.3.2 Experiment 2: Wave Profiles and Particle Trajectories as Functions of Wave Height, Water Depth, and Wave Period; Progressive and Standing Waves

The purpose of this experiment is to compare measured and theoretical profiles and water particle trajectories of progressive and standing waves.

Background. The maximum water particle displacements $|\zeta|$ and $|\xi|$ in the x and z directions, respectively, can be expressed as functions of the incident and reflected wave heights, the mean position of the particle in the waves (both horizontally and vertically), and the wave period and water depth (see Figure 12.4).

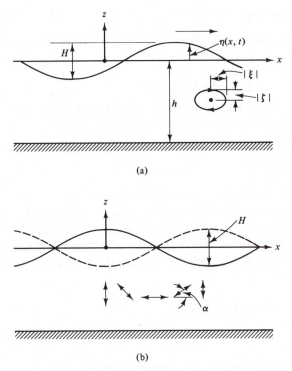

Figure 12.4   Definition sketch for experiment 2: (a) progressive wave; (b) pure standing wave.

$$\eta = \frac{H}{2} \cos(kx - \sigma t) \qquad (12.8)$$

$$|\zeta| = \frac{H}{2} \frac{\cosh k(h + z)}{\sinh kh}$$

$$|\xi| = \frac{H}{2} \frac{\sinh k(h + z)}{\sinh kh}$$

$$\eta = \frac{H}{2} \cos kx \cos \sigma t \qquad (12.9)$$

$$|\zeta| = \frac{H}{2} \frac{\cosh k(h + z)}{\sinh kh} \sin kx$$

$$|\xi| = \frac{H}{2} \frac{\sinh k(h + z)}{\sinh kh} \cos kx$$

$$\alpha = \tan^{-1} \left(\frac{\tanh k(h + z)}{\tan kx} \right)$$

Measurements

Progressive Waves.   With the barrier removed, generate a progressive wave system.

1. Measure the wave characteristics
2. Using approximately neutrally buoyant particles, measure $|\zeta|$ and $|\xi|$ at two depths within the wave.

Standing Waves.   Establish a standing wave system using the vertical barrier as a reflector.

1. Measure the characteristics of the standing wave system.
2. Using approximately neutrally buoyant particles measure the maximum water particle displacement components $|\zeta|$ and $|\xi|$ and inclination of streamlines at any depth at the node and antinode positions and also at a position intermediate to these positions.

Reference: See pp. 80–89.

12.3.3 Experiment 3: Pressure Variations as a Function of Wave Height, Water Depth, and Wave Period; Progressive and Standing Waves

The purpose of this experiment is to compare measured and theoretical pressure variations within progressive and standing waves.

Background.   The pressure deviations from hydrostatic pressure as derived for small amplitude waves is

$$p = \rho g \eta \frac{\cosh k(h + z)}{\cosh kh} \tag{12.10}$$

in which $\eta(x, t)$ can be the water surface displacement for either progressive, standing, or partially standing waves.

Measurements.   Measure the pressure fluctuations near the bottom and at three additional elevations along a tank wall, for a progressive and a standing wave system. Also measure simultaneously the water surface displacement at the longitudinal position (x) of the pressure sensor. Both the amplitudes and phases of these measured pressure fluctuations are to be compared with theory. For the standing wave system, conduct the measurements at two different positions along the standing wave envelope. Waves of two different periods should be used.

Equipment.   The equipment consists of a wave gage, a total head pressure sensor, and a recording oscillograph. If the wave tank is Lucite-walled, it may be worthwhile to drill and tap several permanent pressure taps to be used in conjunction with a manifold. If the tank is glass-walled, a streamlined pressure strut support can be placed along the side of the tank at the desired location (see Figure 12.2).

12.3.4 Experiment 4: Wave Height Transformation in Shoaling Water; Wave Breaking

The purpose of this experiment is to investigate the characteristics of progressive and standing breaking water waves and to compare these results with the available theory.

Theory for breaking waves

Progressive Water Waves.   The breaking characteristics of progressive water waves have been studied theoretically in deep and shallow water. In shallow water, for beaches of mild slope, the relationship is

$$\frac{H}{h} = 0.78 \tag{12.11}$$

and it is remarked that slopes greater than about $1:40$ increase this ratio substantially. For deep water, the deep water steepness (H_b/L_o) at breaking

$$\left(\frac{H_b}{L_o}\right)_{max} = 0.142 \tag{12.12}$$

where $L_o = 1.2(gT^2/2\pi)$ for breaking waves, including nonlinear effects. These asymptotes and some data are presented in Figure 12.5. Additionally, for deep and shallow water, it is predicted that at the inception of breaking, the "interior" angle of the wave is $120°$ as shown in Figure 12.6.

For relatively steep slopes in shallow water, there is considerable scatter of the data, as shown in Figure 12.7.

Standing Waves.   For standing waves the limiting theoretical steepness is

$$\left(\frac{H_b}{L_o}\right)_{max} = 0.218 \tag{12.13}$$

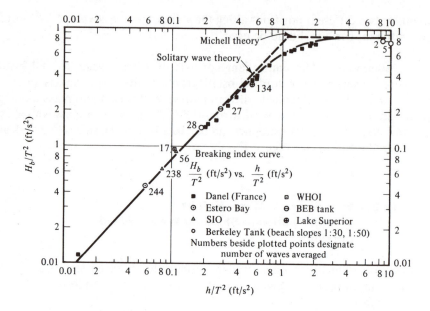

Figure 12.5 Breaking index curve. (From Reid and Bretschneider, 1953.)

and the maximum η_c and minimum η_t water surface displacement at breaking are

$$\eta_c = 0.647H$$
$$\eta_t = 0.353H$$

(12.14)

and the "interior" angle of the wave is 90°.

For shallow water, no theory for the limiting standing wave has been developed, although the experimental results indicate the following ratio:

$$\left(\frac{H}{h}\right)_b = 1.37$$

(12.15)

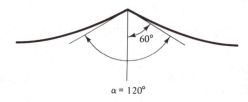

$\alpha = 120°$

Figure 12.6 Crest angle at maximum steepness.

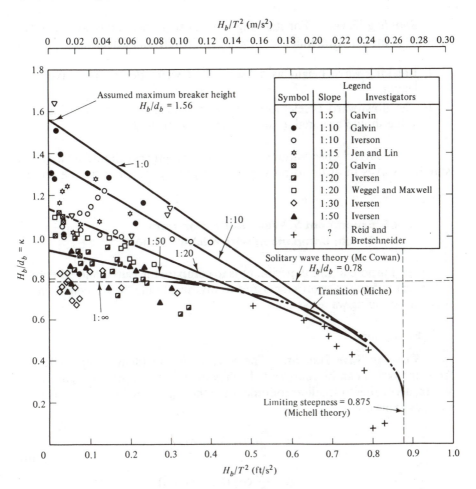

Figure 12.7 Experimental observations of d_b/H_b versus breaker steepness H_b/T^2. (From Weggel, 1972.)

Measurements

Progressive Waves.   Due to the difficulty of measuring kinematics, our experimental study will concentrate on the ratio of wave height to water depth in shallow water. For three wave periods and one water depth (in the uniform depth section), measure the wave height and water depth at which breaking occurs. Also attempt to observe the location of incipient instability. Comment on any extraneous effects in the wave tank, such as reflections from the beach and the effect that these may have on this breaking ratio. You should observe the breaking process carefully to provide a description in your report.

Standing Waves.   For standing waves, using a barrier to provide wave reflection, the experimental efforts will concentrate on measuring:

1. Breaking wave height as a function of water depth and wave length.
2. The downward acceleration at the antinode. At breaking this value should be equal to the gravitational acceleration.

Again careful observations of the breaking process should be made of standing waves in order to provide a perceptive description in the report.

12.3.5 Experiment 5: Wave Reflection from Beach; Comparison with Miche's Theory

The purpose of this experiment is to compare measured and "theoretical" beach reflection coefficients and to investigate the "wave height envelope" for standing wave systems.

Background.

Standing Wave Systems.   The wave system incident on and reflected from the beach can be represented schematically as shown in Figure 12.8, where, according to small-amplitude wave theory, the incident and reflected wave systems are

$$\eta_i = \frac{H_i}{2} \cos (kx - \sigma t)$$

$$\eta_r = \frac{H_r}{2} \cos (kx + \sigma t + \delta)$$

$$(12.16)$$

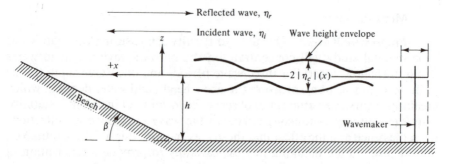

Figure 12.8   Experimental arrangement for experiment 5.

in which H_i and H_r are the incident and reflected wave heights, respectively, and

$$k \equiv \frac{2\pi}{L}$$

$$\sigma \equiv \frac{2\pi}{T}$$

The combined wave system η_c is

$$\eta_c = \eta_i + \eta_r \tag{12.17}$$

The total vertical displacement, $2|\eta_c|$, of the combined wave system can be shown to be

$$2|\eta_c| = \sqrt{H_i^2 + 2H_iH_r \cos{(2kx + \delta)} + H_r^2} \tag{12.18}$$

Equation (12.18) defines a quantity referred to as the "wave height envelope" as a function of distance along the channel. The maximum and minimum of this expression are

$$2|\eta_c|_{max} = H_i + H_r \tag{12.19}$$

and

$$2|\eta_c|_{min} = H_i - H_r \tag{12.19}$$

and occur at positions along the channel separated by $L/4$.

The reflection from the beach can be defined in terms of a reflection coefficient,

$$\kappa_r = \frac{H_r}{H_i} = \frac{2|\eta_c|_{max} - 2|\eta_c|_{min}}{2|\eta_c|_{max} + 2|\eta_c|_{min}} \tag{12.20}$$

The minimum and maximum values of the reflection coefficient are 0 and 1.0, respectively.

Miche's "Theory".   A very approximate "theory" for the reflection coefficient from a plane smooth beach has been developed by A. Miche. Miche defines a critical deep water wave steepness $(H_o/L_o)_{crit}$ in terms of the beach slope β:

$$\left(\frac{H}{L_o}\right)_{crit} = \left(\frac{2\beta}{\pi}\right)^{1/2} \frac{\sin^2{\beta}}{\pi} \tag{12.21}$$

Miche's results predict that the beach reflection coefficient will vary with deep water wave steepness, H_o/L_o, in the following manner:

$$\kappa_r = 1, \qquad\qquad \frac{H_o}{L_o} \leqslant \left(\frac{H_o}{L_o}\right)_{crit} \tag{12.22}$$

$$\kappa_r = \frac{(H_o/L_o)_{crit}}{H_o/L_o}, \qquad \frac{H_o}{L_o} \geqslant \left(\frac{H_o}{L_o}\right)_{crit}$$

The deep water wave height referred to in these equations is that of the incident wave. The relationship between the deep water wave height and the incident wave height is

$$H_o = \sqrt{\frac{2C_G}{C_o}}\, H_i \tag{12.23}$$

where the ratio C_G/C_o is plotted versus h/L_o in Figure 3.9.

Measurement.   For two wave periods, the wave height envelope is to be established by moving a wave gage along the channel over a distance of at least one wave length. From these envelopes, the measured beach reflection coefficients can be determined and compared with those of Miche's theory.

12.3.6 Experiment 6: Wave Reflection from a Partial Vertical Barrier; Comparison with Approximate Theory

The purpose of this experiment is to derive an approximate theory for the wave height transmitted past the vertical partial barrier shown in Figure 12.9 and to test the theory for various wavelengths and a fixed "gap opening" of height Δ.

Background and theory.   A portion of the wave energy incident on the barrier will be reflected as a reflected wave component and a portion will pass beneath the barrier and form a transmitted wave component. As a first approximation to determining the height of the transmitted wave component, one could assume that all the progressive wave energy being propagated at those levels below the lower edge of the barrier is transmitted past the

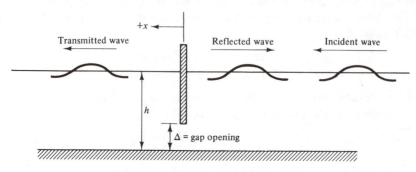

Figure 12.9   Experimental arrangement.

barrier and results in a transmitted wave. Develop an approximate theory on this basis and express the result in the form of transmission coefficient κ_t, where

$$\kappa_t = \text{function of } (k\Delta, \, kh) \tag{12.24}$$

For a ratio $\Delta/h = \frac{1}{2}$, plot κ_t as a function of kh for the range $\pi/10 < kh < \pi$. Also plot the deep and shallow water asymptotes for κ_t. If no energy is lost in the reflection–transmission process, then

$$H_r^2 + H_t^2 = H_i^2 \tag{12.25}$$

or defining a reflection coefficient,

$$\kappa_t = \frac{H_t}{H_i} \tag{12.26}$$

then $\kappa_r^2 + \kappa_t^2 = 1$.

Measurements.   For $\Delta/h = \frac{1}{2}$, measure the wave envelope for $x < 0$ and the transmitted wave height H_t for $x > 0$. From the wave envelope, determine H_i and H_r and compare your experimental values of κ_t with the approximate theory.

Calculate the sum $\kappa_r^2 + \kappa_t^2$ for your individual experiments and determine the percentage energy loss in the reflection–transmission process.

Carry out the measurements and calculations described above for four different wave lengths.

12.3.7 Experiment 7: Wave Forces on Cylinders and Spheres

The purpose of experiment 7a is to measure wave forces and moments on a circular cylinder and to determine the "best fit" drag and inertia coefficients associated with these measurements. Experiment 7b will consist of the measurement of wave forces on a sphere with the *prior* calculation of wave forces based on drag and inertia coefficients obtained from the literature (see, e.g., Grace and Casciano, 1969).

Measurements.   The measurements will be conducted using a portal-type force gage and a cantilever moment gage (see Figure 12.10). In addition, the wave profile near to the object should be measured.

Theory of wave forces.   The Morison equation for horizontal wave forces is written for an elemental length of a cylinder as (see Figure 12.11)

$$dF = C_D \rho A_p \frac{u|u|}{2} \, ds + C_m \rho dV \dot{u} \tag{12.27}$$

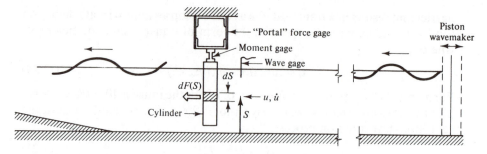

Figure 12.10   Test arrangement for measuring wave forces and moments.

in which

   C_D = drag coefficient
   C_M = inertia coefficient
   ρ = mass density of water
   A_p = cylinder area per unit length projected onto a vertical plane
        perpendicular to the velocity vector
   ds = elemental length of cylinder
   dV = elemental volume in length, ds
   $u, \dot{u}$ = horizontal component of water particle velocity and
        acceleration, respectively

For a *circular* cylinder, Eq. (12.27) becomes

$$dF = C_D \rho D \frac{u|u|}{2} ds + C_M \rho \frac{\pi D^2}{4} \dot{u}\, ds \tag{12.28}$$

which, for linear water wave kinematics, can be integrated to

$$F = \gamma C_D D \frac{H^2}{8} \frac{kh \cos \sigma t\ |\cos \sigma t|}{\sinh 2kh} \tag{12.29}$$

$$\left[\frac{1}{2kh} \sinh 2kh \left(1 + \frac{\eta}{h} \right) + \left(1 + \frac{\eta}{h} \right) \right] - \gamma C_M \frac{\pi D^2}{8} \frac{H \sin \sigma t}{\cosh kh} \left[\sinh kh \left(1 + \frac{\eta}{h} \right) \right]$$

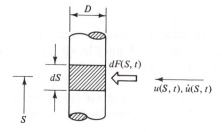

Figure 12.11   Elemental force on a
cylinder.

and the total moment about the bottom of the tank is

$$M = \gamma C_D D \, \frac{H^2}{8} \, \frac{hkh \cos \sigma t \, |\cos \sigma t|}{\sinh 2kh}$$

$$\left[\frac{(1 + \eta/h)^2}{2} + \frac{1 + \eta/h}{2kh} \sinh 2kh \left(1 + \frac{\eta}{h} \right) + \frac{1}{(2kh)^2} \left(1 - \cosh 2kh \left(1 + \frac{\eta}{h} \right) \right) \right]$$

$$\tag{12.30}$$

$$-\gamma C_M \, \frac{\pi D^2}{8} \, \frac{Hh \sin \sigma t}{\cosh kh} \left[\left(1 + \frac{\eta}{h} \right) \sinh kh \left(1 + \frac{\eta}{h} \right) + \frac{1}{kh} \left(1 - \cosh kh \left(1 + \frac{\eta}{h} \right) \right) \right]$$

For a sphere, the equation is

$$F = C_D \rho \, \frac{\pi D^2}{4} \, \frac{u|u|}{2} + C_M \rho \, \frac{\pi D^3}{6} \, \dot{u} \tag{12.31}$$

Scope of measurements.   For a sphere and/or cylinder, measure the waves, wave forces, and moments for two wave periods of approximately 1.0 and 2.5 s. Measure the wave reflection in the tank.

For the two combinations of experimental wave conditions, calculate the waves, wave forces, and wave moments on the object and compare with those measured.

12.3.8 Experiment 8: Plane Wavemaker

The purpose of this experiment is to evaluate the wavemaker theory for the piston-type wavemaker used in our studies. Although the beach is a fairly efficient energy dissipator, the wave envelope should be measured to remove the effect of the reflected wave in the measurements.

Wavemaker theory for a piston-type wavemaker.   The wavemaker theory for a piston-type wavemaker (as presented in Chapter 6) is

$$\frac{H}{S} = \frac{2(\cosh 2kh - 1)}{\sinh 2kh + 2kh} \tag{12.32}$$

See Figure 12.12 for a plot of H/S versus kh.

Measurements.   Measure the wave generated for approximately 10 wave periods (say $0.8 < T < 2.5$ s) for which the waves are well behaved. Evaluate the effect of reflection by measuring the wave envelope.

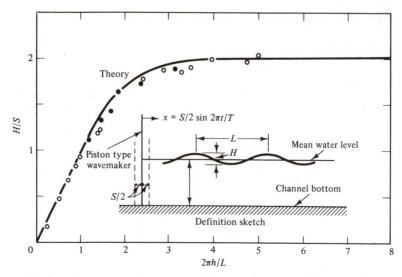

Figure 12.12   Test of wavemaker theory for small wave steepnesses. o, experiments corrected for reflection; •, experiments not corrected for reflection. (From Ursell et al., 1960.)

12.3.9 Experiment 9: Approximate Wavemaker Theory for a Perfectly Reflecting "Beach"

The purpose of this experiment is to develop an approximate theory for the waves in a wave tank with perfectly reflecting boundaries and to conduct measurements to evaluate this theory.

Theory.   The approximate theory will be developed for the case below. Although this problem is for shallow water waves in order to satisfy the boundary condition requirements, in comparing the results with measurements, the actual wave characteristics (particularly the wavelength) appropriate to the water depth and period should be used.

Consider the vertical barrier located at an arbitrary distance l from a piston-type wavemaker (see Figure 12.13). Assuming that shallow water waves are generated, calculate and plot the ratio H/S as a function of l/L. For this problem $\sigma = 1$ rad/s and $h = 1$ ft.

Measurements.   With a rigid vertical barrier located in the tank, conduct sufficient wave height measurements at the barrier over as wide a range of wave periods as possible to verify the approximate theory. Note that it will be helpful (perhaps in locating the barrier) if the theory is developed and incorporated in the planning phase of the experiment.

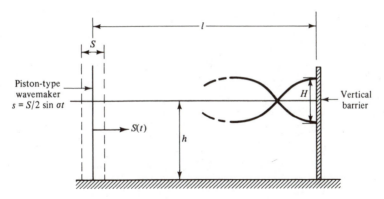

Figure 12.13   Experimental arrangement for experiment 9.

REFERENCES

GRACE, R. A., and F. M. CASCIANO, "Ocean Wave Forces on a Subsurface Sphere," *J. Waterways Harbors Div., ASCE*, Aug., 1969.

REID, R.O., and C. L. BRETSCHNEIDER, "Surface Waves and Offshore Structures," *Texas A. and M. Res. Found. Tech. Rept.*, Oct. 1953.

URSELL, F., R. G. DEAN, and Y. S. YU, "Forced Small Amplitude Water Waves: A Comparison of Theory and Experiment," *J. Fluid Mech.*, Vol. 7, Pt. 1, 1960.

WEGGEL, J. R., "Maximum Breaker Height," *J. Waterways, Harbors Coastal Eng. Div., ASCE*, Vol. 98, WW4, Nov. 1972.

Subject Index

Author Index